Large MIMO Systems

Large MIMO systems, with tens to hundreds of antennas, are a promising emerging communication technology. This book provides a unique overview of this technology, covering the opportunities, engineering challenges, solutions, and state-of-the-art of large MIMO test beds. There is in-depth coverage of algorithms for large MIMO signal processing, based on metaheuristics, belief propagation, and Monte Carlo sampling techniques, and suited for large MIMO signal detection, precoding, and LDPC code designs. The book also covers the training requirement and channel estimation approaches in large-scale point-to-point and multiuser MIMO systems; spatial modulation is also included. Issues like pilot contamination and base station cooperation in multicell operation are addressed. A detailed exposition of MIMO channel models, large MIMO channel sounding measurements in the past and present, and large MIMO test beds is also presented. An ideal resource for academic researchers, next generation wireless system designers and developers, and practitioners in wireless communications.

A. Chockalingam is a Professor in the Department of Electrical Communication Engineering, Indian Institute of Science (IISc), Bangalore, India. He has made pioneering contributions in the area of low complexity near-optimal signal detection in large MIMO systems. He is a recipient of the Swarnajayanti Fellowship from the Department of Science and Technology, Government of India, and a Fellow of the Indian National Academy of Engineering (INAE), the National Academy of Sciences, India (NASI), and the Indian National Science Academy (INSA).

B. Sundar Rajan is a Professor in the Department of Electrical Communication Engineering, Indian Institute of Science (IISc), Bangalore, India. He is a well-known authority in the area of space-time coding for MIMO channels and distributed space-time coding, and a leading expert in the design of space-time codes based on algebraic techniques. He is a recipient of the Professor Rustum Choksi Award from IISc for excellence in research in engineering, and a Fellow of the Indian National Academy of Engineering (INAE), the National Academy of Sciences, India (NASI), the Indian National Science Academy (INSA), and the Indian Academy of Sciences (IASc).

"This cutting-edge portrayal of large-scale MIMO systems provides a shrewd long-term outlook on this salient wireless subject."

Lajos Hanzo
University of Southampton

"This is a very timely and useful book written by authors who are pioneers in the area of large MIMO systems."

Vijay K. Bhargava
The University of British Columbia

"Large MIMO will power our wireless networks before this decade is out and the race is just starting. Chockalingam and Sundar Rajan have compiled an excellent companion for this journey."

Arogyaswami Paulraj
Stanford University

Large MIMO Systems

A. CHOCKALINGAM AND B. SUNDAR RAJAN
Indian Institute of Science, Bangalore

CAMBRIDGE
UNIVERSITY PRESS

University Printing House, Cambridge CB2 8BS, United Kingdom

Published in the United States of America by Cambridge University Press, New York

Cambridge University Press is part of the University of Cambridge.

It furthers the University's mission by disseminating knowledge in the pursuit of education, learning and research at the highest international levels of excellence.

www.cambridge.org
Information on this title: www.cambridge.org/9781107026650

First published 2014

Printed in the United Kingdom by TJ International Ltd. Padstow Cornwall

A catalog record for this publication is available from the British Library

Library of Congress Cataloging in Publication Data

Chockalingam, A., author.
Large MIMO systems / A. Chockalingam, Indian Institute of Science, Bangalore; B. Sundar Rajan, Indian Institute of Science, Bangalore.
pages cm
ISBN 978-1-107-02665-0 (hardback)
1. MIMO systems. I. Rajan, B. Sundar, author. II. Title.
TK5103.4836.C49 2014
621.39′8–dc23 2013041123

ISBN 978-1-107-02665-0 Hardback

To our teachers and students

Contents

Preface

The physical layer capabilities in wireless transmissions are growing. In particular, the growth trajectory of the achieved data transmission rates on wireless channels has followed Moore's law in the past decade and a half. Over a span of 15 years starting mid-1990s, the achieved wireless data transmission rates in several operational scenarios have increased over 1000 times. The data transmission rate in WiFi which was a mere 1 Mbps in 1996 (IEEE 802.11b) had reached 1 Gbps by 2011 (IEEE 802.11ac). During the same span of time, the data rate in cellular communication increased from about 10 kbps in 2G to more than 10 Mbps in 4G (LTE). One of the promising technologies behind such a sustained rate increase is multiantenna technology – more popularly referred to as the multiple-input multiple-output (MIMO) technology, whose beginnings date back to the late 1990s.

The interest shown in the study and implementation of MIMO systems stems from the promise of achieving high data rates as a result of exploiting independent spatial dimensions, without compromising on the bandwidth. Theory has predicted that the greater the number of antennas, the greater the rate increase without increasing bandwidth (in rich scattering environments). This is particularly attractive given that the wireless spectrum is a limited and expensive resource.

More than a decade of sustained research, implementation, and deployment efforts has given MIMO technology the much needed maturity to become commercially viable. More and more wireless products and standards have started adopting MIMO techniques, mainly in the small number of antennas regime (2–8 antennas). However, the promise of achieving very high spectral efficiencies using a much larger number of antennas still remains open to research and subsequent commercial exploitation. We call MIMO systems which achieve spectral efficiencies of tens to hundreds of bps/Hz using tens to hundreds of antennas "large MIMO systems." This book is exclusively about large MIMO systems.

Large MIMO systems, by their very nature, merit special attention and treatment. For example, algorithms and techniques which are known to work well with a small number of antennas may not scale well for a large number of antennas. Therefore, newer and alternative approaches are needed. Also, in addition to increased rate and diversity gains, large dimensionality brings other advantages (e.g., channel hardening, which can be exploited to achieve low complexity signal processing) which do not come with smaller systems. Bringing out such large MIMO centric opportunities, issues, and solution approaches and techniques is one of the key objectives in this book.

A few words about what motivated us to write this book are in order. Our teaching and research interest in space-time coding and multiuser detection in the early- to mid-2000s brought us together to collaborate on MIMO wireless research. Being in the same department and having offices in the same building helped – we could discuss ideas over casual chats during coffee/tea breaks and evening walks. Our first set of results on large MIMO systems were published in mid-2008. Since then, we have continued our research on various signal processing aspects in large MIMO systems, which has led to several of our subsequent publications on large MIMO. The large MIMO idea seems to have caught on, as we can see in the chapter on large MIMO testbeds (Chapter 12). Over these years, we have given tutorial talks on this topic to conferences and industry. We felt that, in the process, we had generated a critical mass of material, enough to write a book on large MIMO systems. Also, we found that a book written exclusively on large MIMO systems was yet to appear at the time of proposing this book to the publisher. We thank the publisher for having accepted our proposal for writing this book, and here we are with our intended book on large MIMO systems.

It is heartening to see that large MIMO systems have become more popular now compared to the days when we first started publishing on this topic in 2008. Large MIMO systems seem to have started to flourish under several names; large-scale MIMO, massive MIMO, hyper-MIMO, higher-order MIMO, to name a few. It is even more heartening to realize that large MIMO technology is one of the key technologies being considered for standardization in 5G and beyond.

We hope that this book will be of interest and use to researchers, graduate students, and wireless system designers and implementers, and will create the interest needed to take large MIMO research, development, and standardization activities to the next level.

Acknowledgments

We would first like to thank our graduate students for their valuable contributions to our large MIMO research. At a time when people started thinking that there is not much of interest left in MIMO research, they took on the challenges of exploring the uncharted area of MIMO systems with tens to hundreds of antennas. Thanks to their dedicated and sustained efforts, we were able to make some of the early contributions to the field of large MIMO systems. This book to a large extent draws on these contributions, and we thank all our students for their commitment, hard work, and help. Our many thanks are due to: K. Vishnu Vardhan, Saif K. Mohammed (currently an Assistant Professor at the Indian Institute of Technology, Delhi), Ahmed Zaki, N. Srinidhi, Suneel Madhekar, P. J. Thomas Sojan, Pritam Som, Tanumay Datta, N. Ashok Kumar, Suresh Chandrasekaran, Yogendra Umesh Itankar, P. M. Chandrakanth, M. Raghavendra Nath Reddy, Harsha Eshwaraiah, T. Lakshmi Narasimhan, Kamal Agarwal Singhal, Manish Mandloi, and Shovik Biswas.

Our research and teaching in multiuser detection and space-time coding had a positive influence on our understanding of and contribution to large MIMO research. We thank all the students who attended our courses on CDMA and multiuser detection, and space-time signal processing and coding. We also thank the students who contributed to our research in these areas. Parts of early drafts of this book were used in the CDMA and multiuser detection course. We thank the students for the valuable feedback on these drafts.

N. Srinidhi, Tanumay Datta, and T. Lakshmi Narasimhan were helpful in the development of the manuscript in many ways (generating figures, proofreading, LaTex help, offering general feedback and comments on the structure and contents of the book). Our special thanks are due to them. We also thank Ms. G. Nithya, our project associate, for her help in the preparation of the manuscript. We appreciate her technical support to our laboratory activities and large MIMO related activities.

We thank Emanuele Viterbo and Yi Hong for their fruitful research collaboration on MIMO precoding and sampling based lattice decoding. We also thank Onkar Dabeer for his collaboration on AdaBoost for MIMO signal detection. We are grateful to Rajesh Sundaresan and Vivek Borkar for useful discussions on MCMC techniques.

We gratefully acknowledge the continuous support of Defence Research and Development Organization (DRDO) in our research. Our special thanks are due to Defence Electronics Application Laboratory (DEAL), Dehradun, for active support for large MIMO system development. We also acknowledge a gift from the Cisco University Research Program, a corporate advised fund of Silicon Valley Community Foundation.

We thank the academic institutions and industries who hosted our talks and discussion meetings on large MIMO systems, on one occasion or another. These interactions stimulated us to continuously engage in and broaden our views and scope of our research in large MIMO.

Our thanks are due to Dheeraj Sreedhar, N. Srinidhi, Tanumay Datta, Sanjay Vishwakarma, T. Lakshami Narasimhan, S. N. Padmanabhan, and S. V. R. Anand for useful discussions on large MIMO system development. We also thank our colleagues K. J. Vinoy, Gaurab Banerjee, and Bharadwaj Amrutur for discussions on RF and hardware related issues. Working with the scientists of DEAL, Dehradun, and the engineers of Tata Elxsi, Bangalore, on large MIMO system development was an exciting experience. We thank them for their zeal and commitment to the development effort.

We thank Cambridge University Press for accepting our proposal to write this book. It was a pleasure working with Philip Meyler and Mia Balashova, Cambridge University Press, on this project. Special thanks are due to Mia for the excellent support she rendered throughout the various stages of writing this book. Her response to our queries and concerns at every stage was always prompt and clear. She made our book writing experience a smooth one.

Finally, we express our sincere gratitude to our families – our indulgence in several of our academic pursuits, including writing of this book, would not have been possible without their patience, understanding, and support.

Abbreviations

2G	Second generation
3G	Third generation
3GPP	Third generation partnership project
4G	Fourth generation
5G	Fifth generation
ADC	Analog-to-digital conversion
AGC	Automatic gain control
AoA	Angle of arrival
AoD	Angle of departure
AP	Access point
APP	A posteriori probability
AS	Angular spread
ASIC	Application specific integrated circuit
AWGN	Additive white Gaussian noise
BC	Broadcast channel
BCJR	Bahl–Cocke–Jelinek–Raviv
BER	Bit error rate
BP	Belief propagation
bpcu	Bits per channel use
BPSK	Binary phase shift keying
BQP	Binary quadratic program
BS	Base station
CCDF	Complementary cumulative distribution function
CDA	Cyclic division algebra
CDF	Cumulative distribution function
CDMA	Code division multiple access
CN	Check node
COMP	Coordinated multipoint
COST	Cooperation in science and technology
CP	Cyclic prefix
CPSC	Cyclic prefixed single-carrier
CRB	Cramer–Rao bound
CRLB	Cramer–Rao lower bound
CSI	Channel state information

CSIR	Channel state information at receiver
CSIT	Channel state information at transmitter
DAC	Digital-to-analog conversion
dB	Decibel
DFT	Discrete Fourier transform
DoA	Direction of arrival
DoD	Direction of departure
DPC	Dirty paper coding
EPA	Extended pedestrian A model
ETU	Extended typical urban model
EVA	Extended vehicular A model
EXIT	Extrinsic information transfer
FDD	Frequency division duplex
FDMA	Frequency division multiple access
FIR	Finite impulse response
FFT	Fast Fourier transform
FGBP	Factor graph belief propagation
FPGA	Field-programmable gate array
GA	Gaussian approximation
GAI	Gaussian approximation of interference
GDL	Generalized distributive law
GPDA	Generalized PDA
GPS	Global positioning system
GSM	Generalized spatial modulation
GTA	Gaussian tree approximation
HDTV	High-definition television
IC	Integrated circuit
ICI	Inter-carrier interference
IDFT	Inverse DFT
IF	Intermediate frequency
IFA	Inverted F antenna
IFFT	Inverse FFT
iid	independent and identically distributed
ILS	Integer least-squares
ISDIC	Iterative soft decision interference cancelation
ISI	Inter-symbol interference
IUI	Inter-user interface
KL	Kullback–Leibler
LAN	Local area network
LAS	Likelihood ascent search
LD	Linear dispersion
LDPC	Low-density parity-check
LHS	Left hand side
LLL	Lenstra–Lenstra–Lovasz

LLR	Log-likelihood ratio
LOS	Line-of-sight
LR	Lattice reduction
LS	Local search
LTE	Long-term evolution
LTE-A	Long-term evolution advanced
LTS	Layered tabu search
MAC	Media access control
MAP	Maximum a posteriori probability
MCMC	Markov chain Monte Carlo
MF	Matched filter
MGS	Mixed-Gibbs sampling
MGS-MR	Mixed-Gibbs sampling with multiple restarts
MIMO	Multiple-input multiple-output
ML	Maximum likelihood
MMSE	Minimum mean square error
MMSE-ISDIC	MMSE based iterative soft-decision interference cancelation
MMSE-SIC	MMSE successive interference cancelation
MRF	Markov random field
MSE	Mean square error
MUBF	Multiuser beamforming
MUD	Multiuser detection
NDS	Norm descent search
NLOS	Non-line-of-sight
NO-STBC	Non-orthogonal space-time block code
OFDM	Orthogonal frequency division multiplexing
OFDMA	Orthogonal frequency division multiple access
OLA	Overlap-and-add
OSTBC	Orthogonal space-time block code
PAM	Pulse amplitude modulation
PAPR	Peak-to-average power ratio
PAS	Power angular spectrum
PC	Personal computer
PDA	Probabilistic data association
pdf	Probability density function
PDP	Power delay profile
PIC	Parallel interference cancelation
PIFA	Planar inverted F antenna
pmf	Probability mass function
PSK	Phase shift keying
QAM	Quadrature amplitude modulation
QPSK	Quadrature phase shift keying
R3TS	Random-restart reactive tabu search
RF	Radio frequency

RFID	Radio-frequency identification
rms	Root mean square
RS	Randomized search
RTS	Reactive tabu search
SA	Seysen's algorithm
SAGE	Space-alternating generalized expectation-maximization
SC-FDMA	Single-carrier frequency division multiple access
SCM	Spatial channel model
SCME	Spatial channel model – extended
SD	Sphere decoder
SDMA	Space division multiple access
SDR	Semi-definite relaxation
SFBC	Space-frequency block code
SGA	Scalar Gaussian approximation
SIC	Successive interference cancelation
SIMO	Single-input multiple-output
SINR	Signal-to-interference plus noise ratio
SISO	Single-input single-output
SM	Spatial modulation
SMSE	Sum mean square error
SNR	Signal-to-noise ratio
spcu	Symbols per channel use
SSK	Space shift keying
STBC	Space-time block code
STTC	Space-time trellis codes
SVD	Singular value decomposition
TCM	Trellis coded modulation
TDD	Time division duplex
TDL	Tapped delay line
TDMA	Time division multiple access
TGn	Task group IEEE 802.11n
THP	Tomlinson–Harashima precoding
TOA	Time of arrival
TS	Tabu search
TV	Television
UCA	Uniform circular array
UE	User equipment
UHF	Ultra high frequency
ULA	Uniform linear array
USB	Universal serial bus
UT	User terminal
UWB	Ultra wideband
V-BLAST	Vertical Bell laboratories layered space-time architecture
VGA	Vector Gaussian approximation

VHF	Very high frequency
VLAN	Virtual local area network
VP	Vector perturbation
VP-SE	Vector perturbation with sphere encoding
WINNER	Wireless world initiative new radio
WiFi	Wireless fidelity
WLAN	Wireless local area network
ZF	Zero forcing
ZF-SIC	Zero forcing successive interference cancelation
ZP	Zero padding
ZPSC	Zero padded single-carrier

Notation

$(\cdot)^*$	Complex conjugation
$(\cdot)^H$	Hermitian transposition
$(\cdot)^T$	Transposition
$\lvert \cdot \rvert$	Absolute value of a complex number (or cardinality of a set)
$\lVert \cdot \rVert$	Euclidean norm of a vector
$\lfloor . \rceil$	Rounding operation to the nearest integer
$\lfloor c \rfloor$	Largest integer less than c
$\odot$	Element-wise multiplication operation
$\otimes$	Kronecker product
$\mathcal{CN}(\mu, \sigma^2)$	Circularly symmetric complex Gaussian distribution with mean μ and variance σ^2
n_t	Number of transmit antennas
n_r	Number of receive antennas
$\mathrm{vec}(\cdot)$	Stack columns of the input matrix into one column vector
$\det(\mathbf{X})$	Determinant of matrix $\mathbf{X}$
$\mathrm{tr}\{\mathbf{X}\}$	Trace of matrix $\mathbf{X}$
$\mathbf{I}_n$	$n \times n$ identity matrix
$\mathbf{x}$	Vector $\mathbf{x}$
$\mathbf{X}$	Matrix $\mathbf{X}$
$\mathbb{C}$	Field of complex numbers
$\mathbb{E}[.]$	Expectation operation
$\mathbb{R}$	Field of real numbers
$\mathbb{R}^+$	Non-negative real numbers
$\mathbb{Z}$	Set of all integers
$\Re(\cdot)$	Real part of the complex argument
$\Im(\cdot)$	Imaginary part of the complex argument

1 Introduction

The practical demonstration of the vertical Bell laboratories layered space-time architecture (V-BLAST) multiantenna wireless system by Bell Labs [1], and the theoretical prediction of very high wireless channel capacities in rich scattering environments by Telatar in [2] and Foschini and Gans in [3] in the late 1990s opened up immense possibilities and created wide interest in multiantenna wireless communications. Since then, multiantenna wireless systems, more commonly referred to as multiple-input multiple-output (MIMO) systems, have become increasingly popular. The basic premise of the popularity of MIMO is its theoretically predicted capacity gains over single-input single-output (SISO) channel capacities. In addition, MIMO systems promise other advantages like increased link reliability and power efficiency. Realizing these advantages in practice requires careful exploitation of large spatial dimensions.

Significant advances in the field of MIMO theory and practice have been made as a result of the extensive research and development efforts carried out in both academia and industry [4]–[7]. A vast body of knowledge on MIMO techniques including space-time coding, detection, channel estimation, precoding, MIMO orthogonal frequency division multiplexing (MIMO-OFDM), and MIMO channel sounding/modeling has emerged and enriched the field. It can be safely argued that MIMO systems using 2 to 4 antennas constitute a fairly mature area now. Technological issues in such small systems are fairly well understood and practical implementations of these systems have become quite common. Indeed, MIMO techniques have found their way into major wireless standards like long-term evolution (LTE) and WiFi (IEEE 802.11n/ac), leading to the successful commercial exploitation of MIMO technology.

At this point in time, when numerous papers and several books on MIMO have already been written and MIMO implementations are increasingly being embedded in wireless products, a natural question that arises is "What is next in MIMO?" One can pose this question a little differently: "Have we exploited the full potential of MIMO?" In addressing this question, one can realize that the main potential of MIMO, which is the feasibility of achieving "very" high spectral efficiencies/sum-rates, has not yet been well exploited in practice. Although the early days of MIMO technology witnessed the practical demonstration by Bell Labs of spectral efficiencies as high as 24 bps/Hz using 8 transmit and 12 receive antennas (the V-BLAST system) in fixed indoor environments, a majority of the

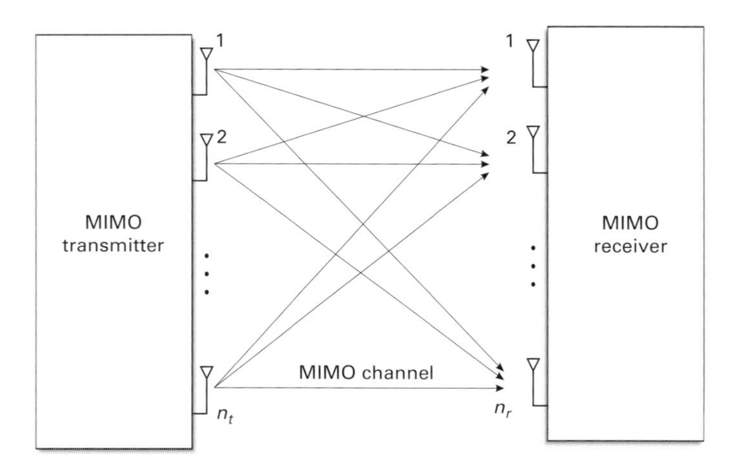

Figure 1.1 Point-to-point MIMO system.

subsequent research and development activities in MIMO seem to have focused on systems with fewer antennas and much lower spectral efficiencies. Except in a few cases (e.g., DoCoMo's 12 × 12 MIMO system [8]), MIMO configurations with 2–4 antennas and lower than 15 bps/Hz spectral efficiency have dominated MIMO research and development efforts since 2000. However, "large MIMO systems" (MIMO systems with a large number of antennas) are now a practical proposition and, hence, interest in them is growing [9]–[13].

The term large MIMO systems refers to systems in which large numbers (tens to hundreds) of antennas are employed in communication terminals. For example, in a point-to-point MIMO wireless link (Fig. 1.1), both the transmitter and receiver sides can be provided with a large number of antennas to achieve increased data rates without increasing bandwidth (i.e., they achieve very high spectral efficiencies). High-speed wireless backhaul connectivity between base stations (BSs) can adopt such a point-to-point MIMO configuration. In point-to-multipoint MIMO communication (e.g., multiuser MIMO downlink), the BS can be provided with a large number of transmit antennas for multiuser precoding and the user terminal can have one or more receive antennas (Fig. 1.2) so that increased sum-rates can be achieved. Likewise, in multipoint-to-point MIMO communication (e.g., multiuser MIMO uplink), each user terminal can transmit using one or more transmit antennas and the BS can receive through a large number of receive antennas and perform multiuser detection (MUD) using spatial signatures of all the users.

1.1 Multiantenna wireless channels

Multiantenna wireless channels are a broad category of channels that include point-to-point and multiuser channels. One of the defining characteristics of a

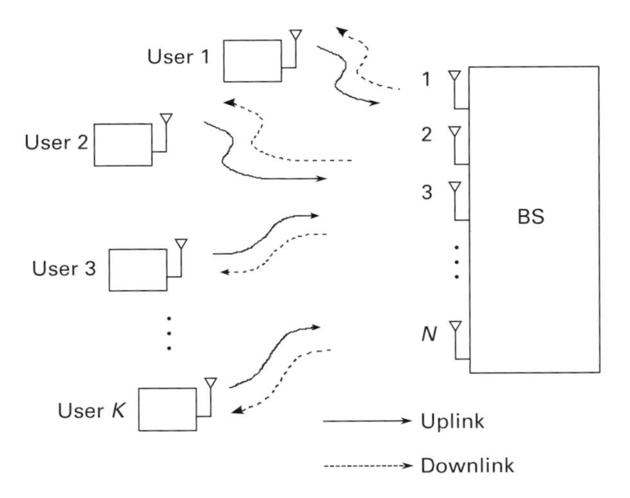

Figure 1.2 Multiuser MIMO system.

wireless channel is the variation of the channel strength over time and over frequency [6],[14]. These variations are typically classified into two types: large scale fading and small scale fading. Large scale fading is due to path loss as a function of distance and shadowing by large objects like buildings, bridges, trees, etc., and is typically frequency independent. Small scale fading, on the other hand, is due to the constructive and destructive interference of the multiple signal paths between the transmitter and receiver. Small scale fading happens at the spatial scale of the order of the carrier wavelength, and is frequency dependent. The channel is then classified as frequency-selective or frequency-flat. When the signaling bandwidth is larger than the coherence bandwidth of the channel (which has an inverse relation with the maximum delay spread of the channel), the channel is frequency-selective [14]. On the other hand, in frequency-flat channels, the signaling bandwidth is much smaller than the coherence bandwidth of the channel. Even when the channel is frequency-selective, techniques like OFDM can convert the channel into multiple frequency-flat channels on which the techniques designed for frequency-flat fading can be employed.

In terms of time variation, wireless channels are further classified as slowly fading or fast fading, depending on the fade rate relative to the signaling rate. If the fade remains constant over the signaling duration, the fading is termed slow (or time-flat) fading, whereas if the fade varies within the signaling duration, it is termed fast (or time-selective) fading. The carrier wavelength and velocity of the communication terminal determine the amount of time selectivity (or Doppler spread) in the channel [14].

Most multiantenna wireless channels with n_t transmit and n_r receive antennas (Fig. 1.1) are modeled as a linear channel with an equivalent baseband channel matrix $\mathbf{H}_c \in \mathbb{C}^{n_r \times n_t}$. The (i, j)th entry of $\mathbf{H}_c$ represents the channel gain from the jth transmit antenna to the ith receive antenna. The channel gains are also referred to as the "channel state information (CSI)." The availability of knowledge of these gains at the receiver and transmitter is an important factor which decides the performance of the communication system. CSI at the receiver (CSIR) refers to the scenario where the receiver has knowledge of the channel gains. Likewise, CSI at the transmitter (CSIT) refers to the scenario where the transmitter has knowledge of the channel gains. In fast fading channels, accurate estimation of the channel gains can become an issue, in which case non-coherent or blind techniques can be considered. In addition, obtaining CSIT through feedback can become ineffective in fast fading. However, in applications where the channel is not varying fast, it is generally possible to estimate the channel gains accurately through pilot-assisted transmission. Also, CSIT based on measured CSI fed back from the receiver is effective in such slow fading channels.

The channel gains can be independent or correlated, depending on various factors including the spacing between antenna elements, the amount of scattering in the environment, pin-hole effects, etc. Mathematical models that characterize the spatial correlation in MIMO channels are used in the performance evaluation of MIMO systems. Spatial correlation at the transmit and/or receive side can affect the rank structure of the MIMO channel resulting in degraded MIMO capacity. The structure of scattering in the propagation environment can also affect the capacity. In addition, transmit correlation in MIMO fading can be exploited by using non-isotropic inputs (precoding) based on knowledge of the channel correlation matrices.

1.2 MIMO system model

A widely studied point-to-point MIMO system model is one which is assumed to be frequency-flat and slow fading. The channel gains are assumed to remain constant over the signaling interval. With these assumptions, the equivalent complex baseband MIMO system model can be written as

$$\mathbf{y}_c = \mathbf{H}_c \mathbf{x}_c + \mathbf{n}_c, \tag{1.1}$$

where $\mathbf{x}_c \in \mathbb{C}^{n_t}$ is the transmitted vector, $\mathbf{y}_c \in \mathbb{C}^{n_r}$ is the received vector, and $\mathbf{n}_c \in \mathbb{C}^{n_r}$ is the additive white Gaussian noise (AWGN) vector. The jth entry of $\mathbf{x}_c$ is the symbol transmitted from the jth transmit antenna, $j = 1, \ldots, n_t$. In a typical communication system, information bits (e.g., the output of some source coder which performs voice or image compression, followed by a channel coder) are grouped into messages, and each message then corresponds to an n_t-dimensional complex vector, whose jth component is transmitted from the jth transmit antenna. In practice, these vectors belong to some codebook $\mathcal{X}$. The transmitter groups $R = \log_2 |\mathcal{X}|$ bits into a message, which is then used to index the codebook. R is often referred to as the rate of the codebook or simply as the rate of transmission. Alternatively, the $\mathbf{x}_c$ vector could be a pilot symbol vector (known to the receiver) during the training phase in a pilot-aided channel estimation scheme. The ith entry of $\mathbf{y}_c$ is the signal received at the ith receive antenna, $i = 1, \ldots, n_r$. Assuming a rich scattering environment, the entries of the channel matrix $\mathbf{H}_c$ are often modeled as independent and identically distributed (iid) $\mathcal{CN}(0, 1)$. Since the transmitter is power constrained, we have $\mathbb{E}[\text{tr}(\mathbf{x}_c\mathbf{x}_c^H)] = P$, where P is the total power available at the transmitter. Also, $\mathbb{E}[\mathbf{n}_c\mathbf{n}_c^H] = \sigma^2\mathbf{I}_{n_r}$, where σ^2 is the noise variance at each receive antenna. The average received signal-to-noise ratio (SNR) at each receive antenna is given by $\gamma = P/\sigma^2$.

The MIMO signal detection problem at the receiver can be stated as: given $\mathbf{y}_c$ and knowledge of $\mathbf{H}_c$, determine $\widehat{\mathbf{x}}_c$, an estimate of the transmitted symbol vector $\mathbf{x}_c$. Likewise, the MIMO channel estimation problem in a training based scheme can be stated as: given knowledge of the transmitted pilot symbol vector $\mathbf{x}_c$, determine $\widehat{\mathbf{H}}_c$, an estimate of the channel gain matrix $\mathbf{H}_c$.

1.3 MIMO communication with CSIR-only

In communication channels, error probability is one of the key performance indicators. Most communication schemes employ channel coding schemes to increase robustness against errors. To achieve an arbitrarily low probability of error, the rate of transmission R must be strictly below the MIMO channel capacity. The MIMO channel capacity is dependent on $\mathbf{H}_c$ and the transmit covariance matrix $\mathbf{K}_x \triangleq \mathbb{E}[\mathbf{x}_c\mathbf{x}_c^H]$, $\text{tr}\{\mathbf{K}_x\} = P$, and is given by

$$C_{MIMO}(\gamma, \mathbf{H}_c, \mathbf{K}_x) = \log_2 \det\left(\mathbf{I}_{n_r} + \frac{1}{\sigma^2}\mathbf{H}_c\mathbf{K}_x\mathbf{H}_c^H\right). \tag{1.2}$$

In the case of availability of CSIR only (i.e., no CSIT), since the transmitter has no knowledge of the channel gains, it cannot adapt its transmission scheme with respect to the channel gains. Therefore, for a fixed γ and rate R, the transmitter uses a fixed codebook, which does not change with changing channel gains. The transmitter codebook selection is very much dependent on whether the channel is slow fading or fast fading.

1.3.1 Slow fading channels

In slow fading channels, where the channel does not change during the length of the codeword, if the channel is such that $C_{MIMO}(\gamma, \mathbf{H}_c, \mathbf{K}_x) < R$, then no detector can recover the transmitted codeword correctly, and the channel is said to be in "outage." Hence, for slow fading channels with CSIR only, outage cannot be avoided and it is impossible to achieve an arbitrary low probability of error. In such scenarios, an appropriate performance indicator of any encoding–decoding scheme is the codeword error probability or codeword error rate. For codewords of large length, the theoretical limit for the codeword error rate of any encoding–decoding scheme is the channel outage probability, which is defined as

$$P_{outage}(\gamma, R) = \min_{\mathbf{K}_x \mid \text{tr}\{\mathbf{K}_x\}=P} p\big(C_{MIMO}(\gamma, \mathbf{H}_c, \mathbf{K}_x) < R\big). \tag{1.3}$$

Any practical encoding–decoding scheme would have a codeword error rate more than the channel outage probability given in (1.3). Therefore, it is important to design transmit schemes and corresponding receivers which can perform very close to the channel outage probability for all values of γ and R. For slow fading channels, there are two important parameters, namely, diversity gain and multiplexing gain. Diversity gain is a measure of reliability, whereas multiplexing gain is a measure of the degrees of freedom in the MIMO channel. These two parameters are usually related by the so called diversity–multiplexing gain tradeoff [15]. The maximum diversity gain achievable is $n_r n_t$ and the maximum multiplexing gain achievable is $\min(n_r, n_t)$. When the rate of transmission R is fixed, the limiting value (as $\gamma \to \infty$) of the negative of the slope of $\log(P_{outage}(\gamma, R))$ wrt $\log\gamma$ can be no more than $n_r n_t$. For a given scheme, we can therefore define the diversity order achievable (with fixed R) as

$$d = -\lim_{\gamma\to\infty} \frac{\log(P_e(\gamma))}{\log\gamma}, \tag{1.4}$$

where $P_e(\gamma)$ is the codeword error rate of the scheme. For simple MIMO schemes like V-BLAST [16], it can be shown that the maximum diversity order achievable is only n_r. This is because symbols transmitted from the antennas in V-BLAST are independent, and each such symbol reaches the receiver only through n_r different paths.

Space-time block coding is a well-known technique which can achieve the full diversity gain of $n_r n_t$ [5]. To achieve full diversity, symbols are coded across both space and time. Orthogonal space-time block codes (STBC) allow simple decoding achieving full diversity [17],[18]. However, they make sacrifices regarding the multiplexing rate, and are therefore not suited for systems with high target spectral efficiencies. Subsequent to orthogonal STBCs, several high-rate and high-diversity STBCs were proposed. One such class of STBCs is non-orthogonal STBCs (NO-STBCs) from cyclic division algebras (CDAs) [19],[20]. STBCs from CDA can achieve the full diversity of $n_r n_t$ without sacrificing rate.

1.3.2 Fast fading channels

In fast fading channels, the channel fade changes multiple times in the duration of the codeword. By spreading portions of the codeword across multiple fades, the reliability of codeword reception can be improved. In such a scenario, if the MIMO channel is ergodic, in the limit of infinitely long codewords, it is possible to achieve error-free communication if the rate of transmission R satisfies

$$R \le C_{ergodic}(\gamma) \triangleq \max_{\mathbf{K}_x \mid \text{tr}\{\mathbf{K}_x\}=P} \mathbb{E}_{\mathbf{H}_c}\big[C_{MIMO}(\gamma, \mathbf{H}_c, \mathbf{K}_x)\big]. \tag{1.5}$$

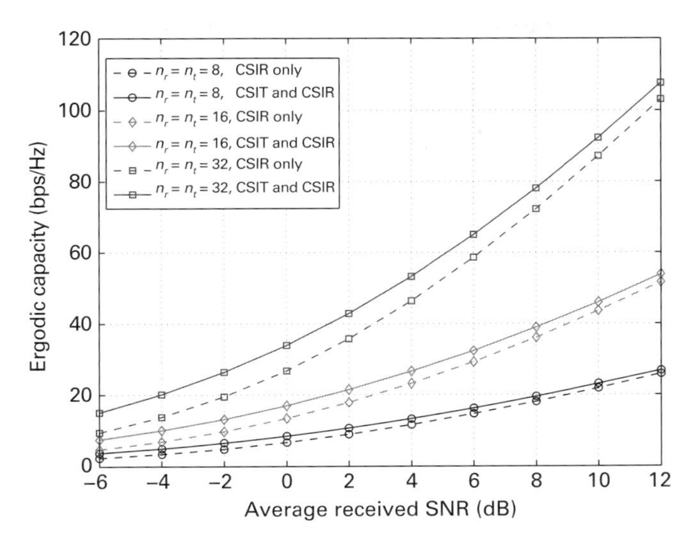

Figure 1.3 Ergodic MIMO capacity for increasing $n_t = n_r$ with (i) CSIR only, and (ii) CSIT and CSIR.

$C_{ergodic}(\gamma)$ is often referred to as the ergodic MIMO capacity, and is achieved with $\mathbf{K}_x = (P/n_t)\mathbf{I}_{n_t}$. This transmit architecture is also known as the V-BLAST scheme, where the symbol streams transmitted from each transmit antenna are uncorrelated. The ergodic MIMO capacity is therefore given by

$$C_{ergodic}(\gamma) = \mathbb{E}_{\mathbf{H}_c}\Big[\log_2 \det\Big(\mathbf{I}_{n_r} + \frac{\gamma}{n_t}\mathbf{H}_c\mathbf{H}_c^H\Big)\Big]. \tag{1.6}$$

The ergodic MIMO capacity increases linearly with increasing $n_t = n_r$ [2]. In Fig. 1.3, the ergodic MIMO capacity is plotted as a function of the average received SNR, γ, for the case with CSIR only as well as for the case with CSIT and CSIR, for different values of $n_t = n_r$. It can be observed that, for a given SNR, the ergodic MIMO capacity increases linearly with $n_t = n_r$. For example, at an SNR of 6.8 dB, the ergodic capacity is 16, 32, and 64 bps/Hz for $n_t = n_r = 8, 16, 32$, respectively. This implies that, at an SNR of 6.8 dB, an $n_t = n_r$ MIMO system would have an ergodic capacity of roughly $2n_t$ bps/Hz.

1.4 MIMO communication with CSIT and CSIR

If MIMO systems are operating in time division duplex (TDD) mode or if MIMO channels are slowly varying, it is possible for both the transmitter as well as the receiver to acquire the CSI. When both CSIT and CSIR are available, the ergodic MIMO capacity is known to be achieved with independent Gaussian inputs beamformed along the right singular vectors of the channel matrix. This transforms the MIMO channel into a set of parallel non-interfering $n = \min(n_t, n_r)$ subchannels. Capacity is then achieved by waterfilling power allocation among these n subchannels [2]. Note that the optimal power allocation is isotropic for the case of CSIR-only.

When CSIT is available, it is possible to use the available power judiciously by allocating more power to the subchannel with higher channel gain. At low SNRs, the availability of CSIT in addition to CSIR has an even higher impact on the ergodic capacity when compared to the CSIR-only scenario. This is because, at low SNRs, capacity is known to increase almost linearly with SNR, and therefore with CSIT the transmitter allocates all available power to the subchannel with the highest channel gain. In contrast to this, with CSIR-only, the available power is equally divided among the subchannels, resulting in a lower achievable capacity when compared to the scenario with CSIT. At high SNRs, waterfilling power allocation distributes roughly equal power to all the subchannels. Therefore, power allocation at high SNRs is almost the same for scenarios with CSIR-only as well as those with CSIR and CSIT. This implies that at high SNRs, both CSIR-only and the CSIR and CSIT scenarios have roughly the same ergodic capacity. This can be seen in Fig. 1.3, where the "CSIT and CSIR" ergodic capacity is plotted, in addition to the "CSIR only" capacity. It can be observed that indeed, for a given $n_t = n_r$ and SNR, γ, the ergodic capacity with "CSIT and CSIR" is more than the ergodic capacity with "CSIR only." Also, the gap between the ergodic capacity of the two scenarios reduces with increasing SNR. Another important fact, which is not highlighted in Fig. 1.3 is that, at low SNRs, the ergodic capacity with "CSIT and CSIR" is more than $n \log_2(1 + \gamma)$, which is the capacity of n parallel, independent SISO non-faded AWGN channels [6].

In slow fading channels, the codewords transmitted are subject to block fading (i.e., the channel remains almost the same for the duration of the transmitted codeword). As pointed out earlier, in such block fading scenarios, if the capacity of the channel is below the rate of transmission, there will always be a codeword error (outage) irrespective of the coding scheme used. With the availability of CSIT, however, it is possible to theoretically achieve zero outage probability by adapting the transmitted codewords (i.e., codeword rate and transmit power) for a given long-term average power constraint [21]. This leads to a variable rate transmission scheme, and also a large peak to average requirement on the transmit radio frequency (RF) amplifiers, which are undesirable in many applications. Therefore, in such applications, it is obvious that outages cannot be avoided. Hence, it is important that encoding and decoding schemes are devised to achieve high diversity and multiplexing gains. The maximum diversity gain is $n_t n_r$ and the maximum multiplexing gain is $\min(n_t, n_r)$. CSIT can be used to encode the information symbols into transmit vectors, a process commonly called "precoding." Several precoding schemes are known in the literature. Most known precoding schemes (or precoders for short) achieve either (*i*) high rate or high diversity at low complexity (e.g., linear precoders and non-linear precoders based on Tomlinson–Harashima precoding) or (*ii*) both high rate and high diversity but at high complexity (e.g., precoders based on lattice reduction techniques and vector perturbation).

Table 1.1. *Reliability and capacity of SISO, SIMO, and MIMO channels*

Number of antennas	Error probability (P_e)	Capacity (C), bps/Hz
SISO ($n_t = n_r = 1$)		
non-fading	$P_e \propto e^{-\gamma}$	$C = \log_2(1+\gamma)$
fading	$P_e \propto \gamma^{-1}$	$C = \log_2(1+\gamma)$
SIMO ($n_t = 1, n_r > 1$)		
fading	$P_e \propto \gamma^{-n_r}$	$C = \log_2(1+\gamma)$
MIMO ($n_t > 1, n_r > 1$)		
fading	$P_e \propto \gamma^{-n_t n_r}$	$C = \min(n_t, n_r) \log_2(1+\gamma)$

1.5 Increasing spectral efficiency: quadrature amplitude modulation (QAM) vs MIMO

The achievable link reliability (in terms of probability of error) and capacity in bps/Hz in SISO ($n_t = n_r = 1$), SIMO ($n_1 = 1, n_r > 1$), and MIMO ($n_t > 1, n_r > 1$) channels are summarized in Table 1.1. In non-fading SISO AWGN channels, the probability of error falls exponentially with increasing SNR, whereas the capacity grows only logarithmically with increasing SNR. With fading, the probability of error degrades to a linear fall with increase in SNR; this is a detrimental effect of fading in SISO channels. This performance degradation in fading can be alleviated by using more receive antennas, which offers receive diversity. That is, in SIMO fading channels, the probability of error falls with SNR as γ^{-n_r}. While this means better error performance in SIMO fading compared to SISO fading, the capacity of SIMO fading, like that of SISO fading, grows only logarithmically with increasing SNR. That is, in SISO and SIMO fading channels, significant power increase is needed to increase capacity. However, MIMO fading channels are attractive in terms of both achievable reliability as well as capacity. The probability of error in MIMO channels falls with SNR as $\gamma^{-n_t n_r}$, which approaches an exponential fall for large n_t, n_r. More importantly, the MIMO channel capacity increases linearly with the minimum of the number of transmit and receive antennas, which is much better than the logarithmic increase in capacity with increasing SNR.

Spectral efficiency in communication systems can be increased by increasing the size of the modulation alphabet (e.g., increasing M in M-QAM), or increasing the number of spatial dimensions for signaling (i.e., increasing n_t), or a combination of both. To achieve a given spectral efficiency, using a small modulation alphabet size and increasing number of antennas is more power efficient than using a small number of antennas and increasing the modulation alphabet size. This can be illustrated as follows. Consider the communication systems in Fig. 1.4. The system in Fig. 1.4(a) uses one transmit antenna and 64-QAM to achieve 6 bps/Hz spectral efficiency. The system in Fig. 1.4(b) achieves the same spectral efficiency using six transmit antennas and binary phase shift keying

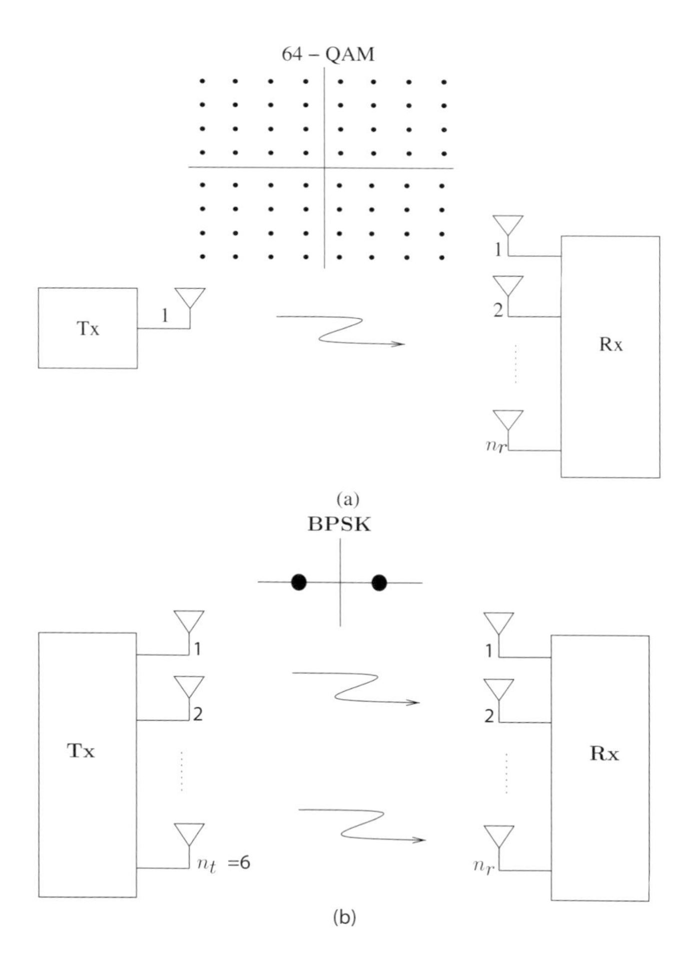

Figure 1.4 Communication systems with 6 bps/Hz spectral efficiency: (a) SISO/SIMO with 64-QAM. (b) MIMO with $n_t = 6$ and BPSK. (Rx: receiver; Tx: transmitter.)

(BPSK). The achieved bit error rates (BERs) (p_e) versus SNR (γ) performances of these systems are compared in Fig. 1.5. The performance of the 64-QAM with $n_t = n_r = 1$ is least power efficient. As mentioned earlier, in this SISO fading case p_e falls as γ^{-1}. This can be seen by noting that for a 10 dB increase in γ, the p_e falls by one order: e.g., p_e improves from 2×10^{-3} at $\gamma = 35$ dB to 2×10^{-4} at $\gamma = 45$ dB. By increasing the number of receive antennas from $n_r = 1$ to $n_r = 6$, keeping $n_t = 1$ and 64-QAM, the performance improves due to receive diversity. However, the MIMO system with $n_t = n_r = 6$ and BPSK significantly outperforms the SIMO and SISO systems with 64-QAM. At a p_e of 10^{-3}, the $n_t = n_r = 6$ MIMO system with BPSK is power efficient by more than 6 dB compared to the $n_t = 1, n_r = 6$ SIMO system with 64-QAM. This illustrates the better power efficiency of MIMO systems compared to SIMO and SISO systems with the same spectral efficiency. Therefore, increasing the num-

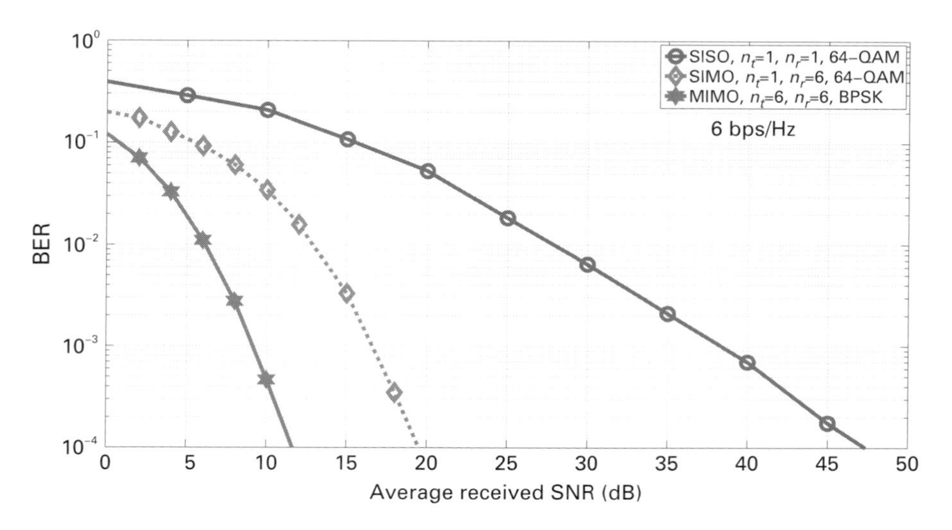

Figure 1.5 Performance of SISO, 1×6 SIMO and 6×6 MIMO systems, all at 6 bps/Hz spectral efficiency.

ber of spatial dimensions for signaling is a preferred choice to increase spectral efficiency.

1.6 Multiuser MIMO communication

Several wireless and mobile applications are cast in multiuser communication settings, where wireless resources like spectrum and power are shared to enable communication between centralized communication terminals (e.g., BSs, access points, etc.) and multiple user terminals (e.g., phones, laptops). Commercial multiuser wireless systems with single-antenna communication/user terminals have been in popular use for quite some time now. Access techniques like frequency division multiple access (FDMA), time division multiple access (TDMA), code division multiple access (CDMA), and orthogonal frequency division multiple access (OFDMA), and coding techniques like convolutional codes, turbo codes, and low density parity-check (LDPC) codes have been used in such systems. MIMO techniques applied to multiuser communications are increasingly being adopted in wireless systems and standards. A key motivation to adopt MIMO techniques in multiuser communications is the high sum capacities (maximum aggregation of all users' rates) that are possible with MIMO.

Commonly studied multiuser channels include the broadcast channel (BC), the multiple access channel (MAC) and relay channels. In a broadcast channel, a transmitter (e.g., a BS) sends data which contains information for all users in the system. The users extract information pertinent to them. In such multiuser communication scenarios, the set of all rates achievable by the users is called the rate region. Given a multiuser channel (which comprises the individual channel gains from the transmitter to each user), there exists a corresponding rate region.

The rate region for the Gaussian BC has been characterized, and it is known that a subset of this region can be achieved by dirty paper coding (DPC) [22],[23]. The sum capacity, which is the maximum aggregation of all users' data rates, grows linearly with the minimum of the number of antennas n_t and the number of single-antenna users n_u, provided the transmitter and receivers all know the channel. The sum capacity was shown to be achievable by DPC [24],[25]. In DPC, the information is coded in such a way that, despite interference from other users, each user can still receive its information perfectly as if there were no interference at all. DPC is, however, not suited to practical implementations due to high complexity. Hence, practical encoding and decoding schemes are required to achieve rates in the BC rate region. A simple idea that seems to give reasonable performance is that of prenulling the interference from users by precoding the information vector with the inverse of the channel matrix. This precoder (also known as the zero-forcing (ZF) precoder) does indeed prevent interference, but the penalty to be paid is in the increase of average transmit power (particularly when the channel is ill conditioned). Due to this, the ZF precoder is known to achieve poor diversity order in fading channels. Most other low complexity precoders also suffer from loss in performance. Vector perturbation (VP) based techniques were proposed as a low complexity alternative to DPC [26],[27]. VP techniques are known to achieve good performance. In VP based techniques, the precoder matrix is still the channel inverse matrix. However, prior to ZF, the information symbol vector is perturbed by an integer vector in such a way that the transmit power requirement is minimized. The optimal integer vector is usually searched using sphere encoding or lattice reduction techniques. Low complexity search techniques can further reduce complexity to suit large multiuser MIMO systems [12].

When the number of antennas at the BS transmitter is made significantly larger than the number of downlink users, simple linear precoders can achieve very good performance because of the additional spatial dimensions available at the transmitter. Likewise, on MAC channels with fading, providing the BS receiver with a large number of receive antennas results in large dimension spatial signatures towards users, which can be exploited to achieve good detection performance using low complexity MUD algorithms at the BS receiver.

1.7 Organization of the book

The main focus of this book is MIMO systems with a large number of antennas. Since several earlier books on MIMO have already treated MIMO systems with a small number of antennas in detail, we will keep our discussions on such small systems only to an extent that is relevant for this book. Many chapters exclusively deal with techniques and performance of large MIMO systems, in line with our intent and motivation to write this book.

In Chapter 2, we introduce large MIMO systems in the context of both point-to-point as well as multiuser scenarios, their potential benefits, and the several technological challenges in realizing them in practice. The channel hardening effect that happens in large dimensional MIMO channels, an effect that can simplify large MIMO signal processing significantly, is presented. Issues related to the placement of a large number of antennas and RF chains, large MIMO signal processing, and multicell operation, and possible solution approaches to address these issues are highlighted.

Chapters 3 and 4 discuss MIMO encoding and detection techniques, respectively. These two chapters mainly summarize various techniques and algorithms well known in the MIMO literature, and hence are kept brief. In addition to the traditional MIMO encoding techniques like spatial multiplexing and space-time coding, we have introduced a section on "spatial modulation," a relatively new modulation technique suited for multiantenna systems. In particular, the generalized spatial modulation scheme exhibits some attributes which are beneficial in large MIMO systems. The MIMO detection algorithms presented in Chapter 4 are well known, but may not be suited for large MIMO systems, on account of either inadequate performance or high complexity.

Chapters 5–8 present a detailed treatment of a set of algorithms suited for large MIMO signal detection. These algorithms play a crucial role in enabling large MIMO signal processing at low complexities in practical systems. A common feature in all these algorithms is scalability and near-optimal performance in large dimensions. While the basic versions of these algorithms achieve near-optimal performance for binary modulations like BPSK and 4-QAM, suitable variants of the basic algorithms need to be devised in order to achieve near-optimal performance in higher-order QAM as well. We have presented the rationale behind some of these variants and their successes in achieving near-optimal performance. Approximations that exploit the channel hardening effect in large dimensions and suitable stopping criteria in these algorithms are instrumental in achieving low complexities. While we have presented these algorithms in the context of point-to-point MIMO systems in Chapters 5–7, they are applicable in multiuser uplink as well. As an illustration of this point, we have presented the algorithm in Chapter 8 in a multiuser uplink setting. In Chapter 5, we present algorithms based on local search techniques and metaheuristics. These include likelihood ascent search (LAS) and reactive tabu search (RTS) algorithms, and their variants. Chapter 6 presents an algorithm based on probabilistic data association (PDA). Chapters 7 and 8 present algorithms based on belief propagation (BP) and Markov chain Monte Carlo (MCMC) techniques, respectively.

Channel knowledge is essential for detection, precoding, and other functions. In Chapter 9, we discuss the training requirement for channel estimation in large point-to-point and multiuser MIMO systems. We also present channel estimation techniques and their performance in large MIMO systems. Chapter 10 covers precoding techniques in large MIMO systems. It discusses the pilot contamination problem in multicell operation, and presents multicell precoding using BS cooperation to address this problem.

Channel models are useful and very much needed for analysis, simulation, development, and testing of any communication system. The topic of MIMO channel models and large MIMO channel sounding measurements is covered in Chapter 11. References are made to some large MIMO channel sounding campaigns and the results obtained from these campaigns. Compact antenna designs using planar inverted F antennas (PIFA) and MIMO cubes are described. Several large MIMO testbeds and prototypes have started to appear in the literature, confirming the view that large MIMO systems are indeed practical. In Chapter 12, we capture the details of some of these testbeds and the achieved results.

References

[1] G. J. Foschini, "Layered space-time architecture for wireless communication in a fading environment when using multi-element antennas," *Bell Labs. Tech. J.*, vol. 1, no. 2, pp. 41–59, 1996.

[2] I. E. Telatar, "Capacity of multi-antenna Gaussian channels," *European Trans. Telecommun.*, vol. 10, no. 6, pp. 585–595, Nov. 1999.

[3] G. J. Foschini and M. J. Gans, "On limits of wireless communications in a fading environment when using multiple antennas," *Wireless Pers. Commun.*, vol. 6, pp. 311–335, Mar. 1998.

[4] A. Paulraj, R. Nabar, and D. Gore, *Introduction to Space-Time Wireless Communications*. Cambridge, UK: Cambridge University Press, 2003.

[5] H. Jafarkhani, *Space-Time Coding: Theory and Practice*. Cambridge, UK: Cambridge University Press, 2005.

[6] D. Tse and P. Viswanath, *Fundamentals of Wireless Communication*. Cambridge, UK: Cambridge University, 2005.

[7] H. Bolcskei, D. Gesbert, C. B. Papadias, and A.-J. van der Veen (Ed), *Space-Time Wireless Systems: From Array Processing to MIMO Communications*. Cambridge, UK: Cambridge University Press, 2006.

[8] H. Taoka and K. Higuchi, "Field experiments on 5-Gbit/s ultra-highspeed packet transmission using MIMO multiplexing in broadband packet radio access," *NTT DoCoMo Tech. J.*, vol. 9, no. 2, pp. 25–31, Sep. 2007.

[9] Y.-C. Liang, S. Sun, and C. K. Ho, "Block-iterative generalized decision feedback equalizers for large MIMO systems: algorithm design and asymptotic performance analysis," *IEEE Trans. Signal Process.*, vol. 54, no. 6, pp. 2035–2048, Jun. 2006.

[10] K. V. Vardhan, S. K. Mohammed, A. Chockalingam, and B. S. Rajan, "A low-complexity detector for large MIMO systems and multicarrier CDMA systems," *IEEE J. Sel. Areas Commun.*, vol. 26, no. 3, pp. 473–485, Apr. 2008.

[11] S. K. Mohammed, A. Zaki, A. Chockalingam, and B. S. Rajan, "High-rate space-time coded large-MIMO systems: low-complexity detection and channel estimation," *IEEE J. Sel. Topics Signal Proc.*, vol. 3, no. 6, pp. 958–974, Dec. 2009.

[12] S. K. Mohammed, A. Chockalingam, and B. S. Rajan, "A low-complexity precoder for large multiuser MISO systems," in *IEEE VTC'2008*, Marina Bay, May 2008, pp. 797–801.

[13] F. Rusek, D. Persson, B. K. Lau, *et al.*, "Scaling up MIMO: opportunities and challenges with very large arrays," *IEEE Signal Proc. Mag.*, vol. 30, no. 1, pp. 40–60, Jan. 2013.

[14] J. G. Proakis, *Digital Communications*. New York, NY: McGraw-Hill, 2000.

[15] L. Zheng and D. Tse, "Diversity and multiplexing: a fundamental tradeoff in multiple-antenna channels," *IEEE Trans. Inform. Theory*, vol. 49, no. 5, pp. 1073–1096, May 2003.

[16] P. W. Wolniansky, G. J. Foschini, G. D. Golden, and R. A. Valenzuela, "V-BLAST: an architecture for realizing very high data rates over the rich-scattering wireless channel," in *URSI Intl. Symp. Signals, Systems and Electronics (ISSSE)*, Pisa, Sept–Oct. 1998, pp. 295–300.

[17] S. M. Alamouti, "A simple transmit diversity technique for wireless communications," *IEEE J. Sel. Areas Commun.*, vol. 16, no. 8, pp. 1451–1458, Oct. 1998.

[18] V. Tarokh, H. Jafarkhani, and A. R. Calderbank, "Space-time block codes from orthogonal designs," *IEEE Trans. Inform. Theory*, vol. 45, no. 5, pp. 1456–1467, Jul. 1999.

[19] B. A. Sethuraman, B. S. Rajan, and V. Shashidhar, "Full-diversity high-rate space-time block codes from division algebras," *IEEE Trans. Inform. Theory*, vol. 49, no. 10, pp. 2596–2616, Oct. 2003.

[20] F. Oggier, J.-C. Belfiore, and E. Viterbo, "Cyclic division algebras: a tool for space-time coding," *Foundations and Trends in Commun. and Inform. Theory*, vol. 4, no. 1, pp. 1–95, Oct. 2007.

[21] G. Caire, G. Taricco, and E. Biglieri, "Optimum power control over fading channels," *IEEE Trans. Commun.*, vol. 51, no. 8, pp. 1389–1398, Aug. 2003.

[22] M. H. M. Costa, "Writing on dirty-paper," *IEEE Trans. Inform. Theory*, vol. 29, no. 3, pp. 439–441, May 1983.

[23] C. B. Peel, "On dirty-paper coding," *IEEE Signal Proc. Mag.*, vol. 20, no. 3, pp. 112–113, May 2003.

[24] S. Vishwanath, N. Jindal, and A. Goldsmith, "Duality, achievable rates and sum-rate capacity of MIMO broadcast channels," *IEEE Trans. Inform. Theory*, vol. 49, no. 10, pp. 2658–2668, Oct. 2003.

[25] P. Viswanath and D. Tse, "Sum capacity of the vector Gaussian broadcast channel and uplink-downlink duality," *IEEE Trans. Inform. Theory*, vol. 49, no. 8, pp. 1912–1921, Aug. 2003.

[26] C. B. Peel, B. M. Hochwald, and A. L. Swindlehurst, "A vector-perturbation technique for near-capacity multi-antenna multiuser communication – part i: channel inversion and regularization," *IEEE Trans. Commun.*, vol. 53, no. 1, pp. 195–202, Jan. 2005.

[27] ——, "A vector-perturbation technique for near-capacity multi-antenna multiuser communication - part ii: perturbation," *IEEE Trans. Commun.*, vol. 53, no. 3, pp. 537–544, Mar. 2005.

2 Large MIMO systems

Large MIMO systems are systems which use tens to hundreds of antennas in communication terminals. Depending on the application scenario, different MIMO system configurations are possible. These include point-to-point MIMO and multiuser MIMO configurations. In multiuser MIMO, point-to-multipoint (e.g., downlink in cellular systems), and multipoint-to-point (e.g., uplink in cellular systems) configurations are common. In a point-to-point communication scenario (Fig. 1.1), the number of transmit antennas n_t at the transmitter and the number of receive antennas n_r at the receiver can be large. A typical application scenario for a point-to-point large MIMO configuration is providing high-speed wireless backhaul connectivity between BSs using multiple antennas at each BS. Since space constraint need not be a major concern at the BSs, a large number of antennas can be used at both the transmit and receive BSs in this application scenario.

In multiuser MIMO (Fig. 1.2), the communication is between a BS and multiple user terminals. These user terminals can be small devices like mobile/smart phones or medium sized terminals like laptops, set-top boxes, TVs, etc. In mobile applications where mobile/smart phones and personal digital assistants are the user terminals, only a limited number of antennas can be mounted on them because of space constraints. However, in applications involving user terminals like TVs, set-top boxes, and laptops a larger number of antennas can be used on the user terminal side as well. Regardless of the size of the user terminals and the number of antennas that can be accommodated in them, use of tens to hundreds of antennas at the BS end in multiuser MIMO is not difficult. In such cases, the greater the number of antennas at the BS, the greater can be the spatial degrees of freedom available to perform precoding on the downlink and detection on the uplink.

2.1 Opportunities in large MIMO systems

There are several benefits in using a large number of antennas. Fundamentally, using more antennas creates more degrees of freedom in the spatial domain, which can be gainfully exploited for several purposes including, for example, increasing the data rate without increasing bandwidth and increasing link reliability through spatial diversity [1]. More specifically, in a point-to-point MIMO system with n_t transmit and n_r receive antennas, the probability of link outage behaves as

$$P_{outage} \propto \mathrm{SNR}^{-n_t n_r},$$

indicating a potential to achieve a diversity order of $n_t n_r$. That is, with large n_t, n_r, the MIMO link performance in terms of error rate can approach an exponential fall in error rate with increase in SNR. Also, the achievable rate scales as

$$\min(n_t, n_r) \log_2(1 + \mathrm{SNR}),$$

which indicates the possibility of achieving very high data rates using large n_t, n_r without increasing the bandwidth. The rate gains in multiuser MIMO with a large number of antennas can also be substantial [1]. For example, space division multiple access (SDMA) in a multiuser uplink becomes quite attractive when a large number of receive antennas are used at the BS. The larger the number of BS receive antennas, the larger can be the number of uplink users supported in the system. Also, a large number of transmit antennas at the BS in multiuser downlink can allow the use of simple precoding methods and flexible user selection and scheduling.

2.2 Channel hardening in large dimensions

Though the most obvious advantages of large MIMO systems are increased data rate and diversity gain, the large dimensionality they offer can also result in a host of other advantages that do not come with smaller systems. As the $n_r \times n_t$ channel matrix $\mathbf{H}$ becomes larger (i.e., both n_t and n_r increase, keeping their ratio fixed), the distribution of its singular values becomes less sensitive to the actual distribution of the entries of the channel matrix (as long as they are independent and identically distributed (iid)). This is a result of the Marčenko–Pastur law, which states that if the entries of an $n_r \times n_t$ matrix $\mathbf{H}$ are zero mean iid with variance $1/n_r$, then the empirical distribution of the eigenvalues of $\mathbf{H}^H\mathbf{H}$ converges almost surely, as $n_t, n_r \to \infty$ with $n_t/n_r \to \beta$, to the density function [2]

$$\mathsf{f}_\beta(x) = \left(1 - \frac{1}{\beta}\right)^+ \delta(x) + \frac{\sqrt{(x-a)^+(b-x)^+}}{2\pi\beta x}, \tag{2.1}$$

where $(z)^+ = \max(z, 0)$, $a = (1 - \sqrt{\beta})^2$, and $b = (1 + \sqrt{\beta})^2$. In a similar way, the empirical distribution of the eigenvalues of $\mathbf{H}\mathbf{H}^H$ converges to

$$\tilde{\mathsf{f}}_\beta(x) = (1 - \beta)\delta(x) + \beta \mathsf{f}_\beta(x). \tag{2.2}$$

Equations (2.1) and (2.2) are plotted in Figs. 2.1(a) and (b), respectively, for different values of β. Note that the mass points at 0 are not plotted.

An effect of the Marčenko–Pastur law is that very tall or very wide channel

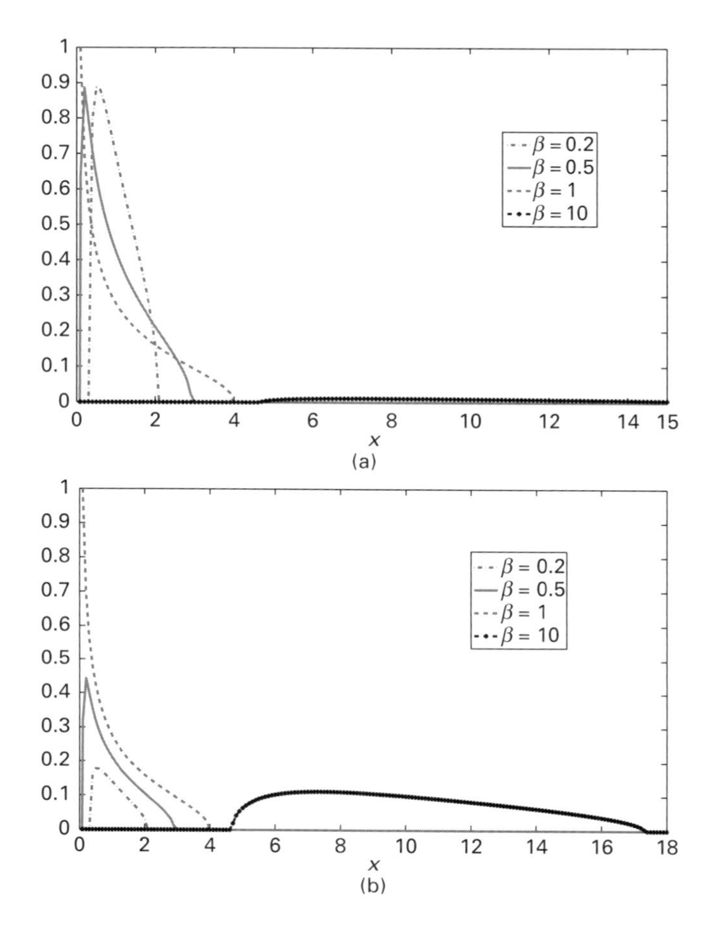

Figure 2.1 Marčenko–Pastur density function for (a) $\mathbf{H}^H\mathbf{H}$ and (b) $\mathbf{H}\mathbf{H}^H$. The mass at zero points are not shown.

matrices are very well conditioned. This can be seen from Figs. 2.1(a) and (b) for $\beta = 0.2$ and $\beta = 10$, as the support of the non-zero eigenvalues of $\mathbf{H}^H\mathbf{H}$ and $\mathbf{H}\mathbf{H}^H$ moves away from zero. The Marčenko–Pastur law also implies that the channel "hardens," meaning that the eigenvalue histogram of a single realization converges to the average asymptotic eigenvalue distribution. In this sense, the channel becomes more and more deterministic as the number of antennas increases. The channel hardening behavior in large dimensions can be seen pictorially in the intensity plots of $\mathbf{H}^H\mathbf{H}$ for $n_t = n_r = 8, 32, 96, 256$ in Fig. 2.2, where the entries of $\mathbf{H}$ are iid Gaussian entries with zero mean and unit variance. Figure 2.2 shows that, as the size of $\mathbf{H}$ increases, the diagonal terms of $\mathbf{H}^H\mathbf{H}$ become increasingly larger in magnitude than the off-diagonal terms.

Channel hardening results in several advantages in large dimensional signal

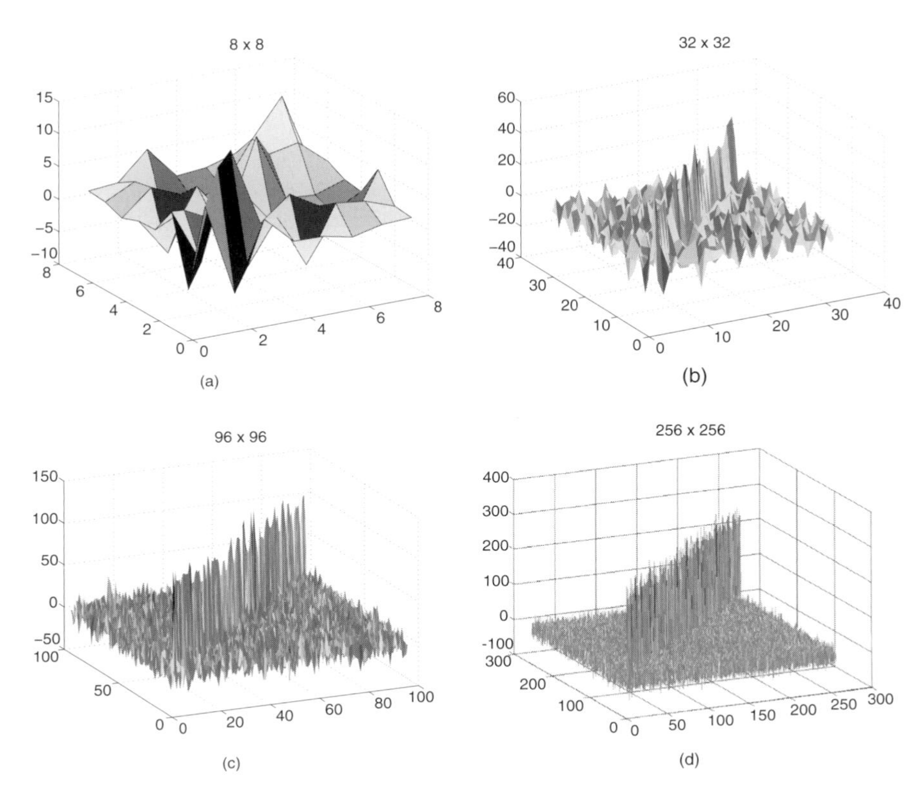

Figure 2.2 Intensity plots of $\mathbf{H}^H\mathbf{H}$ matrices for (a) 8×8 MIMO, (b) 32×32 MIMO, (c) 96×96 MIMO, and (d) 256×256 MIMO channels.

processing. For example, linear detectors like ZF and minimum mean square error (MMSE) detectors need to perform matrix inversions. Inversion of large random matrices can be done quickly, using series expansion techniques [3]. Because of channel hardening, approximate matrix inversions using series expansion and deterministic approximations from the limiting distribution become effective in large dimensions. Also, channel hardening can allow simple detection methods/algorithms to achieve a very good performance in large dimensions. Such low complexity detection algorithms suited for large dimensions are treated in Chapters 5–8.

2.3 Technological challenges and solution approaches

Current wireless standards including IEEE 802.11n/11ac (WiFi) and 3GPP LTE-A have adopted MIMO techniques to achieve increased spectral efficiency and reliability. These standards can harness only some of the potential benefits of MIMO, since they currently use only a small number of antennas (e.g., 2–8 antennas) and achieve spectral efficiencies of only about 15 bps/Hz or less. However, significant benefits can be realized if a larger number of antennas is used; e.g., large MIMO systems with tens to hundreds of antennas can enable multigigabit rate transmissions at high spectral efficiencies of the order of tens to hundreds of bps/Hz. However, several technical challenges need to be addressed in realizing such large MIMO systems. Some of the key challenges include:

- the availability of independent spatial dimensions in real-world channels,
- the placement of a large number of antennas and RF chains,
- large MIMO signal processing at practically implementable low complexities,
- issues in multicell operation.

2.3.1 Availability of independent spatial dimensions

The number of independent spatial dimensions in real-world channels is limited by the amount/richness of scattering around the wireless communication nodes. Even in the presence of rich scattering, things like the pin-hole effect can arise, where all the paths from the transmit to receive antennas go through a common pin-hole [4]. Such effects can result in a reduced number of independent spatial dimensions (i.e., low rank channel matrices). Also, the spacing between the antennas in the communication terminal is crucial in determining the number of independent spatial dimensions. Inadequate spacing between antenna elements causes spatial correlation which can degrade MIMO channel capacity. While medium-/large-sized communication terminals can accommodate many antennas providing adequate spacing between antenna elements, compact antenna array designs are needed to mount more antenna elements in user terminals with space constraints. This point is discussed in the next subsection.

Rich scattering is common in typical indoor and outdoor urban environments. Determination of the rank structure and other statistical characteristics of large MIMO channels requires intensive channel sounding measurements in different physical environments (indoor, outdoor, urban, semi-urban, rural, etc.) and different frequency bands of interest (e.g., 2.5 GHz, 5 GHz, 11 GHz, etc.). Interestingly, some of the large MIMO channel sounding measurement campaigns (e.g., with 128 antenna arrays) have indicated that in spite of significant differences between the theoretical full rank channel models and the experimentally measured channels, a large proportion of the theoretical performance gains of large antenna arrays could be achieved in practice [5]–[7]. More details on early as well as more current large MIMO channel sounding campaigns are presented in Section 11.4.

2.3.2 Placement of a large number of antennas and RF chains

The placement of a large number of antenna elements in communication terminals can be a challenging issue, particularly when the terminals are small. For a given antenna aperture constraint, increasing the number of antenna elements decreases the inter-element spacing which increases the spatial correlation. This, in turn, can degrade MIMO capacity. As a rule of thumb, a spacing of more than $\lambda/2$, where λ is the carrier wavelength, is considered to be necessary to achieve almost no correlation between antennas. The following approaches can help to alleviate the antenna and RF chain placement issue.

- *Use higher carrier frequencies* One approach to enable the placement of more antennas in a given antenna aperture is to use higher carrier frequencies. Since higher carrier frequencies have smaller wavelengths (e.g., $\lambda/2 = 3$ cm in 5 GHz), more antennas can be mounted in a given antenna aperture. Also, operation at 11 GHz, 30 GHz, and 60 GHz carrier frequencies can be attractive in terms of the antenna placement issue, since the carrier wavelengths in these frequencies are in millimeters.
- *Exploit volume* Another approach is to mount antennas not only in one or two dimensions (as in linear and planar arrays) but in a compact volume. Placing multiple antenna elements in cubic structures (referred to as MIMO cubes) is a promising approach. More details on MIMO cube antennas are given in Section 11.5.3.
- *Compact antenna arrays* Compact antenna arrays are antenna arrays with an inter-element spacing of less than half the wavelength. The design of arrays that are compact yet demonstrate acceptable mutual coupling and radiation efficiency is another interesting approach to address the antenna placement issue in large MIMO systems. The use of PIFAs as the basic elements in compact arrays is appealing. The topic of compact antenna arrays is discussed in Section 11.5.
- *Spatial modulation (SM)* The number of transmit RF chains at the MIMO transmitter can be reduced without compromising on the spectral efficiency using SM, a relatively new modulation scheme suitable for multiantenna communications [8],[9]. It reduces RF hardware complexity, size, and cost. SM is introduced in Chapter 3.

2.3.3 Low complexity large MIMO signal processing

Low complexity signal processing algorithms for synchronization, signal detection, precoding, channel estimation, and channel decoding are key to the practical implementation of large MIMO systems.

- *MIMO detection* The MIMO detector at the receiver, whose job is to recover the symbols that are transmitted simultaneously from multiple transmitting antennas, is often a bottleneck in terms of overall performance and complexity. Complexities involved in optimum detectors based on the maximum likelihood (ML) or the maximum a posteriori probability (MAP) criterion are exponential in the number of transmit antennas, and hence are prohibitive for large MIMO systems. Widely known detectors in the literature either perform well but do not scale in complexity (e.g., the sphere decoder (SD) [10] and variants) or scale well in complexity but perform poorly in large systems (e.g., linear detectors like ZF, MMSE detectors). Fortunately, the channel hardening behavior witnessed in large matrices (discussed in Section 2.2) becomes helpful. Several low complexity detection algorithms based on local search, metaheuristics, BP, and sampling techniques have shown promising performance and complexity attributes suited for large MIMO systems. They have the same complexity orders as linear detectors yet exhibit near-optimal performance, particularly when applied for signal detection in large dimensions. This new generation of MIMO detection algorithms is treated in detail in Chapters 5–8.

- *LDPC codes* Another interesting aspect of detection using message passing algorithms like in BP large dimensions is that these graphical model based algorithms can be naturally combined with turbo or LDPC decoding algorithms (which are also graphical model based algorithms) to achieve joint processing of detection and decoding in large MIMO systems. Such a joint detection–decoding approach allows one to design good LDPC codes tailored for large MIMO channels. In particular, LDPC codes matched to large MIMO channels can be designed using the extrinsic information transfer (EXIT) behavior of the joint detection–decoding message passing receiver. In large MIMO channels, such specially designed LDPC codes outperform off-the-shelf LDPC codes designed for other types of channels. LDPC code design for large MIMO systems is covered in Section 7.8.
- *SM* This is a relatively new modulation scheme suited for multiantenna communications [8],[9]. It is a promising technique for large MIMO systems as it allows the use of fewer transmit RF chains than the number of transmit antennas without compromising on the spectral efficiency. This reduces RF hardware complexity, size, and cost in large MIMO systems. A novel aspect of this modulation scheme is that it conveys information in the indices of the chosen antennas for transmission, in addition to information conveyed through conventional modulation alphabets like QAM. Chapter 3 introduces SM as a MIMO encoding scheme. The low complexity detection algorithms in Chapters 5–8 can be used to detect spatial modulation signals.
- *Single-carrier communication* Single-carrier communication techniques are increasingly preferred over multicarrier techniques like OFDM/OFDMA because of the high peak-to-average power ratio (PAPR) in multicarrier systems. The LTE standard already employs single-carrier frequency division multiple access (SC-FDMA) on the uplink. SC-FDMA can be beneficially employed in multiuser MIMO downlink as well [11]. While SC-FDMA offers PAPR and performance advantage over OFDMA, the receiver complexity will be greater in SC-FDMA because of the need to perform equalization. However, SC-FDMA equalization complexity in large MIMO systems can be addressed by using the detection algorithms in Chapters 5–8 for the purpose

of low complexity frequency domain equalization on the large dimension received signal vectors.

- *Channel estimation* Channel estimation schemes that provide CSI at the receiver and transmitter play an important role in MIMO receivers (for signal detection) and MIMO transmitters (for precoding). In pilot-aided channel estimation schemes, the number of training slots needed for pilot transmission grows linearly with the number of transmit antennas. This reduces the data throughput. However, with large channel coherence times, which are typical in slow fading channels (e.g., fixed wireless channels), this throughput loss due to pilot transmission can be less. Channel estimation in high mobility scenarios and the feedback requirements to send a large number of estimated channel coefficients from the receiver to the transmitter in frequency division duplex (FDD) systems are issues that need investigation in the context of large MIMO systems. The topic of channel estimation for large MIMO systems is treated in Chapter 9.
- *Precoding* The availability of CSI at the transmitter allows the precoding of signals at the transmitter. With a large number of antennas at the BS for multiuser communication on the downlink, the available spatial degrees of freedom can be exploited to design simple precoders and to reduce PAPR [12],[13]. Precoding techniques in large point-to-point MIMO and large multiuser MIMO settings are treated in Chapter 10.

2.3.4 Multicell operation

In addition to the various wireless link level issues and challenges discussed above, several system level issues have to be addressed in large MIMO systems. An important area in this regard is multicell operation. Issues related to cell sizing, frequency/resource allocation across cells, interference management in general and managing inter-cell interference in particular can bring new challenges in the deployment of large MIMO systems. One powerful and emerging technique that can address multicell related issues is BS cooperation. For example, multicell precoding through BS cooperation can be an effective inter-cell interference management technique. BS cooperation has already been adopted in wireless standards like LTE – e.g., coordinated multipoint (COMP), which enables dynamic coordination of transmission and reception over a variety of different BSs. BS cooperation is expected to play a crucial role in large MIMO systems. Another issue that arises in multicell operation is pilot contamination. The pilot contamination problem is an inter-cell interference problem encountered when non-orthogonal pilot sequences are used for uplink channel estimation in multicell systems [14]. Here again, BS cooperation can help to alleviate the pilot contamination problem. The pilot contamination problem and addressing it through multicell precoding using BS cooperation are covered in Section 10.3.

Notwithstanding the various technological challenges discussed above, several real-world large MIMO testbeds have started to emerge in the literature. These testbeds adopt different large MIMO architectures/configurations, use different numbers of antennas, frequency bands, and bandwidths, and are deployed and tested in different physical environments (e.g., indoor, outdoor). A summary of these large MIMO testbeds is presented in Chapter 12. A common theme, however, is that in practice all these testbeds demonstrate very high spectral efficiencies using large numbers of antennas, advancing the practical realization of MIMO benefits to the next level.

References

[1] D. Tse and P. Viswanath, *Fundamentals of Wireless Communication.* Cambridge, UK: Cambridge University, 2005.

[2] A. Tulino and S. Verdu, *Random Matrix Theory and Wireless Communications.* Foundations and Trends in Communications and Information Theory. Delft, The Netherlands: Now Publishers, Inc., 2004.

[3] S. Moshavi, E. Kanterakis, and D. Schilling, "Multistage linear receivers for DS-CDMA systems," *Intl. J. Wireless Inf. Netw.*, vol. 3, no. 1, pp. 1–17, Jan. 1996.

[4] D. Gesbert, H. Bolcskei, D. A. Gore, and A. J. Paulraj, "Outdoor MIMO wireless channels: models and performance prediction," *IEEE Trans. Commun.*, vol. 50, no. 12, pp. 1926–1934, Dec. 2002.

[5] X. Gao, O. Edfors, F. Rusek, and F. Tufvesson, "Linear pre-coding performance in measured very-large MIMO channels," in *IEEE VTC'2011 Fall*, San Francisco, CA; Sep. 2011, pp. 1–5.

[6] S. Payami and F. Tufvesson, "Channel measurements and analysis for very large array systems at 2.6 GHz," in *European Conf. Antennas and Prop. (EU-CAP'2012)*, Prague, Mar. 2012, pp. 433–437.

[7] J. Hoydis, C. Hoek, T. Wild, and S. ten Brink, "Channel measurements for large antenna arrays," in *Intl. Symp. on wireless Commun. Sys. (ISWCS)*, Aug. 2012, pp. 811–815.

[8] M. D. Renzo, H. Haas, and P. M. Grant, "Spatial modulation for multiple-antenna wireless systems: a survey," *IEEE Commun. Mag.*, pp. 182–191, Dec. 2001.

[9] R. Y. Mesleh, H. Haas, C. W. Ahn, and S. Yun, "Spatial modulation," *IEEE Trans. Veh. Tech.*, vol. 57, no. 4, pp. 2228–2241, Jul. 2008.

[10] E. Viterbo and J. Boutros, "A universal lattice code decoder for fading channels," *IEEE Trans. Inform. Theory*, vol. 45, no. 5, pp. 1639–1642, Jul. 1999.

[11] H. S. Eshwaraiah and A. Chockalingam, "SC-FDMA for multiuser communication on the downlink," in *Intl. Conf. on Commun. Syst. and Netw. (COMSNETS'2013)*, Bangalore, Jan. 2013, pp. 1–7.

[12] S. K. Mohammed, A. Chockalingam, and B. S. Rajan, "A low-complexity precoder for large multiuser MISO systems," in *IEEE VTC'2008*, Singapore, May 2008, pp. 797–801.

[13] C. Studer and E. G. Larsson, "PAR-aware large-scale multi-user MIMO-OFDM downlink," *IEEE J. Sel. Areas in Commun.*, vol. 31, no. 2, pp. 303–313, Feb. 2013.

[14] J. Jose, A. Ashikhmin, T. Marzetta, and S. Vishwanath, "Pilot contamination and precoding in multi-cell TDD systems," *IEEE Trans. Wireless Commun.*, vol. 10, no. 8, pp. 2640–2651, Aug. 2011.

3 MIMO encoding

The job of MIMO encoding is to map the input symbols, say, from a modulation alphabet, to symbols to be transmitted over multiple transmit antennas. Spatial multiplexing and space-time coding are two well-known MIMO encoding techniques [1],[2]. Spatial modulation (SM) is a more recently proposed scheme for multiantenna communications [3]. These MIMO encoding schemes do not require any knowledge of the CSI at the transmitter, and hence are essentially "open-loop" schemes. MIMO encoding using CSI at the transmitter is referred to as MIMO precoding, which is treated in Chapter 10. Spatial multiplexing is an attractive architecture for achieving high rates. Space-time coding is attractive for achieving increased reliability through transmit diversity. SM serves a different purpose. It allows fewer transmit RF chains than the number of transmit antennas to be used without compromising much on the rate. This reduces the RF hardware complexity, size, and cost. In spatial multiplexing and space-time coding, information is carried on the modulation symbols. In SM, on the other hand, in addition to modulation symbols, the indices of the antennas on which transmission takes place also convey information. This is why SM does not compromise much on the rate. Among the three MIMO encoding schemes, spatial multiplexing is the simplest, and its complexity rests more at the receiver in detecting the transmitted symbol vector. SM, though simple conceptually, needs additional memory to construct the encoding table at the transmitter for selecting the antennas for transmission. Detection is more involved in SM than in spatial multiplexing, since, at the receiver in addition to detecting the modulation symbols, the indices of the transmitting antennas also need to be detected. Space-time coding is the most sophisticated of the three MIMO encoding schemes. Rich and sophisticated mathematical tools (e.g., CDA, Clifford algebra) have been applied to design space-time codes that achieve good diversity and rate. We will describe spatial multiplexing, space-time coding, and SM in some detail in the following sections.

3.1 Spatial multiplexing

Spatial multiplexing is a simple, yet very popular MIMO encoding technique. In the literature, spatial multiplexing is more popularly referred to as the V-BLAST scheme [4]. In an n_t transmit antenna system, it transmits n_t independent data streams simultaneously, one on each transmit antenna. That is, independent data streams are multiplexed in space. There is no coding across the transmit antennas, however. A key advantage of spatial multiplexing is that it utilizes all the available spatial degrees of freedom and achieves the full rate of n_t symbols per channel use. Another advantage is that it applies to systems with any number of transmit antennas. A drawback, however, is that it does not achieve the maximum spatial diversity of $n_t n_r$, where n_r is the number of receive antennas. It achieves receive diversity but not transmit diversity (because there is no coding across transmit antennas).

The transmit signal vector in a given channel use is $\mathbf{x} = [x_1\ x_2\ \cdots\ x_{n_t}]^T$, where the symbols x_k, $k = 1, \ldots, n_t$, come from a modulation alphabet. These transmitted symbols are detected jointly at the receiver using n_r receive antennas. In a V-BLAST system with $n_t = n_r = 2$, an upper bound on the pairwise error probability of a transmit vector $\mathbf{x}_1$ being decoded as $\mathbf{x}_2$ at the receiver can be expressed as [1]

$$\begin{aligned} P(\mathbf{x}_1 \to \mathbf{x}_2) &\leq \left[\frac{1}{1 + \text{SNR}\|\mathbf{x}_1 - \mathbf{x}_2\|^2/4}\right]^2 \\ &\leq \frac{16}{\text{SNR}^2\|\mathbf{x}_1 - \mathbf{x}_2\|^4}. \end{aligned} \tag{3.1}$$

The exponent in the SNR factor is the diversity gain. In the above, the achieved diversity order is only 2, whereas the maximum diversity order is $n_t n_r = 4$.

Spatial multiplexing is a commonly used MIMO encoding scheme in practical wireless systems and standards. Also, a multiuser uplink system with K single-antenna user terminals transmitting simultaneously on the same frequency to a BS with N receive antennas can be viewed as a virtual spatial multiplexing system – also referred to as SDMA [1]. SDMA users' signals are decoded at the BS receiver based on the spatial signatures they establish at the receive antenna array. SDMA is highly spectrally efficient since neither orthogonality among users' transmissions (as in TDMA and FDMA) nor code signatures which involve bandwidth expansion (as in CDMA) are needed in SDMA. All users can transmit at the same time on the same frequency in SDMA without any bandwidth expansion.

Several algorithms for detecting spatially multiplexed signals with varying levels of performance and complexity are known in the literature. Some of the well-known detection algorithms are covered in Chapter 4. These algorithms either scale well in complexity for a large number of antennas but perform poorly, or perform well but do not scale well in complexity. Advanced low complexity approaches and algorithms with good performance are needed for detection in large MIMO systems. Such algorithms suited for large MIMO signal detection are covered in Chapters 5–8.

3.2 Space-time coding

Space-time coding is another popular MIMO encoding technique, where coding is done across the transmit antennas as well as time [2]. While encoding is independent from one channel use to another in spatial multiplexing, encoding is done across multiple channel uses in space-time coding. This allows detection to be carried out on a per-channel-use basis in spatial multiplexing, whereas, in space-time coding, detection has to be carried out using signals received in multiple channel uses, which causes increased decoding latency. Also, since coding is done across transmit antennas, space-time coding can potentially achieve the maximum spatial diversity of $n_t n_r$.

Space-time codes are designed to introduce structured redundancy in space and time so as to achieve good diversity and a good rate. A space-time codeword $\mathbf{C}$ is an $n_t \times p$ array in which the rows are indexed by transmit antennas, the columns are indexed by channel uses in a frame, and the entries are the symbols to be transmitted. In designing space-time codes, a quasi-static fading model is usually assumed. This means that the fading coefficients remain constant over a frame and change independently from one frame to the next. With this assumption, the pairwise error probability of a codeword $\mathbf{C}_1$ being decoded as codeword $\mathbf{C}_2$ at the receiver is given by [2]

$$P(\mathbf{C}_1 \to \mathbf{C}_2) \leq \left(\prod_{i=1}^{n_t} \frac{1}{1 + \lambda_i \gamma_s} \right)^{n_r}, \tag{3.2}$$

where $\gamma_s = E_s/4N_0$, E_s is the average energy per complex symbol, and λ_is are the singular values of the difference matrix $\mathbf{B} = \mathbf{C}_1 - \mathbf{C}_2$. If r denotes the rank of the matrix $\mathbf{A} = \mathbf{B}\mathbf{B}^H$ and $\lambda_1, \lambda_2, \ldots, \lambda_r$ denote the non-zero eigenvalues of $\mathbf{A}$, then

$$P(\mathbf{C}_1 \to \mathbf{C}_2) \leq \left(\prod_{i=1}^{r} \lambda_i \right)^{-n_r} (\gamma_s)^{-rn_r}. \tag{3.3}$$

Thus a diversity gain of rn_r and a coding gain of $(\lambda_1 \lambda_2 \cdots \lambda_r)^{1/r}$ are achieved. The following criteria are generally used to design space-time codes.

Rank criterion Maximize the minimum rank of the matrix $\mathbf{B}(\mathbf{C}_1, \mathbf{C}_2)$ over all distinct pairs of codewords $\mathbf{C}_1$ and $\mathbf{C}_2$. If the minimum rank is r, then a diversity of rn_r is achieved.

Determinant criterion To achieve a better coding gain, the minimum of the determinant of $\mathbf{A}(\mathbf{C}_1, \mathbf{C}_2)$ taken over all pairs of distinct code words $\mathbf{C}_1$ and $\mathbf{C}_2$ must be maximized.

Two types of space-time codes, namely, space-time trellis codes (STTCs) and STBCs, are well known [2]. The rank and determinant criteria apply to the design of both STTCs and STBCs. STTCs combine modulation and trellis coding to transmit information over multiple transmit antennas, and they can be viewed as trellis coded modulation (TCM) for MIMO channels. While STTCs need the vector-Viterbi algorithm for decoding, some STBCs admit very simple decoding.

3.2.1 Space-time block codes

An (n_t, p, k) STBC is represented by a matrix $\mathbf{X}_c \in \mathbb{C}^{n_t \times p}$, where n_t and p denote the number of transmit antennas and number of channel uses, respectively, and k denotes the number of complex data symbols sent in one STBC matrix. The (i, j)th entry in $\mathbf{X}_c$ represents the complex number transmitted from the ith transmit antenna in the jth channel use. The rate of an STBC is given by $r \triangleq k/p$, which is the number of symbols per channel use. A large r is desired since more information is loaded in one STBC matrix.

Orthogonal STBCs (OSTBCs)

A matrix $\mathbf{X}$ is said to be an OSTBC if

$$\mathbf{X}^H\mathbf{X} = \left(|x_1|^2 + |x_2|^2 + \cdots + |x_k|^2\right)\mathbf{I}_{n_t}, \tag{3.4}$$

where $x_1, x_2, \ldots, x_k$ are the information symbols and the elements of $\mathbf{X}$ are linear combinations of $x_1, x_2, \ldots, x_k$ and their conjugates [5]. A well-known orthogonal STBC is the Alamouti code [6], which is a 2×2 code (i.e., $n_t = p = 2$). Two symbols x_1 and x_2 and their conjugates are sent in two time slots using two transmit antennas. The corresponding STBC matrix is given by

$$\mathbf{X} = \begin{bmatrix} x_1 & -x_2^* \\ x_2 & x_1^* \end{bmatrix}. \tag{3.5}$$

It can be seen that $k = 2$, and therefore the rate $r = k/p = 1$.

An advantage of OSTBCs is that they admit simple decoding. This can be explained using the decoding of Alamouti code with one receive antenna ($n_r = 1$) as follows. Let h_1 and h_2 denote the channel gains from transmit antennas 1 and 2, respectively, to the receive antenna. The received signals in two time slots, denoted by y_1 and y_2, are given by

$$\begin{aligned} y_1 &= h_1 x_1 + h_2 x_2 + n_1, \\ y_2 &= -h_1 x_2^* + h_2 x_1^* + n_2, \end{aligned}$$

which can be written in vector form as

$$\mathbf{y} = \begin{bmatrix} y_1 \\ y_2^* \end{bmatrix} = \underbrace{\begin{bmatrix} h_1 & h_2 \\ h_2^* & -h_1^* \end{bmatrix}}_{\mathbf{H}} \underbrace{\begin{bmatrix} x_1 \\ x_2 \end{bmatrix}}_{\mathbf{x}} + \underbrace{\begin{bmatrix} n_1 \\ n_2^* \end{bmatrix}}_{\mathbf{n}}.$$

Note that the $\mathbf{H}$ in the above is orthogonal, i.e.,

$$\mathbf{H}^H\mathbf{H} = \left(|h_1|^2 + |h_2|^2\right)\mathbf{I}_2 = \|\mathbf{h}\|^2\mathbf{I}_2,$$

where

$$\mathbf{h} = \begin{bmatrix} h_1 \\ h_2 \end{bmatrix}.$$

Assuming that the receiver has knowledge of the channel gains, the following receiver operation can be performed

$$\begin{aligned} \widetilde{\mathbf{y}} &= \mathbf{H}^H \mathbf{y} \\ &= \|\mathbf{h}\|^2 \mathbf{x} + \widetilde{\mathbf{n}}. \end{aligned}$$

The noise vector $\widetilde{\mathbf{n}}$ is still white. Therefore, x_1 and x_2 can be decoded separately rather than jointly. Since $\|\mathbf{h}\|^2$ is the amplitude of the symbols, the diversity order is 2, which is the achieved transmit diversity (since $n_r = 1$). If n_r receive antennas are used, then the above receiver operation results in

$$\widetilde{\mathbf{r}} = \sum_{i=1}^{n_r} \mathbf{H}_i^H \mathbf{r}_i = \Big(\sum_{i=1}^{n_r} \|\mathbf{h}\|^2\Big) \mathbf{x} + \widetilde{\mathbf{n}},$$

which achieves the maximum spatial diversity of $2n_r$.

The main advantages of OSTBCs are low decoding complexity and full transmit diversity. However, a major drawback with OSTBCs, particularly in the context of MIMO systems with a large number of antennas, is that the rate of OSTBCs falls linearly with increasing number of transmit antennas [5]. Note that with $n_t = 2$, the rate of OSTBCs (Alamouti code) is 1. Compare this rate with the rate achieved in spatial multiplexing with $n_t = 2$, which is 2. The rate of OSTBCs falls below 1 for $n_t > 2$; this fall is linear in increasing n_t. Quasi-orthogonal STBCs [2], in which the separate decoding property is relaxed (resulting in increased decoding complexity), achieve a rate of 1. This rate is still much less than the maximum rate of n_t symbols per channel use (spcu) achieved by spatial multiplexing. Therefore, STBC designs which can simultaneously achieve both the maximum rate of n_t spcu as well as the maximum spatial diversity of $n_t n_r$ are desired. NO-STBCs presented in the next section achieve these two desired attributes simultaneously for any number of transmit antennas. This makes these NO-STBCs preferred in large MIMO systems.

3.2.2 High-rate NO-STBCs

A well-known 2×2 NO-STBC which achieves the maximum rate of 2 spcu is the Golden code [7]. The Golden code is given by

$$\mathbf{X} = \begin{bmatrix} x_1 + \tau x_2 & x_3 + \tau x_4 \\ \mathbf{j}(x_3 + \mu x_4) & x_1 + \mu x_2 \end{bmatrix},$$

where $\tau = (1 + \sqrt{5})/2$, $\mu = (1 - \sqrt{5})/2$, and $\mathbf{j} = \sqrt{-1}$. Note that $k = 4, p = 2$ in the above code, and therefore its rate $r = k/p = 2$. Also, this code achieves the maximum transmit diversity of $n_t = 2$. NO-STBCs which achieve the attributes of information losslessness, full transmit diversity, and coding gain are called perfect codes [8],[9]. The Golden code is a perfect code. As can be expected, the decoding of these high-rate NO-STBCs is more complex than the decoding of OSTBCs. Low complexity near-optimal decoding of such high rate NO-STBCs with a large number of antennas is dealt in Chapter 5.

3.2.3 NO-STBCs from CDAs

Division algebras have been used as a tool for constructing good NO-STBCs, since they are non-commutative algebras that naturally yield linear fully diverse codes. Their algebraic properties can be further exploited to improve the design of good codes. An introduction to the algebraic tools involved in the design of codes based on CDAs is presented in [10].

An NO-STBC from a CDA is an $n_t \times n_t$ matrix constructed using n_t^2 symbols. The matrix is sent using n_t transmit antennas in n_t channel uses so that the rate of the STBC is n_t symbols per channel use (the same as that of V-BLAST). In addition to full rate, this STBC gives the full transmit diversity order of n_t as well (the same as that of OSTBCs). The number of dimensions is n_t^2, which creates hundreds of dimensions with tens of antennas (e.g., the number of complex dimensions is 256 for a 16×16 NO-STBC from CDAs). The construction of $n \times n$ NO-STBCs from CDAs for arbitrary n (i.e., for arbitrary number of transmit antennas) is given by the matrix [11]

$$\mathbf{X} = \begin{bmatrix} \sum_{i=0}^{n-1} x_{0,i}\, t^i & \delta \sum_{i=0}^{n-1} x_{n-1,i}\, \omega_n^i\, t^i & \cdots & \delta \sum_{i=0}^{n-1} x_{1,i}\, \omega_n^{(n-1)i}\, t^i \\ \sum_{i=0}^{n-1} x_{1,i}\, t^i & \sum_{i=0}^{n-1} x_{0,i}\, \omega_n^i\, t^i & \cdots & \delta \sum_{i=0}^{n-1} x_{2,i}\, \omega_n^{(n-1)i}\, t^i \\ \sum_{i=0}^{n-1} x_{2,i}\, t^i & \sum_{i=0}^{n-1} x_{1,i}\, \omega_n^i\, t^i & \cdots & \delta \sum_{i=0}^{n-1} x_{3,i}\, \omega_n^{(n-1)i}\, t^i \\ \vdots & \vdots & \vdots & \vdots \\ \sum_{i=0}^{n-1} x_{n-2,i}\, t^i & \sum_{i=0}^{n-1} x_{n-3,i}\, \omega_n^i\, t^i & \cdots & \delta \sum_{i=0}^{n-1} x_{n-1,i}\, \omega_n^{(n-1)i} t^i \\ \sum_{i=0}^{n-1} x_{n-1,i}\, t^i & \sum_{i=0}^{n-1} x_{n-2,i}\, \omega_n^i\, t^i & \cdots & \sum_{i=0}^{n-1} x_{0,i}\, \omega_n^{(n-1)i}\, t^i \end{bmatrix}, \tag{3.6}$$

where $\omega_n = e^{\mathbf{j}2\pi/n}$, $\mathbf{j} = \sqrt{-1}$, and $x_{u,v}$, $0 \le u, v \le n-1$ are the data symbols from a QAM alphabet. When $\delta = e^{\sqrt{5}\mathbf{j}}$ and $t = e^{\mathbf{j}}$, the resulting STBC achieves full transmit diversity as well as information losslessness. When $\delta = t = 1$, the code ceases to be of full diversity, but continues to be information lossless.

The main attractive features of the above NO-STBCs from CDAs for large MIMO systems are that (i) they achieve both the maximum rate of n_t spcu as well as the full transmit diversity of n_t, and (ii) they are valid for any number of transmit antennas. However, because of the large dimensions involved, decoding NO-STBCs from CDAs is challenging. Note that a 16×16 NO-STBC from CDAs has 512 real dimensions. Fortunately, detection approaches and algorithms that scale well in complexity and achieve near-optimum performance in such large dimensions are available. One such approach based on local search techniques is treated in Chapter 5. Other detection approaches based on PDA, BP, and Monte-Carlo sampling (treated in Chapters 6–8) can also be used to decode NO-STBCs from CDAs in large dimensions.

3.3 Spatial modulation (SM)

A key issue in the practical realization of large MIMO systems is the need to have a large number of RF chains. This increases the hardware complexity, size, and cost. SM, which was proposed for multiantenna systems, can alleviate this issue by using fewer transmit RF chains than transmit antennas. Space shift keying (SSK) is a special case of SM, and generalized spatial modulation (GSM) is a generalized version of SM. SM, SSK, and GSM schemes for multiantenna communication are described in the following subsections.

3.3.1 SM

SM employs a multiple antenna array at the transmitter but only a single transmit RF chain [3],[12]. This reduces the RF hardware complexity and cost. Though SM employs a multiple antenna array, it uses only one antenna from the array at a time for transmission. The choice of which antenna to activate is made based on a group of m data bits, where $m = \log_2 n_t$, and $n_t = 2^m$ is the number of transmit antennas in the array. On the chosen antenna, a symbol from an M-ary modulation alphabet $\mathbb{A}_M$ (e.g., M-QAM) is sent. The remaining $n_t - 1$ antennas remain silent. Therefore, the achieved rate in SM (number of bits conveyed in one time unit) in SM is

$$m + \log_2 M \quad \text{bits per channel use (bpcu)}.$$

The mapping of data bits to SM signals for $m = 2$, $n_t = 4$ is shown in Table 3.1. The SM signal set for an n_t antenna system, $\mathbb{S}_{n_t,M}$, is given by

$$\begin{aligned} \mathbb{S}_{n_t,M} &= \{\mathbf{x}_{j,l} : j = 1,\ldots,n_t, \quad l = 1,\ldots,M\}, \\ \text{st} \ \ \mathbf{x}_{j,l} &= [0,\ldots,0, \underbrace{x_l}_{j\text{th coordinate}}, 0,\ldots,0]^T, \quad x_l \in \mathbb{A}_M. \end{aligned} \tag{3.7}$$

SM signal detection

The job of signal detection at the receiver in SM involves determining the index of the transmitting antenna and the M-ary modulation symbol sent on it. Let $\mathbf{x} \in \mathbb{S}_{n_t,M}$ denote the $n_t \times 1$ transmitted signal vector. Then $\mathbf{x}$ will have an

Table 3.1. *Data bits to SM signal mapping for $m = 2$, $n_t = 4$. $\mathbb{A}_M$: M-ary modulation alphabet. Achieved rate: $m + \log_2 M$ bpcu*

Antenna sel. bits, $m = 2$	SM Tx signal vector, $\mathbf{x}$	Status of Tx antennas ($n_t = 2^m = 4$) Antenna 1	Antenna 2	Antenna 3	Antenna 4
0 0	$[x, 0, 0, 0]^T$	$x \in \mathbb{A}_M$	OFF	OFF	OFF
0 1	$[0, x, 0, 0]^T$	OFF	$x \in \mathbb{A}_M$	OFF	OFF
1 0	$[0, 0, x, 0]^T$	OFF	OFF	$x \in \mathbb{A}_M$	OFF
1 1	$[0, 0, 0, x]^T$	OFF	OFF	OFF	$x \in \mathbb{A}_M$

M-ary modulation symbol in one of the coordinates and zeros in all the other coordinates. Let $\mathbf{H}$ denote the $n_r \times n_t$ channel gain matrix, whose entries are assumed to be iid complex Gaussian with zero mean and unit variance. Let $\mathbf{n}$ denote the $n_r \times 1$ noise vector at the receiver, whose entries are iid complex Gaussian with zero mean and variance σ^2. Let $\mathbf{y}$ denote the $n_r \times 1$ received signal vector. Assuming equally likely inputs, the ML decision rule is

$$\begin{aligned} \widehat{\mathbf{x}} &= \underset{\mathbf{x} \in \mathbb{S}_{n_t,M}}{\operatorname{argmax}} \; P(\mathbf{y}|\mathbf{H}, \mathbf{x}) \\ &= \underset{\mathbf{x} \in \mathbb{S}_{n_t,M}}{\operatorname{argmin}} \; \|\mathbf{y} - \mathbf{H}\mathbf{x}\|^2. \end{aligned} \tag{3.8}$$

Bounds on the BER performance of ML detection in SM have been obtained in [13]. The BER performance of SM over generalized fading channels is analyzed in [14]. The complexity of ML detection in SM is exponential in the number of antennas and the size of the modulation alphabet. A reduced complexity SD for the detection of SM signals has been reported in [15].

3.3.2 SSK

SSK is a special case of SM [16]. Like SM, SSK also uses a one-to-one mapping between a group of m information bits and the spatial position (index) of the active transmitting antenna, which is chosen among the available $n_t = 2^m$ transmit antennas. But instead of sending an M-ary modulation symbol (e.g., an M-QAM symbol) as is done in SM, in SSK a signal known to the receiver, say +1, is sent on the chosen antenna. The remaining $n_t - 1$ transmit antennas remain silent. By doing so, the problem of SSK signal detection at the receiver becomes one of merely finding out which antenna is transmitting, whereas in SM, demodulation of the M-ary modulation (e.g., QAM) symbol is needed in addition to finding out which antenna is transmitting. So, SSK has a lower detection complexity than SM. Also, the achieved rate in SSK is

$$m \quad \text{bpcu}.$$

Table 3.2. *Data bits to SSK signal mapping for $m = 2$, $n_t = 4$. Achieved rate: m bpcu*

Data bits	SSK Tx. signal	Status of Tx antennas ($n_t = 2^m = 4$)			
($m = 2$)	vector, $\mathbf{x}$	Antenna 1	Antenna 2	Antenna 3	Antenna 4
0 0	$[1, 0, 0, 0]^T$	+1	OFF	OFF	OFF
0 1	$[0, 1, 0, 0]^T$	OFF	+1	OFF	OFF
1 0	$[0, 0, 1, 0]^T$	OFF	OFF	+1	OFF
1 1	$[0, 0, 0, 1]^T$	OFF	OFF	OFF	+1

The mapping of data bits to SSK signals for $m = 2$, $n_t = 4$ is shown in Table 3.2.

SSK signal detection

The SSK signal detector's main function is to determine the index of the transmitting antenna. Let $\mathbf{x}_j$ denote the transmitted signal vector with a one in the jth coordinate and zeros in all the other coordinates, i.e., the jth antenna transmits a +1 and all the other antennas remain silent. Let $\mathbf{h}_j$ denote the jth column of matrix $\mathbf{H}$. Assuming equally likely inputs, the ML rule for finding the transmitting antenna index is given by

$$\begin{aligned} \hat{j} &= \underset{j,\ 1 \le j \le n_t}{\operatorname{argmax}}\ p(\mathbf{y}|\mathbf{x}_j, \mathbf{H}) \\ &= \underset{j}{\operatorname{argmin}}\ \|\mathbf{y} - \mathbf{h}_j\|^2. \end{aligned} \tag{3.9}$$

The estimated antenna index $\hat{j}$ is demapped to the information bits which represent that index. Exact BER expressions for SSK with $m = 1$ ($n_t = 2$) and BER upper bounds based on union bounding for $m > 1$ ($n_t > 2$) have been derived in the literature [17],[18]. The performance of SSK and SM in single-carrier communication on frequency-selective channels is studied in [19].

3.3.3 GSM

Two limitations in SM and SSK are: (*i*) the number of transmit antennas is restricted to powers of 2, and (*ii*) the number of transmit RF chains is restricted to 1 because of which only one antenna can be active at a time. Both these restrictions are relaxed in GSM. In GSM, n_{rf} transmit RF chains, $1 \le n_{rf} \le n_t$, are used, and the number of transmit antennas n_t is not restricted to powers of 2 [20],[21]. By using more than one transmit RF chain, GSM allows multiple transmit antennas to be active simultaneously. This enables GSM to achieve higher spectral efficiencies compared to SM and SSK.

In GSM, the transmitter has n_t transmit antennas and n_{rf} transmit RF chains, $1 \leq n_{rf} \leq n_t$. An $n_{rf} \times n_t$ switch connects the RF chains to the transmit antennas. n_{rf} out of n_t transmit antennas are chosen, and M-ary information symbols are sent on these chosen antennas. The remaining $n_t - n_{rf}$ antennas remain silent (i.e., they can be viewed as transmitting the value zero). Therefore, if $\mathbb{A}_M$ denotes the M-ary alphabet used on the active antennas, the effective GSM alphabet becomes $\mathbb{A}_0 \triangleq \mathbb{A}_M \cup 0$.

Let us define an antenna activation pattern to be an n_t-length vector that indicates which antennas are active (denoted by a "1" in the corresponding antenna index) and which antennas are silent (denoted by a "0"). There are

$$L = \binom{n_t}{n_{rf}}$$

antenna activation patterns possible, and

$$K = \left\lfloor \log_2 \binom{n_t}{n_{rf}} \right\rfloor \quad \text{bits}$$

are used to choose an activation pattern for a given channel use. Note that not all L activation patterns are needed, and any 2^K patterns out of them are adequate. Let us take any 2^K patterns out of L patterns and form a set called the "antenna activation pattern set," $\mathbb{S}$. We illustrate this using the following example. Let $n_t = 4$ and $n_{rf} = 2$. Then, $L = \binom{4}{2} = 6$, $K = \lfloor \log_2 6 \rfloor = 2$, and $2^K = 4$. The $L = 6$ antenna activation patterns are given by

$$\left\{[1,1,0,0]^T, [1,0,1,0]^T, [0,1,0,1]^T, [0,0,1,1]^T, [0,1,1,0]^T, [1,0,0,1]^T\right\}.$$

Out of these six patterns, any $2^K = 4$ patterns can be taken to form the set $\mathbb{S}$. Accordingly, let us take the antenna activation pattern set as

$$\mathbb{S} = \left\{[1,1,0,0]^T, [1,0,1,0]^T, [0,1,0,1]^T, [0,0,1,1]^T\right\}.$$

Table 3.3 shows the mapping of data bits to GSM signals for $n_t = 4$, $n_{rf} = 2$ for the above activation pattern set. Suppose 4-QAM is used to send information on the active antennas. Let $\mathbf{x} \in \mathbb{A}_0^{n_t}$ denote the n_t-length transmit vector. Let 010011 denote the information bit sequence. GSM translates these bits to the transmit vector $\mathbf{x}$ as follows: (i) the first two bits are used to choose the activity pattern, (ii) the second two bits form a 4-QAM symbol, and (iii) the third two bits form another 4-QAM symbol. Using Gray mapping, the transmit vector $\mathbf{x}$ becomes

$$\mathbf{x} = [1 + \sqrt{-1}, \ 0, \ -1 - \sqrt{-1}, \ 0]^T.$$

Note that both SM and spatial multiplexing (i.e., V-BLAST) turn out to be special cases of GSM with $n_{rf} = 1$ and $n_{rf} = n_t$, respectively.

Table 3.3. *Data bits to GSM signal mapping for* $n_t = 4$, $n_{rf} = 2$. $\mathbb{A}_M$: *M-ary alphabet. Achieved rate* $= \left\lfloor \log_2 \binom{n_t}{n_{rf}} \right\rfloor + n_{rf} \log_2 M = 6$ *bpcu for 4-QAM*

Data bits	Ant. activation	Antenna status			
$K = 2$ bits	pattern	Antenna 1	Antenna 2	Antenna 3	Antenna 4
0 0	$[1, 1, 0, 0]^T$	$x_1 \in \mathbb{A}_M$	$x_2 \in \mathbb{A}_M$	OFF	OFF
0 1	$[1, 0, 1, 0]^T$	$x_1 \in \mathbb{A}_M$	OFF	$x_2 \in \mathbb{A}_M$	OFF
1 0	$[0, 1, 0, 1]^T$	OFF	$x_1 \in \mathbb{A}_M$	OFF	$x_2 \in \mathbb{A}_M$
1 1	$[0, 0, 1, 1]^T$	OFF	OFF	$x_1 \in \mathbb{A}_M$	$x_2 \in \mathbb{A}_M$

GSM signal detection

Let $\mathbb{U}$ denote the set of all possible transmit vectors. Then, $\mathbb{U}$ is given by

$$\mathbb{U} = \{\mathbf{x} | \mathbf{x} \in {\mathbb{A}_0}^{n_t \times 1}, \|\mathbf{x}\|_0 = n_{rf}, \mathbf{t}^{\mathbf{x}} \in \mathbb{S}\},$$

where $\mathbb{A}_0 = \mathbb{A}_M \cup 0$ denotes the effective GSM alphabet, $\|\mathbf{x}\|_0$ denotes the zero norm of vector $\mathbf{x}$ (i.e., number of non-zero entries in $\mathbf{x}$), and $\mathbf{t}^{\mathbf{x}}$ denotes the antenna activation pattern vector corresponding to $\mathbf{x}$, where $t_j^{\mathbf{x}} = 1, \text{iff } x_j \neq 0\ , \forall j = 1, 2, \ldots, n_t$. Note that

$$|\mathbb{U}| = 2^R,$$

where

$$R = \left\lfloor \log_2 \binom{n_t}{n_{rf}} \right\rfloor + n_{rf} \log_2 M.$$

The activation pattern set $\mathbb{S}$ and the mapping between elements of $\mathbb{S}$ and antenna selection bits are known at both the transmitter and the receiver. The ML decision rule for GSM signal detection is then given by

$$\widehat{\mathbf{x}} = \underset{\mathbf{x} \in \mathbb{U}}{\text{argmin}} \ \|\mathbf{y} - \mathbf{H}\mathbf{x}\|^2. \tag{3.10}$$

For small n_t and n_{rf}, the set $\mathbb{U}$ may be fully enumerated and ML detection as per (3.10) can be done. But for large n_t and n_{rf}, brute force computation of $\widehat{\mathbf{x}}$ in (3.10) becomes computationally infeasible. A low complexity near-ML detector for GSM which separates the antenna set detection from information bits detection is presented in [22]. A Gibbs sampling based algorithm for detection of GSM signals with large n_t, n_{rf} is presented in [23]. Analytical upper bounds on the BER performance of GSM are derived in [20],[22].

Achievable rates in GSM

In GSM, the transmit vector in a given channel use is formed using (i) antenna activation pattern selection bits, and (ii) M-ary modulation bits. The number

Table 3.4. *Parameters and achievable rates in SSK, SM, GSM*

Modulation	# Tx antennas (n_t)	# Tx RF chains (n_{rf})	Achievable rate (bpcu)
SSK	$n_t = 2^m,\ m \in \{1, 2, \ldots\}$	1	m
SM	$n_t = 2^m,\ m \in \{1, 2, \ldots\}$	1	$m + \log_2 M$
GSM	$n_t \in \{1, 2, \ldots\}$	$1 \leq n_{rf} \leq n_t$	$\left\lfloor \log_2 \binom{n_t}{n_{rf}} \right\rfloor + n_{rf} \log_2 M$

of activation pattern selection bits is

$$\left\lfloor \log_2 \binom{n_t}{n_{rf}} \right\rfloor .$$

The number of M-ary modulation bits is $n_{rf} \log_2 M$. Combining these two parts, the achievable rate in GSM with n_t transmit antennas, n_{rf} transmit RF chains, and M-QAM is given by

$$R = \left\lfloor \log_2 \binom{n_t}{n_{rf}} \right\rfloor + n_{rf} \log_2 M \quad \text{bpcu.} \tag{3.11}$$

The various parameters and the achievable rate expressions for SSK, SM, and GSM are summarized in Table 3.4.

Let us examine the GSM rate R in (3.11) in some detail. In particular, let us examine how R varies as a function of its variables. Figure 3.1 shows the variation of R as a function of n_{rf} for $n_t = 4, 8, 12, 16, 22, 32$ and 4-QAM. The value of n_{rf} in the x-axis is varied from 0 to n_t. As mentioned earlier, $n_{rf} = 1$ corresponds to SM and $n_{rf} = n_t$ corresponds to spatial multiplexing (i.e., V-BLAST). The R versus n_{rf} plot for a given n_t shows an interesting behavior: namely, for a given n_t, there is an optimum n_{rf} that maximizes the achievable rate R. It is interesting to see that the maximum R does not necessarily occur at $n_{rf} = n_t$, but at some $n_{rf} < n_t$. The following interesting observations come to the fore:

1. by choosing the optimum (n_t, n_{rf}) combination (i.e., using fewer RF chains than transmit antennas, $n_{rf} < n_t$), GSM can achieve a higher rate than that of spatial multiplexing (i.e., V-BLAST where $n_{rf} = n_t$); and
2. one can operate GSM at the same rate as that of spatial multiplexing but with even fewer RF chains.

For example, for $n_t = 32$, the optimum n_{rf} that maximizes R is 24 and the corresponding rate R_{opt} is 71 bps/Hz. Compare this rate with 64 bps/Hz which is the rate $n_t \log_2 M$ achieved in V-BLAST. This is an 11% gain in rate in GSM compared to V-BLAST. Interestingly, this rate gain is achieved using fewer RF chains – 24 RF chains in GSM versus 32 RF chains in V-BLAST. This is a 25% savings in transmit RF chains in GSM compared to V-BLAST. Further, if GSM were to achieve the V-BLAST rate of $n_t \log_2 M = 64$ bps/Hz in this case, then it would achieve it with even fewer RF chains, i.e., using just 18 RF

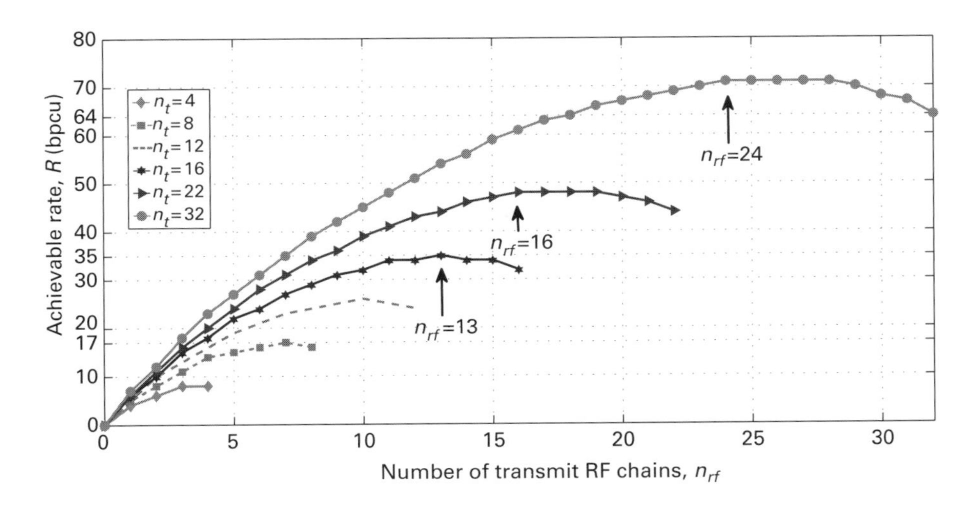

Figure 3.1 Achievable rate R in GSM as a function of n_{rf} for different values of n_t and 4-QAM.

Table 3.5. *Percentage saving in transmit RF chains and percentage increase in rate in GSM compared to spatial multiplexing (V-BLAST) for $n_t = 16, 32$ and BPSK, 4-QAM, 8-QAM, 16-QAM*

M-ary alphabet	% saving in # Tx RF chains at $R = R_{opt}$		% saving in # Tx RF chains at $R = n_t \log_2 M$		% increase in rate at $R = R_{opt}$	
	$n_t = 16$	$n_t = 32$	$n_t = 16$	$n_t = 32$	$n_t = 16$	$n_t = 32$
BPSK	31.25	40.63	68.75	71.88	43.75	46.88
4-QAM	18.75	25	37.5	43.75	9.385	10.94
8-QAM	6.25	12.5	18.75	21.88	2.08	3.13
16-QAM	6.25	3.13	6.25	9.38	0	0.78

chains which is a 43% savings in RF chains compared to V-BLAST. Table 3.5 gives the percentage gains in the number of transmit RF chains at $R = R_{opt}$ and $R = n_t \log_2 M$, and the percentage gains in the rates achieved by GSM compared to V-BLAST for $n_t = 16, 32$ with BSPK, 4-QAM, 8-QAM, and 16-QAM.

The reason why GSM can achieve better rate gains and RF chain savings than V-BLAST can be understood by analyzing the two terms on the right-hand side of the rate expression (3.11). The first term on the right-hand side in (3.11) increases when n_{rf} is increased from 0 to $\lfloor n_t/2 \rfloor$ and then decreases, i.e., it peaks at $n_{rf} = \lfloor n_t/2 \rfloor$. The second term, on the other hand, increases linearly with n_{rf}. These two terms when added can cause a peak at some n_{rf} between $\lfloor n_t/2 \rfloor$ and n_t. While the second term increases with $\log_2 M$, the first term is independent of M. In V-BLAST, $n_t = n_{rf}$, and so the first term is zero and the second term is $n_t \log_2 M$. When $n_{rf} < n_t$, the second term is $(n_t - n_{rf}) \log_2 M$ less than the V-BLAST rate, but there is a positive first term. Therefore, R can exceed the V-BLAST rate of $n_t \log_2 M$ whenever the first term exceeds $(n_t - n_{rf}) \log_2 M$. This explains the rate gains and RF chain savings in GSM. These gains will diminish for large values of M, as the second term will increasingly dominate the first term for increasing values of M. This can be seen in Table 3.5, where the gains are large for BPSK and 4-QAM, but small for 8-QAM and 16-QAM.

References

[1] D. Tse and P. Viswanath, *Fundamentals of Wireless Communication.* Cambridge, UK: Cambridge University, 2005.

[2] H. Jafarkhani, *Space-time coding: Theory and Practice.* Cambridge, UK: Cambridge University Press, 2005.

[3] M. D. Renzo, H. Haas, and P. M. Grant, "Spatial modulation for multiple-antenna wireless systems: a survey," *IEEE Commun. Mag.*, pp. 182–191, Dec. 2001.

[4] P. W. Wolniansky, G. J. Foschini, G. D. Golden, and R. A. Valenzuela, "V-BLAST: an architecture for realizing very high data rates over the rich-scattering wireless channel," in *URSI Intl. Symp. Signals, Systems and Electronics (ISSSE)*, Sept.-Oct. 1998, pp. 295–300.

[5] V. Tarokh, H. Jafarkhani, and A. R. Calderbank, "Space-time block codes from orthogonal designs," *IEEE Trans. Inform. Theory*, vol. 45, no. 7, pp. 1456–1467, Jul. 1999.

[6] S. M. Alamouti, "A simple transmit diversity technique for wireless communications," *IEEE J. Sel. Areas Commun.*, vol. 16, no. 8, pp. 1451–1458, Aug. 1998.

[7] J.-C. Belfiore, G. Rekaya, and E. Viterbo, "The Golden code: a 2 × 2 full-rate space-time code with non-vanishing determinants," *IEEE Tran. Inform. Theory*, vol. 51, no. 4, pp. 1432–1436, Apr. 2005.

[8] F. E. Oggier, G. Rekaya, J.-C. Belfiore, and E. Viterbo, "Perfect space-time block codes," *IEEE Tran. Inform. Theory*, vol. 52, no. 9, pp. 3885–3902, Sep. 2006.

[9] P. Elia, B. A. Sethuraman, and P. V. Kumar, "Perfect space-time codes for any number of antennas," *IEEE Tran. Inform. Theory*, vol. 53, no. 11, pp. 3853–3868, Nov. 2007.

[10] F. Oggier, J.-C. Belfiore, and E. Viterbo, "Cyclic division algebras: a tool for space-time coding," *Foundations and Trends in Commun. and Inform. Theory*, vol. 4, no. 1, pp. 1–95, Oct. 2007.

[11] B. A. Sethuraman, B. S. Rajan, and V. Shashidhar, "Full-diversity high-rate space-time block codes from division algebras," *IEEE Trans. Inform. Theory*, vol. 49, no. 10, pp. 2596–2616, Oct. 2003.

[12] R. Y. Mesleh, H. Haas, C. W. Ahn, and S. Yun, "Spatial modulation," *IEEE Trans. Veh. Tech.*, vol. 57, no. 4, pp. 2228–2241, Jul. 2008.

[13] J. Jeganathan, A. Ghrayeb, and L. Szczecinski, "Spatial modulation: optimal detection and performance analysis," *IEEE Commun. Lett.*, vol. 12, no. 8, pp. 545–547, Aug. 2008.

[14] M. D. Renzo and H. Haas, "Bit error probability of SM-MIMO over generalized fading channels," *IEEE Trans. Veh. Tech.*, vol. 61, no. 3, pp. 1124–1144, Mar. 2012.

[15] A. Younis, R. Mesleh, H. Haas, and P. M. Grant, "Reduced complexity sphere decoder for spatial modulation detection receivers," in *IEEE GLOBECOM'2010*, Miami, FL, Dec. 2010, pp. 1–5.

[16] J. Jeganathan, A. Ghrayeb, L. Szczecinski, and A. Ceron, "Space shift keying modulation for MIMO channels," *IEEE Trans. Wireless Commun.*, vol. 8, no. 7, pp. 3692–3703, Jul. 2009.

[17] M. D. Renzo and H. Haas, "A general framework for performance analysis of space shift keying (SSK) modulation for MISO correlated Nakagami-m fading channels," *IEEE Trans. Commun.*, vol. 58, no. 9, pp. 2590–2603, Sep. 2009.

[18] ——, "Space shift keying (SSK-) MIMO over correlated Rician fading channels: performance analysis and a new method for transmit-diversity," *IEEE Trans. Commun.*, vol. 59, no. 1, pp. 116–129, Jan. 2011.

[19] P. Som and A. Chockalingam, "Spatial modulation and space shift keying in single carrier communication," in *IEEE PIMRC'2012*, Sydney, Sep. 2012, pp. 1962–1967.

[20] A. Younis, N. Serafimovski, R. Mesleh, and H. Haas, "Generalised spatial modulation," in *Asilomar Conf. on Signals, Systems and Computers*, Nov. 2010, pp. 1498–1502.

[21] J. Fu, C. Hou, W. Xiang, L. Yan, and Y. Hou, "Generalised spatial modulation with multiple active transmit antennas," in *IEEE GLOBECOM'2010 Workshops*, Miami, FL, Dec. 2010, pp. 839–844.

[22] J. Wang, S. Jia, and J. Song, "Generalised spatial modulation system with multiple active transmit antennas and low complexity detection scheme," *IEEE Trans. Wireless Commun.*, vol. 11, no. 4, pp. 1605–1615, Apr. 2012.

[23] T. Datta and A. Chockalingam, "On generalized spatial modulation," in *IEEE WCNC'2013*, Shanghai, Apr. 2013, pp. 2716–2721.

4 MIMO detection

Detection of MIMO encoded signals, be it for spatial multiplexing or space-time coding or SM, is one of the crucial receiver functions in MIMO wireless communication [1]. Compared to detection in SISO or SIMO communication in fading channels, detection in MIMO communication is more involved. This is because, in addition to fading, the receive antennas encounter spatial interference due to simultaneous transmission from multiple transmit antennas. Efficient detection of signals in the presence of this spatial interference is therefore a demanding task, and sophisticated signal processing algorithms are needed for this purpose. Consequently, design, analysis, and implementation of efficient algorithms for MIMO detection continues to attract the attention of researchers and system developers.

Often, the roots of several MIMO detection algorithms in the literature can be traced to algorithms for multiuser detection (MUD) in CDMA which have been studied since the mid-1980s [2]. This is because CDMA systems and MIMO systems are both described by a linear vector channel model with the same structural format. In the case of a CDMA system the channel matrix is defined by the normalized cross-correlations between the signature sequences of the active users, whereas the channel matrix in a MIMO system is defined by the spatial signatures between the transmit and receive antennas. Specifically, several real-life communication systems can be characterized by the following linear vector channel model, where the received signal vector ($\mathbf{y}$) is given by the transmit vector ($\mathbf{x}$) transformed by a "channel" matrix ($\mathbf{H}$) plus a noise vector ($\mathbf{n}$):

$$\mathbf{y} = \mathbf{H}\mathbf{x} + \mathbf{n}, \tag{4.1}$$

where $\mathbf{x} \in \mathbb{C}^{d_t}$, $\mathbf{H} \in \mathbb{C}^{d_r \times d_t}$, $\mathbf{y} \in \mathbb{C}^{d_r}$, $\mathbf{n} \in \mathbb{C}^{d_r}$, and d_t and d_r are the number of transmit and receive dimensions, respectively. In CDMA, $d_t = d_r = K$, where K is the number of active users, $\mathbf{x}$ is the transmit vector consisting of the transmitted bits from each active user, $\mathbf{H}$ is the cross-correlation matrix, and $\mathbf{y}$ is the received vector at the output of the K matched filters (matched to the signature sequences of the active users). In MIMO with spatial multiplexing, $d_t = n_t$, $d_r = n_r$, $\mathbf{x}$ is the vector transmitted from n_t transmit antennas, $\mathbf{H}$ is the channel gain matrix, and $\mathbf{y}$ is the received vector at the n_r receive antennas. Because of this structural similarity, approaches and algorithms that were investigated for MUD in CDMA are natural candidates for MIMO detection.

The job of any detection algorithm is to obtain an estimate of the transmit vector $\mathbf{x}$, given knowledge of the received vector $\mathbf{y}$ and the channel matrix $\mathbf{H}$ [3]. The elements of $\mathbf{x}$ often come from a predecided modulation alphabet with discrete-valued symbols. Certain detection algorithms naturally produce soft outputs, e.g., BP based algorithms, where the output will be the soft values of the likelihood of the transmitted symbols. These soft output values can be fed to channel decoders in coded systems, which can offer improved performance compared to feeding hard inputs to channel decoders. Several detection algorithms, on the other hand, produce only hard outputs, e.g., search based algorithms which test a set of discrete-valued candidate vectors and choose one among them as the output. While these hard decision outputs can be fed to the channel decoder input as such, in order to improve performance futher, suitable methods have to be devised to generate soft values from the detector's hard outputs for feeding to the channel decoder.

The early breakthrough in MIMO system implementation was due to the laboratory prototype of the V-BLAST system demonstrated by Bell Labs in the late 1990s [4],[5]. This prototype system used 8 transmit antennas and 12 receive antennas, employed spatial multiplexing, and achieved a spectral efficiency of 25 bps/Hz. The detection algorithm employed in the V-BLAST system was the zero forcing successive interference cancelation (ZE-SIC) algorithm, which was then a popularly studied detection algorithm in the CDMA multiuser detection literature [2],[6],[7]. Subsequently, it has become common to refer to the ZF-SIC algorithm as the V-BLAST detection algorithm in the MIMO literature. The basic idea in the V-BLAST detection algorithm is to detect the symbols in a layered manner. In each layer, one symbol is detected. Detection in each layer is done using ZF (a well-known linear detection method). Interference due to the detected symbol in the first layer is estimated and subtracted from the received signal. From the first layer's interference canceled signal, another symbol is detected in the second layer. The interference due to the second layer's detected symbol is estimated and subtracted. These detection and interference cancelation steps are carried out in each layer, till all the symbols are detected.

Since the days of Bell Lab's V-BLAST system demonstration, MIMO detection research has grown in two main directions. One direction is along the lines of linear and non-linear detection approaches adopted from CDMA MUD literature. Well-known algorithms along these lines are the linear detectors, including matched filter (MF), ZF, and MMSE detectors, and non-linear detectors including multistage interference cancelers like ZF-SIC and MMSE-SIC detectors [1]. These are suboptimum algorithms whose key advantage is their polynomial complexity. The complexity of the ZF-SIC algorithm can be further reduced by the square-root algorithm proposed in [8] by efficiently deducing the pseudo-inverse of the deflated channel matrix in a given layer from the pseudo-inverse computed in the previous layer, using array algorithm ideas in linear estimation theory. Also, the performance of linear detection methods can be improved upon using lattice reduction techniques [9]–[12].

The other main direction of MIMO detection research is along the lines of sphere decoding [13]. This line of research became quite dominant because the sphere decoding algorithm is an ML decoding algorithm whose average complexity over a wide range of SNRs is polynomial in the number of dimensions [14], which is significantly less than the exponential complexity of ML detection. Sphere decoding is based on a bounded distance search among the lattice points falling inside a sphere centered at the received point. A drawback, however, is that its complexity in low and medium SNRs is still exponential in the number of dimensions, making it unsuitable for more than 32 dimensions [13],[14]. Numerous variants of sphere decoding aimed at complexity reduction while retaining the ML decodability have appeared in the literature [15],[16]. Some later variants of sphere decoding have compromised on the ML decodability while reducing or fixing the complexity for implementation ease [17].

Apart from the progress made in the above two main directions, another general trend (as happened in CDMA MUD research) has been to look for promising algorithms and tools from optimization, heuristics, machine learning, and artificial intelligence. For example, detectors based on semi-definite relaxation (SDR) [18]–[20], PDA [21],[22], and BP [23],[24] have exhibited good potential to achieve close-to-optimal performance.

Often, the twin objectives of achieving good performance and low complexity do not seem to be met simultaneously. One seems to come at the expense of the other. For example, linear methods scale well in complexity but suffer significant performance loss compared to optimum detection. Sphere decoding, on the other hand, achieves ML performance but does not scale well in complexity. So, the traditional view has been that complexity needs to be traded off to achieve good performance, and that this tradeoff will render large MIMO systems (with tens to hundreds of antennas) impractical due to the high complexities that may be involved for signal detection in large dimensions. However, quite contrary to this view, the "channel hardening" [25] that happens when the number of dimensions increases can be exploited to achieve near-optimal performance in large MIMO systems at low complexities that are affordable in practice.

Channel hardening (discussed in Section 2.2) is the phenomenon in which the variance of the mutual information (capacity) grows very slowly relative to its mean or even shrinks as the number of antennas grows [25]. In particular, consider an $n_r \times n_t$ MIMO channel with large n_r and fixed n_t. This, in practice, may correspond to an uplink scenario with n_r receive antennas at the BS and n_t synchronized uplink users each having one transmit antenna. n_r can be in the order of hundreds and n_t can be in tens. Since the loading factor $\beta \stackrel{\Delta}{=} n_t/n_r \ll 1$, the system is over-determined. As n_r increases, the diagonal entries of $\mathbf{H}^H\mathbf{H}$ become increasingly more prominent than the off-diagonal entries. Specifically, $\mathbf{H}^H\mathbf{H}/n_r$ converges to $\mathbf{I}_{n_t}$ as $n_r \to \infty$ and the n_t eigenvalues of $\mathbf{H}^H\mathbf{H}/n_r$ approach 1 [25]. Because of this, in large systems with $\beta \ll 1$, simple linear detectors like MF and

MMSE detectors themselves tend to be quite attractive for signal detection at the BS [26],[27]. Their appealing advantages are that the performance achieved with linear detection methods in this setting is close to optimum performance and the complexities involved are small enough for practical implementation.

Though the use of linear detection methods in over-determined systems with $n_t \ll n_r$ ($\beta \ll 1$) is attractive from an implementation point of view, the spectral efficiency potential is limited by the smaller n_t because capacity is proportional to $\min(n_t, n_r)$. On the other hand, systems with large n_t and β admit more uplink users in the system and achieve high spectral efficiency. The performance of linear detectors is severely degraded when used in systems with large n_t and β. Interestingly, certain detection algorithms have been shown to achieve near-optimal performance in large MIMO systems with large n_t, n_r (in the range of tens to hundreds) even when β is large. Also, such good performance is achieved at the same order of complexity as that of linear detectors like the MMSE detector. These algorithms are based on heuristic search, PDA, BP, and MCMC techniques [28]–[32]. In the rest of this chapter, well-known detection algorithms, including MF, ZF, MMSE, ZF-SIC, lattice reduction (LR) aided ZF/MMSE, and sphere decoding algorithms, and their performance and complexities are summarized. Detection algorithms specifically well suited for large MIMO systems are treated in detail in Chapters 5–8.

4.1 System model

In this chapter, we will present several well-known MIMO detection algorithms considering a spatially multiplexed (V-BLAST) MIMO system with n_t transmit and n_r receive antennas, $n_t \leq n_r$. The n_t symbols are transmitted simultaneously in one channel use from n_t transmit antennas. Let $\mathbf{x}$ denote the transmitted symbol vector. The elements of $\mathbf{x}$ come from a known modulation alphabet $\mathbb{A}$ (say, QAM), i.e., $\mathbf{x} \in \mathbb{A}^{n_t}$, and $\mathbb{E}\{\mathbf{x}\mathbf{x}^H\} = \mathbf{I}_{n_t}$. The received signal vector $\mathbf{y} \in \mathbb{C}^{n_r}$ in each channel use is given by

$$\mathbf{y} = \mathbf{H}\mathbf{x} + \mathbf{n}, \tag{4.2}$$

where $\mathbf{H} \in \mathbb{C}^{n_r \times n_t}$ is the $n_r \times n_t$ channel matrix whose entries are modeled as iid complex Gaussian with zero mean and unit variance, and $\mathbf{n}$ is a complex AWGN vector where $\mathbb{E}\{\mathbf{n}\mathbf{n}^H\} = \sigma^2\mathbf{I}_{n_r}$. It is assumed that $\mathbf{H}$ is known perfectly at the receiver but is unknown at the transmitter. With the above assumptions, the output likelihood function for the system model in (4.2) is given by

$$p(\mathbf{y}|\mathbf{H}, \mathbf{x}) = \frac{1}{(\pi\sigma^2)^{n_r}} \exp\left(\frac{1}{\sigma^2}\|\mathbf{y} - \mathbf{H}\mathbf{x}\|^2\right). \tag{4.3}$$

Several detection algorithms work directly with the complex-valued system model in (4.2). Certain other detection algorithms, for computational reasons, may work with an equivalent real-valued system model corresponding to (4.2), which is given by

$$\mathbf{y}_r = \mathbf{H}_r\mathbf{x}_r + \mathbf{n}_r, \tag{4.4}$$

where

$$\mathbf{H}_r \triangleq \begin{bmatrix} \Re(\mathbf{H}) & -\Im(\mathbf{H}) \\ \Im(\mathbf{H}) & \Re(\mathbf{H}) \end{bmatrix} \in \mathbb{R}^{2n_r \times 2n_t}, \qquad \mathbf{y}_r \triangleq [\Re(\mathbf{y})^T \ \Im(\mathbf{y})^T]^T \in \mathbb{R}^{2n_r},$$

$$\mathbf{x}_r \triangleq [\Re(\mathbf{x})^T \ \Im(\mathbf{x})^T]^T \in \mathbb{R}^{2n_t}, \qquad \mathbf{n}_r \triangleq [\Re(\mathbf{n})^T \ \Im(\mathbf{n})^T]^T \in \mathbb{R}^{2n_r}.$$

Note that the elements of vector $\mathbf{x}_r$ in the real-valued system model (4.4) come from the underlying PAM alphabet corresponding to the QAM alphabet employed in (4.2).

4.2 Optimum detection

At the receiver, the detector forms an estimate of the transmitted symbol, $\widehat{\mathbf{x}}$. The optimal detector minimizes the average probability of error, $p(\widehat{\mathbf{x}} \neq \mathbf{x})$. This is achieved by the ML detector which solves the non-linear optimization problem of minimizing the squared Euclidean distance between the actual received vector $\mathbf{y}$ and the hypothesized received signal $\mathbf{Hx}$ with the vector $\mathbf{x}$ constrained to the set $\mathbb{A}^{n_t}$, i.e., the ML solution is given by

$$\widehat{\mathbf{x}}_{ML} = \underset{\mathbf{x} \in \mathbb{A}^{n_t}}{\operatorname{argmin}} \ \|\mathbf{y} - \mathbf{Hx}\|^2. \tag{4.5}$$

Computing the exact solution to the above optimization problem through an exhaustive search requires exponential complexity in n_t. Therefore, this computation is possible only for small n_t. Knowing the exact ML solution is desired since it serves as a benchmark to assess how various detectors perform relative to the optimum solution. When n_t is large (tens to hundreds), computing the exact ML solution becomes infeasible due to the exponential complexity. Low complexity bounds on ML performance can help to address this issue. A simple, yet useful bound for large n_t is the non-faded SISO AWGN performance which is a lower bound on the ML performance. In the CDMA MUD literature, this is referred to as the "single-user" bound. This bound is easy to compute and is tight for large n_t at high SNRs. Another approach is to obtain an upper bound on the ML performance through union bounding of pairwise error probability [33]. However, it turns out that computation of this bound also has exponential complexity [33]. Sphere decoding gives the ML solution. But again, its complexity is exponential in low and medium SNRs, making it impractical for large n_t. More recently, it has been proposed that low complexity large MIMO detection algorithms that are based on local search techniques can be used to compute lower bounds on ML performance [34]. This bounding technique is described in

Chapter 5. Popular suboptimum solutions to (4.5) using linear and non-linear methods are presented next.

4.3 Linear detection

Linear detection methods generate soft estimates of transmitted symbols through a linear transformation of the received vector. To obtain hard estimates, these methods take the form of $\widehat{\mathbf{x}} = f(\mathbf{G}\mathbf{y})$, where $\mathbf{G}$ is a transformation matrix and $f(.)$ is a slicer, which quantizes each entry of $\mathbf{G}\mathbf{y}$ to the nearest symbol in the modulation alphabet to obtain $\widehat{\mathbf{x}}$. Linear detection methods possess the advantage of low (polynomial) complexity.

MF detector

The MF detector is a simple linear detector. In detecting the symbol in a given stream, the MF detector treats the interference from other streams as merely noise. Defining $\mathbf{h}_i, i = 1, 2, \ldots, n_t$, to be the ith column of the channel matrix $\mathbf{H}$, (4.2) can be written in the form

$$\begin{aligned} \mathbf{y} = \mathbf{H}\mathbf{x} + \mathbf{n} &= \sum_{i=1}^{n_t} \mathbf{h}_i x_i + \mathbf{n} \\ &= \mathbf{h}_k x_k + \sum_{i=1, i \neq k}^{n_t} \mathbf{h}_i x_i + \mathbf{n}, \end{aligned} \tag{4.6}$$

where the first term in (4.6) is the component due to the kth stream, and the second term is due to all streams other than the kth stream, i.e., the second term is the interference term as far as the kth stream is concerned. In detecting the kth stream symbol x_k, the MF detector simply ignores the second term as noise and obtains a soft estimate of x_k as

$$\tilde{x}_k = \mathbf{h}_k^* \mathbf{y}, \tag{4.7}$$

and a hard estimate $\widehat{x}_k$ is obtained by mapping $\tilde{x}_k$ to be nearest symbol in the alphabet in terms of Euclidean distance. In vector form, the MF solution can be written as

$$\tilde{\mathbf{x}}_{MF} = \mathbf{H}^H \mathbf{y}, \tag{4.8}$$

i.e., the transformation matrix $\mathbf{G}_{MF} = \mathbf{H}^H$. Computing (4.8) requires a complexity of order $n_t n_r$, which is very attractive. As mentioned earlier, for lightly loaded systems with $\beta \ll 1$ (i.e., $n_t \ll n_r$), the performance of the MF detector is near to optimum. However, its performance severely degrades with increasing n_t in systems with moderate to full loading, due to increased levels of uncanceled interference from other streams.

ZF detector

The ZF detector is a linear detector in which the linear transformation on the received vector is carried out using the pseudo-inverse of the $\mathbf{H}$ matrix. Let $\mathbf{Q}$ denote the $n_t \times n_r$ matrix which is the pseudo-inverse of $\mathbf{H}$, i.e.,

$$\mathbf{Q} = (\mathbf{H}^H\mathbf{H})^{-1}\mathbf{H}^H. \tag{4.9}$$

Since $\mathbf{QH} = \mathbf{I}_{n_t}$, the transformation $\mathbf{Qy}$ completely cancels the interference from other streams (hence the name zero-forcing or interference-nulling detector). A drawback, however, is that noise is enhanced in the process of eliminating the interference completely.

Let $\mathbf{q}_k$, $k = 1, 2, \ldots, n_t$, denote the kth row of $\mathbf{Q}$. Then, $\mathbf{q}_k\mathbf{H}$ is a row vector of length n_t whose entries are all zero except for a 1 in the kth coordinate. A soft estimate of the symbol x_k can be obtained as

$$\begin{aligned} \tilde{x}_k &= \mathbf{q}_k\mathbf{y} = \mathbf{q}_k\mathbf{H}\mathbf{x} + \mathbf{q}_k\mathbf{n} \\ &= x_k + \mathbf{q}_k\mathbf{n}, \end{aligned} \tag{4.10}$$

and a hard estimate is obtained through Euclidean distance mapping to the nearest symbol in the modulation alphabet. Note that the first term in (4.10) is the kth stream symbol without any interference, and that the SNR at the kth stream zero-forced output is given by

$$SNR_k = \frac{|x_k|^2}{||\mathbf{q}_k||^2\sigma^2}. \tag{4.11}$$

Note that the ZF operation, in addition to nulling the interference, has enhanced the noise variance by a factor of $||\mathbf{q}_k||^2$. Because of this, at low SNRs (large σ), the noise enhancement effect dominates and the ZF detector may end up performing worse than the MF detector. At high SNRs, however, the interference nulling effect dominates and the performance of the ZF detector is better than that of the MF detector. The ZF solution, in vector form, can be written as

$$\tilde{\mathbf{x}}_{ZF} = \mathbf{Qy}, \tag{4.12}$$

i.e., the transformation matrix $\mathbf{G}_{ZF} = \mathbf{Q}$. The computation complexity in (4.12) is cubic in n_t because of the computation of the matrix inverse in (4.9). Therefore, the per-symbol complexity is n_t^2, which is one order more than that of the MF detector. While this quadratic per-symbol complexity of the ZF detector is still attractive for large MIMO systems, its performance also degrades severely for large n_t at moderate to full loads.

MMSE detector

The MMSE detector is a linear detector whose transformation matrix is that matrix which minimizes the mean square error between the transmit vector and the estimated vector (i.e., the transformed received vector). That is, the transformation matrix $\mathbf{G}_{MMSE}$ is given by the solution to the following minimization problem:

$$\min_{\mathbf{G}} \mathbb{E}\left[\|\mathbf{x} - \mathbf{G}\mathbf{y}\|^2\right]. \tag{4.13}$$

The solution to (4.13) is given by

$$\mathbf{G}_{MMSE} = \left(\mathbf{H}^H\mathbf{H} + \sigma^2\mathbf{I}_{n_t}\right)^{-1}\mathbf{H}^H, \tag{4.14}$$

and the MMSE solution is given by

$$\tilde{\mathbf{x}}_{MMSE} = \mathbf{G}_{MMSE}\,\mathbf{y}. \tag{4.15}$$

The MMSE detector combines the best performance attributes of MF and ZF detectors. At high SNRs (i.e., small σ), MMSE behaves like ZF since the second term inside the inverse operation in (4.14) becomes negligible. At low SNRs, it behaves like MF because of the prominence of the diagonal entries of $\mathbf{H}^H\mathbf{H}$ as $\sigma \to \infty$. The MMSE detector strictly performs better than both the MF and the ZF detector over the entire range of SNRs. Note that the MMSE solution needs knowledge of the noise variance σ^2, which MF and ZF solutions do not need. Like the ZF detector, because of the matrix inversion involved in (4.14), the per-symbol complexity of the MMSE detector is n_t^2. Like the MF and ZF performances, the MMSE performance is also severely degraded for increasing n_t at medium to full loading.

4.4 Interference cancelation

Detectors based on interference cancelation belong to the class of non-linear detectors which perform interference estimation and removal in multiple stages. Popular interference cancelation techniques include successive interference cancelation (SIC) and parallel interference cancelation (PIC). SIC is known for its simplicity. The steps involved in SIC based detection can be summarized as follows.

1. Initially, the symbol transmitted in a data stream (preferably in the strongest data stream) is detected using a detector (e.g., using any of the MF, ZF, MMSE detectors). SIC is referred to as MF-SIC, ZF-SIC, and MMSE-SIC if the component detector used is MF, ZF, and MMSE, respectively. With ZF-SIC, the data streams can be ordered based on their received SNRs given by (4.11), or, equivalently, based on the norm of $\mathbf{q}_k$, $k = 1, \ldots, n_t$.
2. Using the detected data symbol and knowledge of the channel matrix, an estimate of the interference contributed by it is estimated.
3. The estimated interference is subtracted (canceled) from the received signal.
4. The interference canceled output signal is used to detect the symbol in the next strongest data stream. This stream's interference contribution is then estimated and canceled. This procedure is continued till the last (weakest) data stream is detected.

V-BLAST detector

The detector used in the Bell Lab's V-BLAST system was the ZF-SIC detector, which uses ZF for symbol detection in each stage. The ZF-SIC algorithm for V-BLAST MIMO is summarized below.

Set $\mathbf{y}^{(1)} = \mathbf{y}$, $\mathbf{H}^{(1)} = \mathbf{H}$, where the superscript denotes the stage index. $\mathbf{Q}^{(m)}$ is the pseudo-inverse of the channel matrix $\mathbf{H}^{(m)}$, and $\mathbf{q}_l^{(m)}$ is the lth row of $\mathbf{Q}^{(m)}$. Set stage index $m = 1$. Let k denote the index of the user detected in a given stage.

1. *Symbol detection (ZF)* Detect the symbol of the kth data stream using the ZF detector, i.e.,
$$\widehat{x}_k = f(\mathbf{q}_k^{(m)} \mathbf{y}^{(m)}), \tag{4.16}$$
where $f(.)$ is the slicing function. End the algorithm if $m = n_t$.
2. *Interference estimation* Using $\widehat{x}_k$, obtain an estimate of the interference vector due to the kth stream $\mathbf{a}_k$ as
$$\widehat{\mathbf{a}}_k = \mathbf{h}_k \widehat{x}_k. \tag{4.17}$$
3. *Interference cancelation* Subtract (4.17) from $\mathbf{y}^{(m)}$ to get the canceled output $\mathbf{y}^{(m+1)}$ as
$$\mathbf{y}^{(m+1)} = \mathbf{y}^{(m)} - \widehat{\mathbf{a}}_k = \mathbf{y}^{(m)} - \mathbf{h}_k \widehat{x}_k. \tag{4.18}$$
4. Obtain $\mathbf{H}^{(m+1)}$ by setting the kth column of $\mathbf{H}^{(m)}$ to zero, i.e., $\mathbf{H}^{(m+1)} =$ [$\mathbf{H}^{(m)}$ with kth column set to zero]. $m \leftarrow m + 1$. Go to Step 1.

The algorithm needs to do a matrix inversion in each stage and the resulting complexity is $O(n_t^4)$, an order more than the complexity of the ZF detector. The ZF-SIC performance, however, is better than the ZF performance. This is illustrated in Fig. 4.1, which shows the BER performance of MF, ZF, and ZF-SIC detectors in a V-BLAST MIMO system with $n_t = 8$, $n_r = 12$, and 4-QAM. Though ZF-SIC performs better than linear detectors, its performance is far from optimum. Also, it does not scale well for large MIMO systems.

4.5 LR-aided linear detection

Low complexity receiver structures can be devised based on LR techniques [35]. LR-aided linear detection can achieve better performance than the underlying linear detection (without LR aid) at the same order of complexity. The basic idea is that, instead of directly applying the linear transformation on the received signal model

$$\mathbf{y} = \mathbf{H}\mathbf{x} + \mathbf{n},$$

the transformation is applied on an equivalent system model

$$\mathbf{y} = \tilde{\mathbf{H}}\mathbf{z} + \mathbf{n},$$

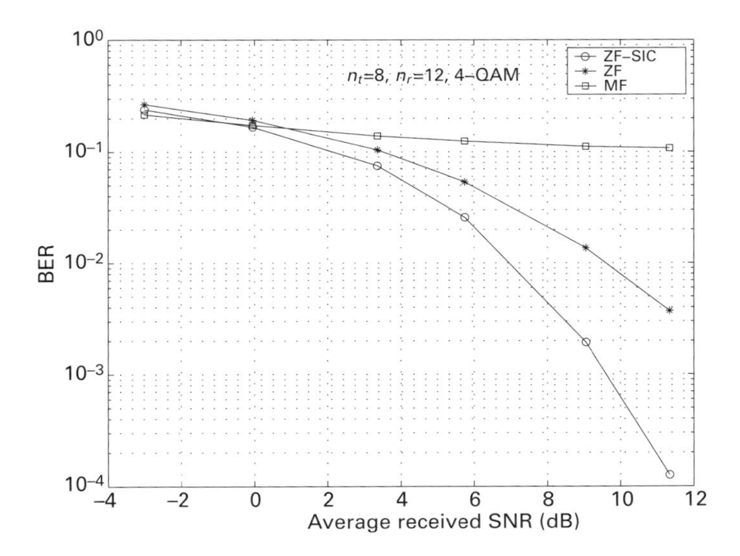

Figure 4.1 BER performance comparison between MF, ZF, and ZF-SIC detectors.

obtained using LR techniques. The channel matrix $\tilde{\mathbf{H}}$ in the equivalent system model is made to be more orthogonal than the channel matrix $\mathbf{H}$ in the original system model. Also, the data vector $\mathbf{z}$ in the equivalent system model is related to the data vector $\mathbf{x}$ in the original system model through a transformation using a unimodular matrix $\mathbf{T}$. Seysen's algorithm (SA) [10] is a low complexity iterative algorithm to obtain $\tilde{\mathbf{H}}$ and $\mathbf{T}$ from $\mathbf{H}$. The Lenstra–Lenstra–Lovasz (LLL) algorithm is another iterative algorithm for finding a good new basis $\tilde{\mathbf{H}}$ [36]. ZF or MMSE operation on the equivalent system model gives estimates in the $\mathbf{z}$ domain, which can be mapped back to the $\mathbf{x}$ domain through $\mathbf{T}$. By doing so, better performance is achieved due to the near-orthogonality of $\tilde{\mathbf{H}}$. Also, since the SA that generates $\tilde{\mathbf{H}}$ and $\mathbf{T}$ has one order less complexity than the ZF and MMSE operation, the order of complexity of LR-aided ZF/MMSE detection is the same as that of ZF/MMSE. Details of the above LR-aided approach are given in the following subsections.

4.5.1 LR-aided detection

The columns of the matrix $\mathbf{H}$, i.e., $\mathbf{h}_i, i = 1, 2, \ldots, n_t$ can be interpreted as the basis of a lattice. Assume that the possible transmit vectors are given by $\mathbb{C}_{\mathbb{Z}}^{n_t}$, where $\mathbb{C}_{\mathbb{Z}}$ represents the set of Gaussian integers, i.e., if $s = a + jb$ and $s \in \mathbb{C}_{\mathbb{Z}}$, then $a \in \mathbb{Z}$ and $b \in \mathbb{Z}$. Consequently, the set of all possible received signals undisturbed by noise is given by the lattice

$$\mathbb{L}(\mathbf{H}) \triangleq \sum_{i=1}^{n_t} \mathbf{h}_i \mathbb{C}_{\mathbb{Z}}. \tag{4.19}$$

The matrix $\tilde{\mathbf{H}} = \mathbf{HT}$ generates the same lattice as $\mathbf{H}$ if and only if the $n_t \times n_t$ matrix $\mathbf{T}$ is unimodular [9], i.e., $\mathbf{T}$ and $\mathbf{T}^{-1}$ contain only Gaussian integers and $\det(\mathbf{T}) = \pm 1$ or $\pm j$. That is,

$$\mathbb{L}(\tilde{\mathbf{H}}) = \mathbb{L}(\mathbf{H}) \Longleftrightarrow \tilde{\mathbf{H}} = \mathbf{HT}. \tag{4.20}$$

Since the inverse of a unimodular matrix always exists, the relation $\mathbf{H} = \tilde{\mathbf{H}}\mathbf{T}^{-1}$ holds. The aim of the lattice reduction is to transform a given basis $\mathbf{H}$ into a new basis $\tilde{\mathbf{H}}$ with vectors of shortest length, or, equivalently, into a basis consisting of roughly orthogonal basis vectors. Usually, $\tilde{\mathbf{H}}$ will be much better conditioned than $\mathbf{H}$.

Linear detection is optimal for an orthogonal channel matrix. With $\tilde{\mathbf{H}} = \mathbf{HT}$ and defining $\mathbf{z} = \mathbf{T}^{-1}\mathbf{x}$, the received signal vector can be written as

$$\begin{aligned} \mathbf{y} &= \mathbf{Hx} + \mathbf{n} \\ &= \mathbf{HTT}^{-1}\mathbf{x} + \mathbf{n} \\ &= \tilde{\mathbf{H}}\mathbf{z} + \mathbf{n}. \end{aligned} \tag{4.21}$$

Note that $\mathbf{Hx}$ and $\tilde{\mathbf{H}}\mathbf{z}$ denote the same point in the lattice, but the reduced matrix $\tilde{\mathbf{H}}$ is much better conditioned than $\mathbf{H}$. For $\mathbf{x} \in \mathbb{C}_{\mathbb{Z}}^{n_t}$, we also have $\mathbf{z} \in \mathbb{C}_{\mathbb{Z}}^{n_t}$. So $\mathbf{x}$ and $\mathbf{z}$ come from the same set. However, for the QAM alphabet, i.e., $\mathbf{x} \in \mathbb{A}$, the lattice is finite and the domain of $\mathbf{z}$ differs from $\mathbb{A}$.

The idea behind LR-aided linear detection is to consider the equivalent system model in (4.21) and perform the slicing on $\mathbf{z}$ instead of $\mathbf{x}$. For LR-aided ZF, this means that

$$\tilde{\mathbf{z}}_{LR-ZF} = \tilde{\mathbf{H}}^{+}\mathbf{y} = \mathbf{z} + \tilde{\mathbf{H}}^{+}\mathbf{n} = \mathbf{T}^{-1}\tilde{\mathbf{x}}_{ZF} \tag{4.22}$$

is calculated, where $(.)^{+}$ denotes the pseudo-inverse operation. The multiplication of $\mathbf{y}$ with $\tilde{\mathbf{H}}^{+}$ usually causes less noise enhancement than the multiplication with $\mathbf{H}^{+}$ in (4.10) due to the roughly orthogonal columns of $\tilde{\mathbf{H}}$. Therefore, a hard decision based on $\tilde{\mathbf{z}}_{LR-ZF}$ is in general more reliable than the hard decision based on $\tilde{\mathbf{x}}_{ZF}$. However, the elements of the transformed vector $\mathbf{z}$ are not independent of each other. A straightforward solution then is to perform an unconstrained element-wise slicing $\widehat{\mathbf{z}}_{LR-ZF} = f(\tilde{\mathbf{z}}_{LR-ZF})$, calculate $\widehat{\mathbf{x}}_{LR-ZF} = \mathbf{T}\,\widehat{\mathbf{z}}_{LR-ZF}$, and finally restrict this result to the set $\mathbb{A}$.

The MMSE solution may be applied instead of the ZF solution in order to get an improved estimate of $\mathbf{z}$, i.e., to obtain the MMSE solution of the lattice-reduced system in (4.21) as

$$\tilde{\mathbf{z}}_{LR-MMSE} = \left(\tilde{\mathbf{H}}^{H}\tilde{\mathbf{H}} + \sigma^{2}\mathbf{TT}^{-1}\right)^{-1}\tilde{\mathbf{H}}^{H}\mathbf{y} = \mathbf{T}^{-1}\,\tilde{\mathbf{x}}_{MMSE}. \tag{4.23}$$

How $\tilde{\mathbf{H}}$ and $\mathbf{T}$ are obtained using SA is described next.

4.5.2 SA

SA is an iterative method of lattice reduction [10]. The metric that SA uses to quantify the orthogonality of the channel matrix is

$$\mathrm{q}(\tilde{\mathbf{H}}) = \sum_{n=1}^{n_t} \|\tilde{\mathbf{h}}_n\|^2 \|\tilde{\mathbf{h}}_n^{'}\|^2, \tag{4.24}$$

where $\tilde{\mathbf{h}}_n^{'}$ is the nth basis vector of the dual lattice $\mathbb{L}^{'}$, i.e., $\tilde{\mathbf{H}}^{'H}\tilde{\mathbf{H}} = \mathbf{I}$, where $\tilde{\mathbf{H}}^{'} \triangleq [\tilde{\mathbf{h}}_1^{'} \cdots \tilde{\mathbf{h}}_{n_t}^{'}]$ denotes the dual basis. $\mathrm{q}(\tilde{\mathbf{H}})$ assumes its minimum, i.e., $\mathrm{q}(\tilde{\mathbf{H}}) = n_t$, if and only if the basis $\tilde{\mathbf{H}}$ is orthogonal. SA finds a (local) minimum of $\mathrm{q}(\tilde{\mathbf{H}}) = \mathrm{q}(\mathbf{HT})$ in an iterative manner. In view of (4.24), it can be said that the basis and its dual are reduced simultaneously.

Basic principle of SA

For any matrix $\mathbf{D}$, let $\mathbf{d}_i$ denote the ith column and $\mathbf{d}_i^u$ denote the updated ith column. Initializing $\mathbf{T} = \mathbf{I}$ and $\tilde{\mathbf{H}} = \mathbf{H}$, SA repeats the following steps until $\tilde{\mathbf{H}}$ is "SA-reduced" [12].

1. Based on $\tilde{\mathbf{H}} = \mathbf{HT}$, an index pair (k, l) with $k, l \in \{1, \ldots, n_t\}$ is selected and a corresponding complex scalar update value $\lambda_{k,l} \in \mathbb{C}_{\mathbb{Z}}$ is calculated.
2. *Basis update*

$$\mathbf{T} = [\mathbf{t}_1 \cdots \mathbf{t}_{k-1}\ \mathbf{t}_k^u\ \mathbf{t}_{k+1} \cdots \mathbf{t}_{n_t}] \text{ with } \mathbf{t}_k^u = \mathbf{t}_k + \lambda_{k,l}\mathbf{t}_l, \tag{4.25}$$

or, equivalently,

$$\tilde{\mathbf{H}} = [\tilde{\mathbf{h}}_1 \cdots \tilde{\mathbf{h}}_{k-1}\ \tilde{\mathbf{h}}_k^u\ \tilde{\mathbf{h}}_{k+1} \cdots \tilde{\mathbf{h}}_{n_t}] \text{ with } \tilde{\mathbf{h}}_k^u = \tilde{\mathbf{h}}_k + \lambda_{k,l}\tilde{\mathbf{h}}_l. \tag{4.26}$$

Note that $\tilde{\mathbf{h}}_k^u = \mathbf{H}\mathbf{t}_k^u$. In each iteration, $\tilde{\mathbf{H}}$ is again a valid basis for $\mathbb{L}$. In fact, any basis for $\mathbb{L}$ can be achieved by a sequence of updates according to (4.25) and (4.26).

SA-reduced basis

Consider a basis vector update (4.26) for a given index pair (i, j) – not necessarily the selected index pair (k, l):

$$\tilde{\mathbf{H}}_{i,j} \triangleq [\tilde{\mathbf{h}}_1 \cdots \tilde{\mathbf{h}}_{i-1}\ \tilde{\mathbf{h}}_i^u\ \tilde{\mathbf{h}}_{i+1} \cdots \tilde{\mathbf{h}}_{n_t}] \text{ with } \tilde{\mathbf{h}}_i^u = \tilde{\mathbf{h}}_i + \lambda_{i,j}\tilde{\mathbf{h}}_j. \tag{4.27}$$

The best update value $\lambda_{i,j}$ such that $\mathrm{q}(\tilde{\mathbf{H}}_{i,j})$ is minimized is obtained as [10]

$$\lambda_{i,j} = \left\lfloor \frac{\tilde{\mathbf{h}}_j^{'H}\tilde{\mathbf{h}}_i^{'}}{2\|\tilde{\mathbf{h}}_i^{'}\|^2} - \frac{\tilde{\mathbf{h}}_j^{H}\tilde{\mathbf{h}}_i}{2\|\tilde{\mathbf{h}}_j^{'}\|^2} \right\rceil. \tag{4.28}$$

It can be shown that $\mathrm{q}(\tilde{\mathbf{H}}_{i,j}) \leq \mathrm{q}(\tilde{\mathbf{H}})$ if and only if $\lambda_{i,j} \neq 0$. We call the basis $\tilde{\mathbf{H}}$ "SA-reduced" if no decrease of $\mathrm{q}(\tilde{\mathrm{H}})$ can be achieved for any (i, j), i.e., $\lambda_{i,j} = 0$ for all possible (i, j). Thus, to obtain an SA-reduced basis, one simply has to repeat

Step 1 and Step 2 until no decrease of $\mathrm{q}(\tilde{\mathbf{H}}_{i,j})$ is observed. This corresponds to a local minimum of Seysen's orthogonality measure.

For the determination of the index pair (k,l), a greedy selection procedure is adopted [10]. During each iteration, one selects (k,l) such that the decrease in the Seysen's orthogonality measure is maximized, i.e.,

$$(k,l) = \underset{(i,j)}{\operatorname{argmax}}\ \Delta_{i,j}, \tag{4.29}$$

where $\Delta_{i,j} \triangleq \mathrm{q}(\tilde{\mathbf{H}}) - \mathrm{q}(\tilde{\mathbf{H}}_{i,j})$. That is, in each iteration $n_t^2 - n_t$ basis updates with respect to their achieved reduction of Seysen's orthogonality measure are calculated and the best basis update is retained. If $\lambda_{k,l} = 0$, a local minimum is found and SA ends.

Implementation of SA

The inputs to SA are the channel matrix $\mathbf{H}$, i.e., the original basis of $\mathbb{L}$, the basis of the dual lattice $\mathbb{L}^{'}$, i.e., $\mathbf{H}^{'} = \mathbf{Q}^H$, where $\mathbf{Q} \triangleq (\mathbf{H}^H\mathbf{H})^{-1}\mathbf{H}^H$ is the pseudo-inverse of $\mathbf{H}$, and the corresponding Gram matrices $\mathbf{S} = \mathbf{H}^H\mathbf{H}$ and $\mathbf{S}^{'} = \mathbf{H}^{'H}\mathbf{H}^{'} = (\mathbf{H}^H\mathbf{H})^{-1}$. Let $s_{i,j}$ and $s_{i,j}^{'}$ denote the (i,j)th elements of the matrices $\mathbf{S}$ and $\mathbf{S}^{'}$, respectively.

Initialization

Set $\tilde{\mathbf{H}} = \mathbf{H}$ and $\tilde{\mathbf{H}}^{'} = \mathbf{H}^{'}$, and calculate all possible update values $\lambda_{i,j}$ with their corresponding reduction $\Delta_{i,j}$ of Seysen's orthogonality measure. The update values are calculated as

$$\lambda_{i,j} = \lfloor v_{i,j} \rceil, \text{ where } v_{i,j} \triangleq \frac{s_{j,i}^{'}}{2s_{i,i}^{'}} - \frac{s_{j,i}}{2s_{j,j}}. \tag{4.30}$$

Using this expression, $\Delta_{i,j}$ can be efficiently calculated as follows. The update of the ith basis vector of $\tilde{\mathbf{H}}$ according to (4.25) corresponds to the update of the jth basis vector of $\tilde{\mathbf{H}}^{'}$ according to

$$\tilde{\mathbf{H}}_{i,j}^{'} = [\tilde{\mathbf{h}}_1^{'} \cdots \tilde{\mathbf{h}}_{j-1}^{'}\ \tilde{\mathbf{h}}_j^{'u}\ \tilde{\mathbf{h}}_{j+1}^{'} \cdots \tilde{\mathbf{h}}_{n_t}^{'}] \text{ with } \tilde{\mathbf{h}}_j^{'u} = \tilde{\mathbf{h}}_j^{'} - \lambda_{i,j}^{*}\tilde{\mathbf{h}}_i^{'}. \tag{4.31}$$

We then have

$$\begin{aligned} \Delta_{i,j} &= \mathrm{q}(\tilde{\mathbf{H}}) - \mathrm{q}(\tilde{\mathbf{H}}_{i,j}) \\ &= \|\tilde{\mathbf{h}}_i\|^2\ \|\tilde{\mathbf{h}}_i^{'}\|^2 + \|\tilde{\mathbf{h}}_j\|^2\ \|\tilde{\mathbf{h}}_j^{'}\|^2 - \|\tilde{\mathbf{h}}_i^{u}\|^2\ \|\tilde{\mathbf{h}}_i^{'}\|^2 - \|\tilde{\mathbf{h}}_j\|^2\ \|\tilde{\mathbf{h}}_j^{'u}\|^2. \end{aligned} \tag{4.32}$$

Substituting $\tilde{\mathbf{h}}_i^{u}$, $\tilde{\mathbf{h}}_j^{'u}$, we obtain

$$\begin{aligned} \Delta_{i,j} &= -2\left(|\lambda_{i,j}|^2\|\tilde{\mathbf{h}}_j\|^2\|\tilde{\mathbf{h}}_i^{'}\|^2 + \|\tilde{\mathbf{h}}_i^{'}\|^2\Re\{\lambda_{i,j}\tilde{\mathbf{h}}_i^H\tilde{\mathbf{h}}_j\} - \|\tilde{\mathbf{h}}_j\|^2\Re\{\lambda_{i,j}^{*}\tilde{\mathbf{h}}_j^{'H}\tilde{\mathbf{h}}_i^{'}\}\right) \\ &= 2\left(|\lambda_{i,j}|^2\, s_{j,j}s_{i,i}^{'} + s_{i,i}^{'}\Re\{\lambda_{i,j}s_{i,j}\} - s_{j,j}\Re\{\lambda_{i,j}^{*}s_{j,i}^{'}\}\right) \\ &= 2s_{j,j}s_{i,i}^{'}\left(\Re\left\{\lambda_{i,j}^{*}\left(\frac{s_{j,i}^{'}}{s_{i,i}^{'}} - \frac{s_{i,j}}{s_{j,j}}\right)\right\} - |\lambda_{i,j}|^2\right) \\ &= 2s_{j,j}s_{i,i}^{'}\left(\Re\left\{2\lambda_{i,j}^{*}v_{i,j} - |\lambda_{i,j}|^2\right\}\right). \end{aligned} \tag{4.33}$$

Iteration
Set $\mathbf{T} = \mathbf{I}$ and repeat the following steps until $\tilde{\mathbf{H}}$ is SA-reduced, i.e., $\lambda_{i,j} = 0$ for all (i, j).

1. Select (k, l) according to (4.29) and update $\mathbf{T}$ (4.25), $\tilde{\mathbf{H}}$ (4.26), and $\tilde{\mathbf{H}}^{'}$ (4.31) using $\lambda_{k,l}$ (4.28).
2. Compute corresponding updates of $\mathbf{S}$ and $\mathbf{S}^{'}$.
3. Calculate new $\lambda_{i,j}$ values (4.30) and $\Delta_{i,j}$ values (4.33) for all index pairs corresponding to the updated elements of $\mathbf{S}$ and $\mathbf{S}^{'}$.

Output
The output of SA is given by the unimodular transformation matrix $\mathbf{T}$, the SA-reduced basis $\tilde{\mathbf{H}} = \mathbf{HT}$, and the associated reduced dual basis $\tilde{\mathbf{H}}^{'} = \tilde{\mathbf{Q}}^{'}$.

Performance and complexity
The effectiveness of the near-orthogonalization of the channel matrix achieved by SA for 4×4, 6×6, and 10×10 MIMO channels is illustrated in Fig. 4.2. The entries of the channel matrix $\mathbf{H}$ are independent complex Gaussian variables with zero mean and unit variance. In Fig. 4.2, the complementary cumulative distribution functions (CCDFs) of the normalized orthogonalization measure, given by

$$\overline{\mathrm{q}}(\mathbf{H}) = 1 - \frac{n_t}{\mathrm{q}(\mathbf{H})}, \tag{4.34}$$

are plotted. CCDFs are plotted for the input channel matrix $\mathbf{H}$ (i.e., before SA) and the output matrix $\tilde{\mathbf{H}}$ (i.e., after the SA). These CCDFs for $\overline{\mathrm{q}}(\mathbf{H})$ and $\overline{\mathrm{q}}(\tilde{\mathbf{H}})$ are obtained through simulation of 10^8 channel realizations. It is seen that the performance in terms of orthogonalization measure improves significantly for $\tilde{\mathbf{H}}$ compared to that for $\mathbf{H}$. This improved orthogonality of $\overline{\mathrm{q}}(\tilde{\mathbf{H}})$ results in significantly improved BER performance of LR-MMSE detection compared to MMSE detection. This is illustrated in Fig. 4.3 for a V-BLAST MIMO system with $n_t = n_r = 4$ and 4-QAM.

The complexity of SA is $O(n_t^2)$. This can be explained as follows. The initialization of SA requires the calculation of $n_t^2 - n_t$ different λ values according to (4.30) and at most (if the corresponding $\lambda_{i,j}$s are all non-zero) $n_t^2 - n_t$ different Δ values according to (4.32). Therefore, the initialization step has a complexity of $O(n_t^2)$. At each iteration, the update of $\mathbf{T}$, $\tilde{\mathbf{H}}$, $\tilde{\mathbf{H}}^{'}$ has a complexity of $O(n_t)$, resulting in a per-iteration complexity that is linear in n_t. The total complexity of SA is therefore dominated by the initialization step, which is $O(n_t^2)$. This complexity is one order less than the complexity of the matrix inversion in the ZF/MMSE operation (cubic in n_t). Therefore, the orders of total complexity of LR-aided ZF/MMSE detection and ZF/MMSE detection are both cubic in n_t.

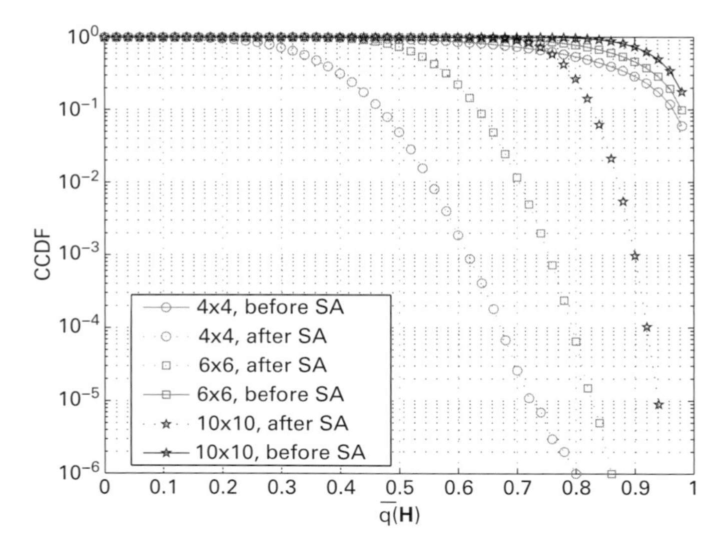

Figure 4.2 CCDF of $\overline{q}(\mathbf{H})$ (before SA) and $\overline{q}(\tilde{\mathbf{H}})$ (after SA).

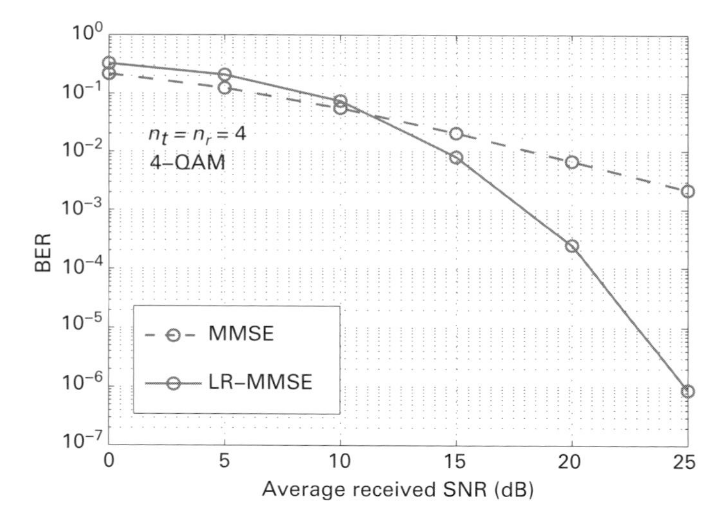

Figure 4.3 BER performance comparison between MMSE detection and LR-MMSE detection. $n_t = n_r = 4$, 4-QAM.

4.6 Sphere decoding

Sphere decoding is a detection method that obtains the exact ML solution, generally faster than a brute-force exhaustive search. The ML optimization problem in (4.5) formulated using the complex-valued system model (4.2) can be written using the real-valued system model (4.4) as

$$\min_{\mathbf{x}_r \in \mathbb{B}^{2n_t}} \|\mathbf{y}_r - \mathbf{H}_r \mathbf{x}_r\|^2, \tag{4.35}$$

where $\mathbb{B}$ is the underlying pulse amplitude modulation (PAM) alphabet corresponding to the QAM alphabet $\mathbb{A}$. The above problem is referred to as an integer least-squares (ILS) problem. Since the elements of $\mathbf{x}_r$ are points from a PAM constellation, $\mathbf{x}_r$ spans a rectangular $2n_t$-dimensional lattice. For any lattice-generating matrix $\mathbf{H}_r$, the $2n_r$-dimensional vector $\mathbf{H}_r\mathbf{x}_r$ spans a skewed lattice. Given the skewed lattice $\mathbf{H}_r\mathbf{x}_r$ and the vector $\mathbf{y}_r \in \mathbb{R}^{2n_r}$, the ILS problem is to find the closest lattice point to $\mathbf{y}_r$ in terms of the Euclidean distance. The idea behind sphere decoding is that, instead of an exhaustive search over the entire lattice, the search is over only lattice points that lie in a $2n_t$-dimensional sphere of radius d around $\mathbf{y}_r$. This reduces the search space/effort and hence the complexity. The closest lattice point inside the sphere will also be the closest lattice point for the whole lattice.

A key question here is how to choose the radius d. A large d means more points to test and hence more complexity. A small d, however, can lead to no or too few points inside the sphere. Another key question is how to tell which lattice points are inside the sphere. Sphere decoding addresses this second question by determining all lattice points inside a $2n_t$-dimensional sphere of radius d by successively determining all lattice points in spheres of lower dimensions and the same radius d. This is motivated by the observation that finding the lattice points in a one-dimensional lattice is easy; a one-dimensional sphere reduces to the end points of an interval, and so the lattice points will be the integer points in this interval. Suppose all one-dimensional lattice points within a radius of sphere d are already determined. It is then easy to see that for any such one-dimensional point, the set of admissible values of the second dimensional coordinate that lie in the two-dimensional sphere of the same radius d forms an interval. Thus it is possible to go successively from dimension k to dimension $k+1$. The algorithm exploiting the above observation is described below.

Algorithm

Assume that $2n_t \leq 2n_r$ so that there are at least as many equations as unknowns. The lattice point $\mathbf{H}_r\mathbf{x}_r$ lies inside a sphere of radius d if and only if

$$d^2 \geq \|\mathbf{y}_r - \mathbf{H}_r\mathbf{x}_r\|^2. \tag{4.36}$$

In order to decompose the $2n_t$-dimensional problem into multiple one-dimensional subproblems, consider the QR decomposition of the matrix $\mathbf{H}_r$ given by

$$\mathbf{H}_r = [\mathbf{Q}_1 \ \mathbf{Q}_2] \begin{bmatrix} \mathbf{R} \\ \mathbf{0}_{(2n_r - 2n_t)\times 2n_t} \end{bmatrix}, \tag{4.37}$$

where $\mathbf{R}$ is a $2n_t \times 2n_t$ upper triangular matrix, $\mathbf{Q}_1$ is a $2n_r \times 2n_t$ matrix, $\mathbf{Q}_2$ is a $2n_r \times (2n_r - 2n_t)$ matrix, and $\mathbf{Q} = [\mathbf{Q}_1 \ \mathbf{Q}_2]$ is a $2n_r \times 2n_r$ orthogonal matrix.

Using (4.37), the condition in (4.36) can be written as

$$
\begin{aligned}
d^2 &\geq \left\| \mathbf{y}_r - [\mathbf{Q}_1 \ \mathbf{Q}_2] \begin{bmatrix} \mathbf{R} \\ \mathbf{0}_{(2n_r-2n_t)\times 2n_t} \end{bmatrix} \mathbf{x}_r \right\|^2 \\
&= \left\| \begin{bmatrix} \mathbf{Q}_1^H \\ \mathbf{Q}_2^H \end{bmatrix} \mathbf{y}_r - \begin{bmatrix} \mathbf{Q}_1^H \\ \mathbf{Q}_2^H \end{bmatrix} [\mathbf{Q}_1 \ \mathbf{Q}_2] \begin{bmatrix} \mathbf{R} \\ \mathbf{0} \end{bmatrix} \mathbf{x}_r \right\|^2 \\
&= \|\mathbf{Q}_1^H \mathbf{y}_r - \mathbf{R}\mathbf{x}_r\|^2 + \|\mathbf{Q}_2^H \mathbf{y}_r\|^2. \qquad (4.38)
\end{aligned}
$$

Defining $\tilde{d}^2 \triangleq d^2 - \|\mathbf{Q}_2^H \mathbf{y}_r\|$, (4.38) can be written as

$$\tilde{d}^2 \geq \|\mathbf{Q}_1^H \mathbf{y}_r - \mathbf{R}\mathbf{x}_r\|^2. \qquad (4.39)$$

Further, defining $\mathbf{z} \triangleq \mathbf{Q}_1^H \mathbf{y}_r$, (4.39) can be written as

$$\tilde{d}^2 \geq \sum_{i=1}^{2n_t} \left(z_i - \sum_{j=1}^{2n_t} r_{i,j} x_j \right)^2, \qquad (4.40)$$

where z_i is the ith element in $\mathbf{z}$, x_j is the jth element in $\mathbf{x}_r$, and $r_{i,j}$ is the (i, j)th element of $\mathbf{R}$. Because of the upper triangular nature of $\mathbf{R}$, the right-hand side of the inequality in (4.40) can be expanded as

$$
\begin{aligned}
&(z_{2n_t} - r_{2n_t,2n_t} \ x_{2n_t})^2 \\
&+ \ (z_{2n_t-1} - r_{2n_t-1,2n_t} \ x_{2n_t} - r_{2n_t-1,2n_t-1} \ x_{2n_t-1})^2 \\
&+ \ (z_{2n_t-2} - r_{2n_t-2,2n_t} \ x_{2n_t} - r_{2n_t-2,2n_t-1} \ x_{2n_t-1} - r_{2n_t-2,2n_t-2} \ x_{2n_t-2})^2 \\
&+ \ \cdots, \qquad (4.41)
\end{aligned}
$$

where the first term depends only on x_{2n_t}, the second term depends only on (x_{2n_t}, x_{2n_t-1}), the third term depends only on $(x_{2n_t}, x_{2n_t-1}, x_{2n_t-2})$, and so on. Therefore, a necessary (not sufficient) condition for $\mathbf{H}_r\mathbf{x}_r$ to lie inside the sphere is that

$$\tilde{d}^2 \geq (z_{2n_t} - r_{2n_t,2n_t} \ x_{2n_t})^2, \qquad (4.42)$$

which is equivalent to x_{2n_t} belonging to the interval

$$\left\lceil \frac{-\tilde{d} + z_{2n_t}}{r_{2n_t,2n_t}} \right\rceil^{\circledR} \leq x_{2n_t} \leq \left\lfloor \frac{\tilde{d} + z_{2n_t}}{r_{2n_t,2n_t}} \right\rfloor^{\circledS}, \qquad (4.43)$$

where $[.]^{\circledR}$ and $[.]^{\circledS}$ denote rounding to the nearest larger element and smaller element, respectively, in the PAM constellation that spans the lattice. For every x_{2n_t} satisfying (4.43), defining

$$\tilde{d}^2_{2n_t-1} \triangleq \tilde{d}^2 - (z_{2n_t} - r_{2n_t,2n_t} \ x_{2n_t})^2, \qquad (4.44)$$

and

$$z_{2n_t-1|2n_t} \triangleq z_{2n_t-1} - r_{2n_t-1,2n_t} \ x_{2n_t}, \qquad (4.45)$$

Algorithm 1. Sphere decoding algorithm

1. **input:** $\mathbf{y}_r$, $\mathbf{H}_r$, and d.
   From $\mathbf{H}_r$ and $\mathbf{y}_r$, obtain $\mathbf{Q} = [\mathbf{Q}_1\ \mathbf{Q}_2]$, $\mathbf{R}$, and $\mathbf{z} = \mathbf{Q}_1^H \mathbf{y}_r$.
2. Set $k = 2n_t$; $\quad \tilde{d}_{2n_t}^2 = d^2 - \|\mathbf{Q}_2^H \mathbf{y}_r\|^2$; $\quad z_{2n_t|2n_t+1} = z_{2n_t}$.
3. Set $UB(x_k) = \left[\dfrac{\tilde{d}_k + z_{k|k+1}}{r_{k,k}}\right]^{Ⓢ}$; $\quad x_k = \left[\dfrac{-\tilde{d}_k + z_{k|k+1}}{r_{k,k}}\right]^{Ⓡ} - 1.$
4. $x_k = x_k + 1$;
   if $x_k \le UB(x_k)$ **then** go to Step 6; **else** go to Step 5.
5. $k = k + 1$;
   if $k = 2n_t + 1$ **then** terminate algorithm; **else** go to Step 4.
6. **if** $k = 1$ **then** go to Step 7;
   else
   $k = k - 1$; $\quad z_{k|k+1} = z_k - \sum_{j=k+1}^{2n_t} r_{k,j} x_j$;
   $\tilde{d}_k^2 = \tilde{d}_{k+1}^2 - (z_{k+1|k+2} - r_{k+1,k+1}\ x_{k+1})^2$;
   go to Step 3;
7. Solution found. Save $\mathbf{x}_r$ and its distance from $\mathbf{y}_r$, $\tilde{d}_{2n_t}^2 - \tilde{d}_1^2 + (z_1 - r_{1,1}x_1)^2$;
   go to Step 4.

a stronger condition can be found by looking at the first two terms in (4.41), which leads to x_{2n_t-1} belonging to the interval

$$\left[\frac{-\tilde{d}_{2n_t-1} + z_{2n_t-1|2n_t}}{r_{2n_t-1,2n_t-1}}\right]^{Ⓡ} \le x_{2n_t-1} \le \left[\frac{\tilde{d}_{2n_t-1} + z_{2n_t-1|2n_t}}{r_{2n_t-1,2n_t-1}}\right]^{Ⓢ}. \tag{4.46}$$

We can continue in a similar manner for x_{2n_t-2}, x_{2n_t-3} and so on until x_1, thereby all points belonging to (4.36) are obtained. The algorithm listing is given in Algorithm 1 [3].

The search radius d can be chosen to be proportional to σ^2 (the noise variance), i.e., choose large d for large σ^2 (low SNRs) and small d for small σ^2 (high SNRs). Specifically, d can be chosen to be a scaled variance of the noise, i.e., $d^2 = 2\alpha n_r \sigma^2$, in such a way that a lattice point is found inside the sphere with high probability. Since $\|\mathbf{n}^2\|$ is χ^2 distributed with $2n_r$ degrees of freedom, this can be achieved by

$$\int_0^{\alpha n_r} \frac{\lambda^{n_r-1} e^{-\lambda}}{\Gamma(n_r)} d\lambda = 1 - \epsilon, \tag{4.47}$$

where $1 - \epsilon$ is set to a value close to 1, e.g., $\epsilon = 0.01$. If the point is not found, increase the probability $1 - \epsilon$, adjust the radius, and search again. For a point $\hat{\mathbf{x}}_r$ found inside the sphere, $\mathbf{H}_r\hat{\mathbf{x}}_r$ need not be the closest point to $\mathbf{y}_r$. So, whenever the algorithm finds a point $\hat{\mathbf{x}}_r$ inside the sphere, set the new radius as $d^2 = \|\mathbf{y}_r - \mathbf{H}_r \mathbf{x}_r\|^2$ and restart the algorithm. Such radius updating may be useful in low SNRs, where the number of points inside the initial sphere can be large.

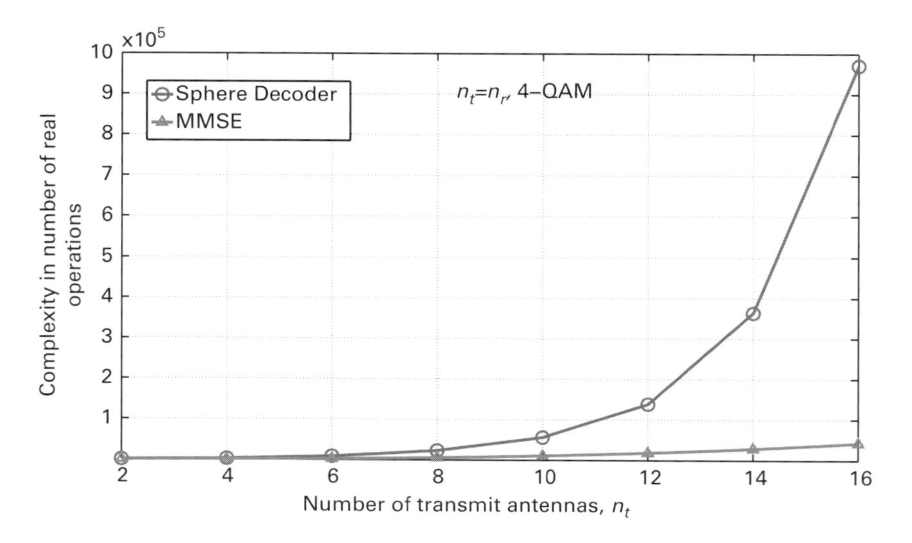

Figure 4.4 Sphere decoding and MMSE complexity at 10^{-2} BER for V-BLAST MIMO with $n_t = n_r = 4$ and 4-QAM.

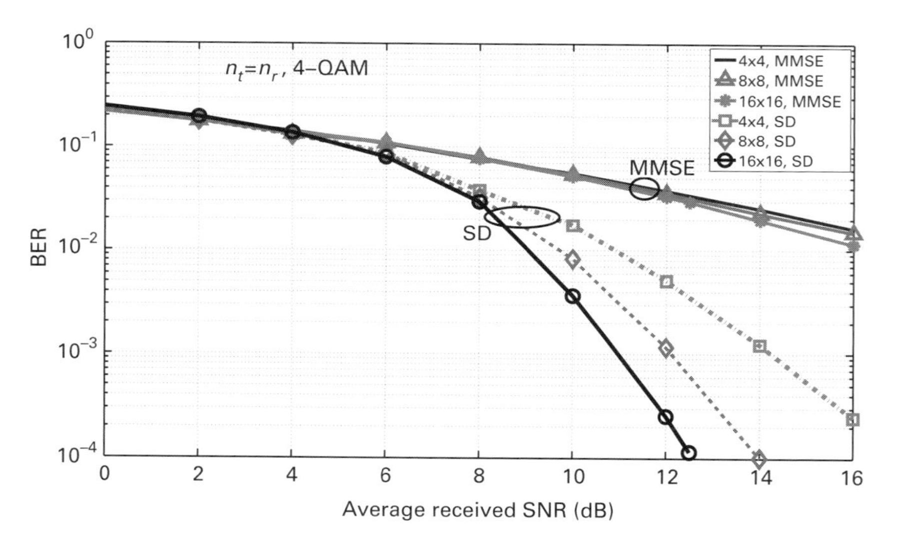

Figure 4.5 BER performance of sphere decoding and MMSE detection for V-BLAST MIMO with $n_t = n_r = 4, 8, 16$ and 4-QAM.

The choice of radius d being statistical in nature, and $\mathbf{H}$ and $\mathbf{n}$ being random, the computational complexity of sphere decoder is a random variable, whose mean and variance can be computed [14]. Sphere decoding is very efficient in terms of complexity at high SNRs due to the small search radius. However, it is inefficient at low to moderate SNRs due to the increased search radius [3]. This is illustrated in Fig. 4.4, which shows the complexity of SD in the number of real operations at a BER target of 10^{-2}. It is seen that at this target BER (which is not as small as it would be at high SNRs), the complexity grows exponentially in n_t. LR techniques can be used as preprocessors to sphere decoding in order to reduce complexity [37]. Several variants of the SD have also been proposed to reduce complexity. Still, although the SD and several of its low complexity variants achieve ML performance (Fig. 4.5 shows the BER performance of sphere decoder in V-BLAST MIMO with $n_t = n_r = 4, 8, 16$ and 4-QAM), their complexity in low to moderate SNRs becomes prohibitive beyond 32 real dimensions [13],[14], making them inadequate for large MIMO systems.

References

[1] D. Tse and P. Viswanath, *Fundamentals of Wireless Communication.* Cambridge, UK: Cambridge University, 2005.

[2] S. Verdu, *Multiuser Detection.* Cambridge, UK: Cambridge University Press, 1998.

[3] T. Kailath, H. Vikalo, and B. Hassibi, "MIMO receive algorithms," in *Space-Time Wireless Systems: From Array Processing to MIMO Communications*, H. Bolcskei, D. Gesbert, C. B. Papadias, and A. jan van der veen, Eds. Cambridge, UK: Cambridge University Press, 2006, ch. 15.

[4] G. J. Foschini, "Layered space-time architecture for wireless communication in a fading environment when using multi-element antennas," *Bell Labs. Tech. J.*, vol. 1, no. 2, pp. 41–59, 1996.

[5] P. W. Wolniansky, G. J. Foschini, G. D. Golden, and R. A. Valenzuela, "V-BLAST: an architecture for realizing very high data rates over the rich-scattering wireless channel," in *URSI Intl. Symp. Signals, Systems and Electronics (ISSSE)*, Sept.-Oct. 1998, pp. 295–300.

[6] R. Kohno, "Pseudo-noise sequences and interference cancellation techniques for spread spectrum systems – spread spectrum theory and techniques in Japan," *IEICE Trans. Commun.*, vol. E74-B, no. 5, pp. 1083–1092, Jun. 2003.

[7] P. Patel and J. Holtzman, "Analysis of a simple successive interference cancellation scheme in a DS/CDMA system," *IEEE J. Sel. Areas in Commun.*, vol. 12, no. 5, pp. 796–807, Jun. 1994.

[8] B. Hassibi, "An efficient square-root algorithm for BLAST," in *IEEE ICASSP'2000*, Istanbul, vol. 2, Jun. 2000, pp. 737–740.

[9] C. P. Schnoor and M. Euchner, "Lattice basis reduction: improved practical algorithms and solving subset sum problems," *Mathematical Programming*, vol. 66, pp. 181–191, 1994.

[10] B. LaMacchia, "Basis reduction algorithms and subset sum problems," Master's Thesis, MIT, May 1991.

[11] M. Seysen, "Simultaneous reduction of a lattice basis and its reciprocal basis," *Combinatorica*, vol. 13, pp. 363–376, 1993.

[12] D. Seethaler, G. Matz, and F. Hlawatsch, "Low complexity MIMO data detection using Seysen's lattice reduction algorithm," in *IEEE ICASSP'2007*, Honolulu, HI, Apr. 2007, pp. 53–57.

[13] E. Viterbo and J. Boutros, "A universal lattice code decoder for fading channels," *IEEE Trans. Inform. Theory*, vol. 45, no. 5, pp. 1639–1642, Jul. 1999.

[14] B. Hassibi and H. Vikalo, "On the sphere decoding algorithm I. Expected complexity," *IEEE Trans. Signal Process.*, vol. 53, no. 8, pp. 2806–2818, Aug. 2005.

[15] H. Vikalo and B. Hassibi, "On the sphere-decoding algorithm II. Generalizations, second-order statistics, and applications to communications," *IEEE Trans. Signal Process.*, vol. 53, no. 8, pp. 2819–2834, Aug. 2005.

[16] Y. Wang and K. Roy, "A new reduced complexity sphere decoder with true lattice boundary awareness for multi-antenna systems," in *IEEE ISCAS'2005*, Kobe, vol. 5, May 2005, pp. 4963–4966.

[17] L. G. Barbero and J. S. Thompson, "Fixing the complexity of the sphere decoder for MIMO detection," *IEEE Trans. Wireless Commun.*, vol. 7, no. 6, pp. 2131–2142, Jun. 2008.

[18] P. H. Tan and L. K. Rasmussen, "The application of semidefinite programming for detection in CDMA," *IEEE J. Sel. Areas Commun.*, vol. 19, no. 8, pp. 1442–1449, Aug. 2001.

[19] ——, "Multiuser detection in CDMA – a comparison of relaxations, exact, and heuristic search methods," *IEEE Trans. Wireless Commun.*, vol. 3, no. 5, pp. 1802–1809, Sep. 2004.

[20] N. D. Sidiropoulos and Z.-Q. Luo, "A semidefinite relaxation approach to MIMO detection for high-order QAM constellations," *IEEE Signal Process. Lett.*, vol. 13, no. 9, pp. 525–528, Sep. 2006.

[21] J. Luo, K. R. Pattipati, P. K. Willett, and F. Hasegawa, "Near-optimal multiuser detection in synchronous CDMA using probabilistic data association," *IEEE Commun. Lett.*, vol. 5, no. 9, pp. 361–363, Sep. 2001.

[22] D. Pham, K. R. Pattipati, P. K. Willet, and J. Luo, "A generalized probabilistic data association detector for multiantenna systems," *IEEE Commun. Lett.*, vol. 8, no. 4, pp. 205–207, Apr. 2004.

[23] Y. Kabashima, "A CDMA multiuser detection algorithm on the basis of belief propagation," *J. Phys. A: Math. Gen.*, vol. 36, pp. 11 111–11 121, Oct. 2003.

[24] M. N. Kaynak, T. M. Duman, and E. M. Kurtas, "Belief propagation over SISO/MIMO frequency selective fading channels," *IEEE Trans. Wireless Commun.*, vol. 6, no. 6, pp. 2001–2005, Jun. 2007.

[25] B. M. Hochwald, T. L. Marzetta, and V. Tarokh, "Multiple-antenna channel hardening and its implications for rate feedback and scheduling," *IEEE Trans. Inform. Theory*, vol. 50, no. 9, pp. 1893–1909, Sep. 2004.

[26] H. Q. Ngo, E. G. Larsson, and T. L. Marzetta, "Uplink power efficiency of multiuser MIMO with very large antenna arrays," in *Allerton Conf. on Commun., Contr., and Comput.*, Monticello, IL, Sep. 2011, pp. 1272–1279.

[27] J. Hoydis, S. ten Brink, and M. Debbah, "Massive MIMO in the UL/DL of cellular networks: how many antennas do we need?" *IEEE J. Sel. Areas in Commun.*, vol. 31, no. 2, pp. 160–171, Feb. 2013.

[28] K. V. Vardhan, S. K. Mohammed, A. Chockalingam, and B. S. Rajan, "A low-complexity detector for large MIMO systems and multicarrier CDMA systems," *IEEE J. Sel. Areas Commun.*, vol. 26, no. 3, pp. 473–485, Apr. 2008.

[29] S. K. Mohammed, A. Zaki, A. Chockalingam, and B. S. Rajan, "High-rate space-time coded large-MIMO systems: low-complexity detection and channel estimation," *IEEE J. Sel. Topics Signal Process.*, vol. 3, no. 6, pp. 958–974, Dec. 2009.

[30] S. K. Mohammed, A. Chockalingam, and B. S. Rajan, "Low-complexity near-MAP decoding of large non-orthogonal STBCs using PDA," in *IEEE ISIT'2009*, Seoul, Jun.-Jul. 2009, pp. 1998–2002.

[31] P. Som, T. Datta, N. Srinidhi, A. Chockalingam, and B. S. Rajan, "Low-complexity detection in large-dimension MIMO-ISI channels using graphical models," *IEEE J. Sel. Topics Signal Process.*, vol. 5, no. 8, pp. 1497–1511, Dec. 2011.

[32] T. Datta, N. A. Kumar, A. Chockalingam, and B. S. Rajan, "A novel MCMC algorithm for near-optimal detection in large-scale uplink mulituser MIMO systems," in *ITA'2012*, San Diego, CA, Feb. 2012, pp. 69–77.

[33] X. Zhu and R. D. Murch, "Performance analysis of maximum-likelihood detection in a MIMO antenna system," *IEEE Trans. Commun.*, vol. 50, no. 2, pp. 187–191, Feb. 2002.

[34] N. Srinidhi, T. Datta, A. Chockalingam, and B. S. Rajan, "Layered tabu search algorithm for large-MIMO detection and a lower bound on ml performance," *IEEE Trans. Commun.*, vol. 59, no. 11, pp. 2955–2963, Nov. 2011.

[35] H. Yao and G. W. Wornell, "Lattice-reduction-aided detectors for MIMO communication systems," in *IEEE GLOBECOM'2002*, Taipei, vol. 1, Nov. 2002, pp. 424–428.

[36] A. K. Lenstra, H. W. Lenstra, and L. Lovasz, "Factoring polynomials with rational coefficients," *Math. Ann.*, vol. 261, pp. 515–534, 1982.

[37] M. O. Damen, H. El Gamal, and G. Caire, "On maximum-likelihood detection and the search for the closest lattice point," *IEEE Trans. Inform. Theory*, vol. 49, no. 10, pp. 2389–2402, Oct. 2003.

5 Detection based on local search

Local search has grown from a simple heuristic idea into an important and mature field of research in combinatorial optimization [1]. When confronted with NP-hard problems, one can resort to (*i*) an enumerative method that is guaranteed to produce an optimal solution, or (*ii*) an approximation algorithm that runs in polynomial time, or (*iii*) some kind of heuristic technique without any guarantee on the quality of the solution and running time. The first approach of true optimization algorithms may become prohibitive due to the problem of size or the lack of insight into the problem structure. The second approach of polynomial-time approximation algorithms, though characterizable by performance bounds, may give inferior solutions. The third approach of heuristics is the preferred choice for NP-hard problems, as it provides a robust means to obtain good solutions to problems of large size in a reasonable time. Local search techniques come under the third approach. Optimum signal detection in MIMO systems involves the minimization of a certain cost over a discrete signal space, where the exhaustive enumerative approach becomes prohibitive when the number of signaling dimensions becomes large. Therefore, the local search approach can be a considered choice for signal detection in MIMO systems with a large number of antennas.

An important characteristic of a local search algorithm is its neighborhood function/definition which guides the search to a good solution. Typically, a local search algorithm starts with an initial solution (often generated by some other low complexity algorithm or just generated randomly) and then continually attempts to find better solutions by searching in the neighborhoods defined by the neighborhood function. A basic version of local search is based on iterative improvement, where the algorithm starts with some initial solution and searches its neighborhood for a solution of lesser cost. If such a solution is found, it replaces the current solution and the search continues. Otherwise, it returns the current solution which is a locally optimal solution. Variants of this iterative algorithm with different neighborhood definitions and stopping criteria offer a rich tradeoff between performance and complexity. Neighborhood definitions depend on the problem under consideration, and finding suitable and efficient neighborhood functions/definitions that can lead to high-quality local optima can be viewed as one of the crucial requirements in local search. Also, the quality of the local optima reached has been found to depend a lot on the initial solution chosen.

A simple neighborhood definition is one where a given solution's neighborhood consists of all solutions which differ from the given solution in only one coordinate (referred to as a "1-coordinate away" neighborhood). A key advantage of this neighborhood definition is that the size of this neighborhood grows only linearly in the problem size. Let n denote the problem size and $\mathcal{N} \triangleq \{\pm 1\}^n$ denote the entire solution space. The size of $\mathcal{N}$ grows exponentially in n, i.e., $|\mathcal{N}| = 2^n$ which includes the enumeration of all possible n-length binary vectors. However, the size of the 1-coordinate away neighborhood of a solution $\mathbf{x}$, denoted by $\mathcal{N}_1(\mathbf{x})$, is $|\mathcal{N}_1(\mathbf{x})| = n$. For example, if $n = 4$ and $\mathbf{x} = [-1, 1, 1, -1]$, then $|\mathcal{N}_1(\mathbf{x})| = 4$ where

$$\mathcal{N}_1(\mathbf{x}) = \Big\{ [\,\textcircled{1}, 1, 1, -1], [-1, \textcircled{-1}, 1, -1], [-1, 1, \textcircled{-1}, -1], [-1, 1, 1, \textcircled{1}] \Big\}.$$

Hence, complexity-wise, the 1-coordinate away neighborhood definition is very attractive for large n. This neighborhood is also referred to as a *1-opt* neighborhood in the literature [2]. A more general neighborhood, where the neighbors differ from the given solution in *at most* k coordinates, $1 \leq k \leq n$, is referred to as a *k-opt* neighborhood [2]. While the solution quality improves for large k under this neighborhood definition, it comes at the cost of increased complexity which grows exponentially in k. Note that $k = n$ under this definition results in the entire solution space $\mathcal{N}$. Since it becomes computationally expensive to search the full *k-opt* neighborhood, one can search only a fraction of the *k-opt* neighborhood [3],[4]. Several variants are possible in choosing the fraction of neighbors. A low complexity neighborhood could be a "k-coordinate away" neighborhood, where the neighbors differ from the given solution in *only* k coordinates, leading to a neighborhood size of

$$|\mathcal{N}_k(\mathbf{x})| = \binom{n}{k}.$$

For example, for $n = 4$ and $\mathbf{x} = [-1, 1, 1, -1]$, the size of the 2-coordinate away neighborhood is $|\mathcal{N}_2(\mathbf{x})| = 6$, where

$$\begin{aligned} \mathcal{N}_2(\mathbf{x}) = \Big\{ & [\textcircled{1}, \textcircled{-1}, 1, -1], [-1, \textcircled{-1}, \textcircled{-1}, -1], [-1, 1, \textcircled{-1}, \textcircled{1}], \\ & [\textcircled{1}, 1, \textcircled{-1}, -1], [-1, \textcircled{-1}, 1, \textcircled{1}], [\textcircled{1}, 1, 1, \textcircled{1}] \Big\}. \end{aligned}$$

Local search performance can be improved by executing *escape strategies* when the algorithm becomes trapped in local optima. The idea is not to stop at the local optima encountered, but to continue the search beyond them. Alternatively put, the idea is to direct the search away from the local optima so that other parts of the search space can be explored for better solutions. Changing the neighborhood definition (e.g., 1-coordinate away neighborhood to 2-coordinate away neighborhood) on reaching a local optimum is a typical escape strategy. Another strategy is to move to the local optimum's best neighboring solution (although this best neighboring solution will be worse than the local optimum solution) and continue the search. Tabu search [5],[6] adopts this second strategy. Escape strategies must be devised along with suitable stopping criteria so that the resulting complexity is not unduly high. In addition to the escape strategy, tabu search uses other features like avoiding already visited solutions using memory structures (e.g., a tabu list) and allowing revisits after a certain tabu period to enhance the efficiency of the search. Another common way to improve local search performance is *multistart* search, where the local search procedure is run several times, each time starting with a different random initial solution and declaring the best solution among the multiple runs.

Analyzing the performance and complexity of particular local search algorithms is generally not a simple task. It is often difficult to get non-trivial bounds on the solution quality (in terms of the amount by which the local optimum cost differs from the optimal cost) and running time of local search. Nevertheless, in practice, many local search algorithms are known to converge quickly and find high-quality solutions. The flexibility and ease of implementation of local search algorithms have resulted in the successful handling of many complex real-world problems.

The local search approach has been adopted in communication problems as well. MUD in CDMA [7] is one prominent area where local search techniques have been widely adopted [8]–[17]. The problem is to jointly detect the binary symbols sent by multiple users to the BS. This can be considered as a problem of optimizing a quadratic objective function (ML cost) with binary constraints on decision variables – referred to as a binary quadratic program (BQP) in the area of combinatorial optimization. Local search algorithms including *1-opt* and *k-opt* local search have been proposed for the BQP problem and were found to be capable of quickly finding near-optimal solutions for large problem sizes [2]. *1-opt* and *k-opt* searches along with the multistart strategy are found to be effective for large problems. Since k-coordinate away neighborhoods are obtained by flipping one or more bits, algorithms using them are also referred to as bit-flipping algorithms.

In CDMA MUD, *1-opt* and *k-opt* local search have been widely studied and have been shown to achieve better performance than other detectors [8]–[15]. MF, decorrelating (ZF), and decision feedback detectors have been used to generate the initial solution for the search leading to better solutions than the initial solutions themselves [10],[12]. Near-single-user performance in CDMA systems with a large number of users has been reported using LAS, a greedy *1-opt* search [15]. CDMA MUD using tabu search has been reported in [16],[17]. In [18], k-opt local search has been used for reduced complexity turbo equalization in coded ISI channels. With the growing popularity of MIMO systems, local search is being adopted for MIMO detection [19]–[26].

The rest of this chapter is devoted to illustrating how local search methods can be effectively exploited to achieve near-optimum signal detection in MIMO systems with a large number of antennas at affordable complexities. In particular, the focus will be on local search algorithms that have been established to be effective for large MIMO systems. These algorithms include variants of k-coordinate away neighborhood search, randomized search, and variants of tabu search. While the basic principles behind these algorithms are known in the general local search literature, the details regarding their suitable adoption in the detection problem in the large MIMO context, defining suitable neighborhoods for higher-order modulation (e.g., M-QAM), choosing suitable memory structures for tabooing solutions in tabu search, devising efficient stopping criteria, achieving near-optimal performance in higher-order QAM, generating soft output from discrete output, etc. need careful attention, without which the desired performance and complexity goals may not be met.

5.1 LAS

In this section, variants of 1-, 2-, ... , K-coordinate away neighborhood based local search algorithms that work well for large MIMO systems are presented.

5.1.1 System model

Consider a V-BLAST system with n_t transmit antennas and n_r receive antennas, $n_t \leq n_r$, where n_t symbols are transmitted from n_t transmit antennas simultaneously. Let $\mathbf{x}_c \in \mathbb{C}^{n_t \times 1}$ be the symbol vector transmitted. Each element of $\mathbf{x}_c$ is an M-PAM or M-QAM symbol. M-PAM symbols take discrete values from $\{A_m, m = 1, 2, \ldots, M\}$, where $A_m = (2m - 1 - M)$, and M-QAM is nothing but two PAMs in quadrature. Let $\mathbf{H}_c \in \mathbb{C}^{n_r \times n_t}$ be the channel gain matrix, such that its (p, q)th entry $h_{p,q}$ is the complex channel gain from the qth transmit antenna to the pth receive antenna. Assuming rich scattering, the entries of $\mathbf{H}_c$ are modeled as iid $\mathcal{CN}(0, 1)$. Let $\mathbf{y}_c \in \mathbb{C}^{n_r \times 1}$ and $\mathbf{n}_c \in \mathbb{C}^{n_r \times 1}$ denote the received signal vector and the noise vector, respectively, at the receiver, where the entries of $\mathbf{n}_c$ are modeled as iid $\mathcal{CN}(0, \sigma^2)$. The received signal vector can then be written as

$$\mathbf{y}_c = \mathbf{H}_c \mathbf{x}_c + \mathbf{n}_c. \tag{5.1}$$

Let $\mathbf{y}_c$, $\mathbf{H}_c$, $\mathbf{x}_c$, and $\mathbf{n}_c$ be decomposed into real and imaginary parts as follows:

$$\mathbf{y}_c = \mathbf{y}_I + j\mathbf{y}_Q, \quad \mathbf{x}_c = \mathbf{x}_I + j\mathbf{x}_Q, \quad \mathbf{n}_c = \mathbf{n}_I + j\mathbf{n}_Q, \quad \mathbf{H}_c = \mathbf{H}_I + j\mathbf{H}_Q.$$

Further, define $\mathbf{H}_r \in \mathbb{R}^{2n_r \times 2n_t}$, $\mathbf{y}_r \in \mathbb{R}^{2n_r \times 1}$, $\mathbf{x}_r \in \mathbb{R}^{2n_t \times 1}$, and $\mathbf{n}_r \in \mathbb{R}^{2n_r \times 1}$ as

$$\mathbf{H}_r = \begin{pmatrix} \mathbf{H}_I & -\mathbf{H}_Q \\ \mathbf{H}_Q & \mathbf{H}_I \end{pmatrix}, \quad \mathbf{y}_r = [\mathbf{y}_I^T \ \mathbf{y}_Q^T]^T, \quad \mathbf{x}_r = [\mathbf{x}_I^T \ \mathbf{x}_Q^T]^T, \quad \mathbf{n}_r = [\mathbf{n}_I^T \ \mathbf{n}_Q^T]^T.$$

Now, the complex-valued system model (5.1) can be written in an equivalent real-valued system model as

$$\mathbf{y}_r = \mathbf{H}_r \mathbf{x}_r + \mathbf{n}_r. \tag{5.2}$$

Dropping the subscript r for notational simplicity, the real-valued system model is written as

$$\mathbf{y} = \mathbf{H}\mathbf{x} + \mathbf{n}, \tag{5.3}$$

where $\mathbf{H} = \mathbf{H}_r \in \mathbb{R}^{2n_r \times 2n_t}$, $\mathbf{y} = \mathbf{y}_r \in \mathbb{R}^{2n_r \times 1}$, $\mathbf{x} = \mathbf{x}_r \in \mathbb{R}^{2n_t \times 1}$, and $\mathbf{n} = \mathbf{n}_r \in \mathbb{R}^{2n_r \times 1}$.

With the above real-valued system model, the real part of the original complex data symbols will be mapped to $[x_1, \ldots, x_{n_t}]$ and the imaginary part of these symbols will be mapped to $[x_{n_t+1}, \ldots, x_{2n_t}]$. For M-PAM, $[x_{n_t+1}, \ldots, x_{2n_t}]$ will be zeros since M-PAM symbols take only real values. In the case of M-QAM, $[x_1, \ldots, x_{n_t}]$ can viewed to be from an underlying PAM signal set and so is $[x_{n_t+1}, \ldots, x_{2n_t}]$. Let $\mathbb{A}_i$ denote the PAM signal set from which x_i takes values $i = 1, 2, \ldots, 2n_t$. For example, for 4-PAM, $\mathbb{A}_i = \{-3, -1, 1, 3\}$ for $i = 1, 2, \ldots, n_t$ and $\mathbb{A}_i = \{0\}$ for $i = n_t + 1, \ldots, 2n_t$. Similarly, for 4-QAM, after transforming the system into an equivalent real-valued system, $\mathbb{A}_i = \{1, -1\}$ for $i = 1, 2, \ldots, 2n_t$. Now, the $2n_t$-dimensional signal space $\mathcal{S}$ is the Cartesian product of $\mathbb{A}_1, \mathbb{A}_2, \ldots, \mathbb{A}_{2n_t}$. Assuming that all elements in $\mathbf{x}$ take values from the same PAM alphabet $\mathbb{A}$, $\mathcal{S} = \mathbb{A}^{2n_t}$. The ML detection rule is then given by

$$\mathbf{x}_{ML} = \underset{\mathbf{x} \in \mathbb{A}^{2n_t}}{\text{argmin}} \; \|\mathbf{y} - \mathbf{H}\mathbf{x}\|^2 \; = \; \underset{\mathbf{x} \in \mathbb{A}^{2n_t}}{\text{argmin}} \; f(\mathbf{x}), \tag{5.4}$$

where $f(\mathbf{x}) \stackrel{\triangle}{=} \mathbf{x}^T\mathbf{H}^T\mathbf{H}\mathbf{x} - 2\mathbf{y}^T\mathbf{H}\mathbf{x}$ is the ML cost.

5.1.2 Multistage LAS algorithm

The LAS algorithm is a greedy variant of k-coordinate away neighborhood search. In its basic version, referred to as the 1-LAS algorithm, a 1-coordinate away neighborhood search is carried out to reach a local optimum which is declared as the final solution. Low complexity escape strategies are devised by additionally invoking limited 2- and 3-coordinate neighborhood searches. On escaping from a local optimum, a 1-LAS search is initiated again. A LAS algorithm that carries out a search in 1- to K-coordinate away neighborhoods is referred to as a K-LAS algorithm.

The K-LAS algorithm considered here consists of a sequence of LAS search stages, where the likelihood of the solution increases monotonically with every search stage. Each search stage consists of several substages, each substage comprising one or more iterations. The first substage can have one or more iterations, whereas all the other substages can have at most one iteration (to limit complexity). In the first substage, the algorithm updates one symbol per iteration (1-coordinate away neighborhood) such that the likelihood monotonically increases from one iteration to the next until a local optimum is reached. Upon reaching this local optimum, the algorithm initiates the second substage.

In the second substage, a two-symbol update (2-coordinate away neighborhood) is tried to further increase the likelihood. If the algorithm succeeds in increasing the likelihood by a two-symbol update, it starts the next search stage again with one-symbol updates. If the algorithm does not succeed in increasing the likelihood in the second substage, it goes to the third substage where a three-symbol update (3-coordinate away neighborhood) is tried. Essentially, in the Kth substage, a K-symbol update (K-coordinate away neighborhood) is tried to further increase the likelihood. This goes on until either (a) the algorithm succeeds in increasing the likelihood in the Kth substage (in which case a new search stage is initiated with 1-symbol update), or (b) the algorithm terminates.

The K-LAS algorithm starts with an initial solution $\mathbf{x}^{(0)}$, given by $\mathbf{x}^{(0)} = \mathcal{Q}(\mathbf{By})$, where $\mathbf{B}$ is the initial solution filter, which can be an MF or ZF or MMSE filter, and $\mathcal{Q}(.)$ refers to the nearest-neighbor quantizer. The index m in $\mathbf{x}^{(m)}$ denotes the iteration number in a substage of a given search stage. The ML cost function after the kth iteration in a given search stage is given by

$$C^{(k)} = \mathbf{x}^{(k)^T}\mathbf{H}^T\mathbf{H}\mathbf{x}^{(k)} - 2\mathbf{y}^T\mathbf{H}\mathbf{x}^{(k)}. \tag{5.5}$$

One-symbol update

Assume that the pth coordinate in the $(k+1)$th iteration is updated; p can take values from $1, \ldots, n_t$ for M-PAM and $1, \ldots, 2n_t$ for M-QAM. The update rule can be written as

$$\mathbf{x}^{(k+1)} = \mathbf{x}^{(k)} + \lambda_p^{(k)}\mathbf{e}_p, \tag{5.6}$$

where $\mathbf{e}_p$ denotes the unit vector with its pth entry only as 1, and all other entries as zero. Also, for any iteration k, $\mathbf{x}^{(k)}$ should belong to the space $\mathcal{S}$, and therefore $\lambda_p^{(k)}$ can take only certain integer values. For example, in the case of 4-PAM or 16-QAM (both have the same signal set $\mathbb{A}_p = \{-3, -1, 1, 3\}$), $\lambda_p^{(k)}$ can take values only from $\{-6, -4, -2, 0, 2, 4, 6\}$. Using (5.5) and (5.6), and defining a matrix $\mathbf{G}$ as

$$\mathbf{G} \triangleq \mathbf{H}^T\mathbf{H}, \tag{5.7}$$

the cost difference can be written as

$$\Delta C_p^{k+1} \triangleq C^{(k+1)} - C^{(k)} = \lambda_p^{(k)^2}(\mathbf{G})_{p,p} - 2\lambda_p^{(k)} z_p^{(k)}, \tag{5.8}$$

where $\mathbf{z}^{(k)} = \mathbf{H}^T(\mathbf{y} - \mathbf{H}\mathbf{x}^{(k)})$, $z_p^{(k)}$ is the pth entry of the $\mathbf{z}^{(k)}$ vector, and $(\mathbf{G})_{p,p}$ is the (p,p)th entry of the $\mathbf{G}$ matrix. Also, define a_p and $l_p^{(k)}$ as

$$a_p = (\mathbf{G})_{p,p}, \qquad l_p^{(k)} = |\lambda_p^{(k)}|. \tag{5.9}$$

With the above variables defined, (5.8) can be rewritten as

$$\Delta C_p^{k+1} = l_p^{(k)^2} a_p - 2 l_p^{(k)} |z_p^{(k)}| \operatorname{sgn}(\lambda_p^{(k)}) \operatorname{sgn}(z_p^{(k)}), \tag{5.10}$$

where sgn(.) denotes the signum function. For the ML cost function to reduce from the kth to the $(k+1)$th iteration, the cost difference should be negative.

Using this fact and that a_p and $l_p^{(k)}$ are non-negative quantities, we can conclude from (5.10) that the sign of $\lambda_p^{(k)}$ must satisfy

$$\mathrm{sgn}(\lambda_p^{(k)}) = \mathrm{sgn}(z_p^{(k)}). \tag{5.11}$$

Using (5.11) in (5.10), the ML cost difference can be rewritten as

$$\mathcal{F}(l_p^{(k)}) \triangleq \Delta C_p^{k+1} = l_p^{(k)^2} a_p - 2 l_p^{(k)} |z_p^{(k)}|. \tag{5.12}$$

For $\mathcal{F}(l_p^{(k)})$ to be non-positive, the necessary and sufficient condition from (5.12) is

$$l_p^{(k)} < \frac{2|z_p^{(k)}|}{a_p}. \tag{5.13}$$

The value of $l_p^{(k)}$ which satisfies (5.13) and at the same time gives the largest descent in the ML cost function from the kth to the $(k+1)$th iteration when symbol p is updated (i.e., greedy choice) can be found. Also, $l_p^{(k)}$ is constrained to take only certain integer values, and therefore the brute-force way to get the optimum $l_p^{(k)}$ is to evaluate $\mathcal{F}(l_p^{(k)})$ at all possible values of $l_p^{(k)}$. This becomes computationally expensive as the constellation size M increases. However, for the case of one-symbol update, a closed-form expression for the optimum $l_p^{(k)}$ that minimizes $\mathcal{F}(l_p^{(k)})$ can be found, which is given by

$$l_{p,opt}^{(k)} = 2 \left\lfloor \frac{|z_p^{(k)}|}{2a_p} \right\rceil, \tag{5.14}$$

where $\lfloor . \rceil$ denotes the rounding operation, i.e., for a real number x, $\lfloor x \rceil$ is the integer closest to x. If the pth symbol in $\mathbf{x}^{(k)}$, i.e., $x_p^{(k)}$, were indeed updated, then the new value of the symbol would be given by

$$\tilde{x}_p^{(k+1)} = x_p^{(k)} + l_{p,opt}^{(k)} \mathrm{sgn}(z_p^{(k)}). \tag{5.15}$$

However, $\tilde{x}_p^{(k+1)}$ can take values only in the set $\mathbb{A}_p$, and therefore the possibility of $\tilde{x}_p^{(k+1)}$ being greater than $(M-1)$ or less than $-(M-1)$ needs to be checked. If $\tilde{x}_p^{(k+1)} > (M-1)$, then $l_{p,opt}^{(k)}$ is adjusted so that the new value of $\tilde{x}_p^{(k+1)}$ with the adjusted value of $l_p^{(k)}$ using (5.15) is $(M-1)$. Similarly, if $\tilde{x}_p^{(k+1)} < -(M-1)$, then $l_{p,opt}^{(k)}$ is adjusted so that the new value of $\tilde{x}_p^{(k+1)}$ is $-(M-1)$. Let $\tilde{l}_{p,opt}^{(k)}$ be obtained from $l_{p,opt}^{(k)}$ after these adjustments. It can be shown that if $\mathcal{F}(l_{p,opt}^{(k)})$ is non-positive, then $\mathcal{F}(\tilde{l}_{p,opt}^{(k)})$ is also non-positive. Compute $\mathcal{F}(\tilde{l}_{p,opt}^{(k)})$, $\forall\ p = 1, \ldots, 2n_t$. Now, let

$$s = \underset{p}{\mathrm{argmin}}\ \mathcal{F}(\tilde{l}_{p,opt}^{(k)}). \tag{5.16}$$

If $\mathcal{F}(\tilde{l}_{s,opt}^{(k)}) < 0$, the update for the $(k+1)$th iteration is

$$\mathbf{x}^{(k+1)} = \mathbf{x}^{(k)} + \tilde{l}_{s,opt}^{(k)}\, \mathrm{sgn}(z_s^{(k)})\, \mathbf{e}_s, \tag{5.17}$$

$$\mathbf{z}^{(k+1)} = \mathbf{z}^{(k)} - \tilde{l}_{s,opt}^{(k)}\, \mathrm{sgn}(z_s^{(k)})\, \mathbf{g}_s, \tag{5.18}$$

where $\mathbf{g}_s$ is the sth column of $\mathbf{G}$. The update in (5.18) follows from the definition of $\mathbf{z}^{(k)}$ in (5.8). If $\mathcal{F}(\tilde{l}_{s,opt}^{(k)}) \geq 0$, then the one-symbol update search terminates. The data vector at this point is referred to as "1-symbol update local minimum," which is the final output in the 1-LAS algorithm. In the K-LAS algorithm, after reaching a one-symbol update local minimum, a further decrease in the cost function is sought by updating multiple symbols simultaneously.

Multiple-symbol updates

The motivation for trying out multiple-symbol updates is as follows. Let $\mathcal{L}_K \subseteq \mathcal{S}$ denote the set of data vectors such that for any $\mathbf{x} \in \mathcal{L}_K$, if a K-symbol update is performed on $\mathbf{x}$ resulting in a vector $\mathbf{x}'$, then $||\mathbf{y} - \mathbf{H}\mathbf{x}'|| \geq ||\mathbf{y} - \mathbf{H}\mathbf{x}||$. Note that $\mathbf{x}_{ML} \in \mathcal{L}_K, \forall\, K = 1, 2, \ldots, 2n_t$, because any number of symbol updates on $\mathbf{x}_{ML}$ will not decrease the cost function. Define another set $\mathcal{M}_K = \bigcap_{j=1}^{K} \mathcal{L}_j$. Note that $\mathbf{x}_{ML} \in \mathcal{M}_K, \forall\, K = 1, 2, \ldots, 2n_t$, and $\mathcal{M}_{2n_t} = \{\mathbf{x}_{ML}\}$, i.e., $\mathcal{M}_{2n_t}$ is a singleton set with $\mathbf{x}_{ML}$ as the only element. It is noted that if the updates are done optimally, then the output of the K-LAS algorithm converges to a vector in $\mathcal{M}_K$. Also, $|\mathcal{M}_{K+1}| \leq |\mathcal{M}_K|$, $K = 1, 2, \ldots, 2n_t - 1$. For any $\mathbf{x} \in \mathcal{M}_K$, $K = 1, 2, \ldots, 2n_t$ and $\mathbf{x} \neq \mathbf{x}_{ML}$, it can be seen that $\mathbf{x}$ and $\mathbf{x}_{ML}$ will differ in $K+1$ or more locations. The probability that $\mathbf{x}_{ML} = \mathbf{x}$ increases with increasing SNR, and so the separation between $\mathbf{x} \in \mathcal{M}_K$ and $\mathbf{x}$ will monotonically increase with increasing K. Since $\mathbf{x}_{ML} \in \mathcal{M}_K$, and $|\mathcal{M}_K|$ decreases monotonically with increasing K, there are fewer non-ML data vectors to which the algorithm can converge for increasing K. Therefore, the probability of the noise vector $\mathbf{n}$ inducing an error decreases with increasing K. This indicates that K-symbol updates with large K could approach ML performance with increasing complexity for increasing K.

K-symbol update

K-symbol updates can be done in $\binom{2n_t}{K}$ ways, among which the update that gives the largest reduction in the ML cost is of interest. Assume that in the $(k+1)$th iteration, K symbols at the indices $i_1, i_2, \ldots, i_K$ of $\mathbf{x}^{(k)}$ are updated. Each i_j, $j = 1, 2, \ldots, K$, can take values from $1, 2, \ldots, n_t$ for M-PAM and $1, 2, \ldots, 2n_t$ for M-QAM. Further, define the set of indices, $\mathcal{U} \triangleq \{i_1, i_2, \ldots, i_K\}$. The update rule for the K-symbol update can then be written as

$$\mathbf{x}^{(k+1)} = \mathbf{x}^{(k)} + \sum_{j=1}^{K} \lambda_{i_j}^{(k)} \mathbf{e}_{i_j}. \tag{5.19}$$

For any iteration k, $\mathbf{x}^{(k)}$ belongs to the space $\mathcal{S}$, and therefore $\lambda_{i_j}^{(k)}$ can take only certain integer values. In particular, $\lambda_{i_j}^{(k)} \in \mathcal{A}_{i_j}^{(k)}$, where $\mathcal{A}_{i_j}^{(k)} \triangleq \{x | (x + x_{i_j}^{(k)}) \in \mathcal{A}_{i_j}, x \neq 0\}$. For example, for 16-QAM, $\mathcal{A}_{i_j} = \{-3, -1, 1, 3\}$, and if $x_{i_j}^{(k)}$ is -1, then $\mathcal{A}_{i_j}^{(k)} = \{-2, 2, 4\}$. Using (5.5), the cost difference function $\Delta C_{\mathcal{U}}^{k+1}(\lambda_{i_1}^{(k)}, \lambda_{i_2}^{(k)}, \ldots, \lambda_{i_K}^{(k)}) \triangleq C^{(k+1)} - C^{(k)}$ can be written as

$$\Delta C_{\mathcal{U}}^{k+1}(\lambda_{i_1}^{(k)}, \lambda_{i_2}^{(k)}, \ldots, \lambda_{i_K}^{(k)}) = \sum_{j=1}^{K} \lambda_{i_j}^{(k)^2} (\mathbf{G})_{i_j,i_j}$$
$$+ 2\sum_{q=1}^{K} \sum_{p=q+1}^{K} \lambda_{i_p}^{(k)} \lambda_{i_q}^{(k)} (\mathbf{G})_{i_p,i_q} - 2\sum_{j=1}^{K} \lambda_{i_j}^{(k)} z_{i_j}^{(k)}, \tag{5.20}$$

where $\lambda_{i_j}^{(k)} \in \mathcal{A}_{i_j}^{(k)}$, which can be compactly written as $(\lambda_{i_1}^{(k)}, \lambda_{i_2}^{(k)}, \ldots, \lambda_{i_K}^{(k)}) \in \mathcal{A}_{\mathcal{U}}^{(k)}$, where $\mathbb{A}_{\mathcal{U}}^{(k)}$ denotes the Cartesian product of $\mathcal{A}_{i_1}^{(k)}$, $\mathcal{A}_{i_2}^{(k)}$ through to $\mathcal{A}_{i_K}^{(k)}$.

For a given $\mathcal{U}$, in order to decrease the ML cost, it is desired to choose the value of the K-tuple $(\lambda_{i_1}^{(k)}, \lambda_{i_2}^{(k)}, \ldots, \lambda_{i_K}^{(k)})$ such that the cost difference given by (5.20) is negative. If multiple K-tuples exist for which the cost difference is negative, then the K-tuple which gives the most negative cost difference is chosen.

Unlike for one-symbol updates for a K-symbol update a closed-form expression for $(\lambda_{i_1,opt}^{(k)}, \lambda_{i_2,opt}^{(k)}, \ldots, \lambda_{i_K,opt}^{(k)})$ which minimizes the cost difference over $\mathbb{A}_{\mathcal{U}}^{(k)}$ is difficult to obtain, since the cost difference is a function of K discrete-valued variables. Consequently, a brute-force method is to evaluate $\Delta C_{\mathcal{U}}^{k+1}(\lambda_{i_1}^{(k)}, \lambda_{i_2}^{(k)}, \ldots, \lambda_{i_K}^{(k)})$ over all possible values of $(\lambda_{i_1}^{(k)}, \lambda_{i_2}^{(k)}, \ldots, \lambda_{i_K}^{(k)})$. Approximate methods can be adopted to solve this problem using less complexity. One method based on ZF is as follows. The cost difference function in (5.20) can be rewritten as

$$\Delta C_{\mathcal{U}}^{k+1}(\lambda_{i_1}^{(k)}, \lambda_{i_2}^{(k)}, \ldots, \lambda_{i_K}^{(k)}) = \mathbf{\Lambda}_{\mathcal{U}}^{(k)^T} \mathbf{F}_{\mathcal{U}} \mathbf{\Lambda}_{\mathcal{U}}^{(k)} - 2\mathbf{\Lambda}_{\mathcal{U}}^{(k)^T} \mathbf{z}_{\mathcal{U}}^{(k)}, \tag{5.21}$$

where $\mathbf{\Lambda}_{\mathcal{U}}^{(k)} \triangleq [\lambda_{i_1}^{(k)} \lambda_{i_2}^{(k)} \ldots \lambda_{i_K}^{(k)}]^T$, $\mathbf{z}_{\mathcal{U}}^{(k)} \triangleq [z_{i_1}^{(k)} z_{i_2}^{(k)} \ldots z_{i_K}^{(k)}]^T$, and $\mathbf{F}_{\mathcal{U}} \in \mathbb{R}^{K \times K}$, where $(\mathbf{F}_{\mathcal{U}})_{p,q} = (\mathbf{G})_{i_p,i_q}$ and $p, q \in \{1, 2, \ldots, K\}$. Since $\Delta C_{\mathcal{U}}^{k+1}(\lambda_{i_1}^{(k)}, \lambda_{i_2}^{(k)}, \ldots, \lambda_{i_K}^{(k)})$ is a strictly convex quadratic function of $\mathbf{\Lambda}_{\mathcal{U}}^{(k)}$ (the Hessian $\mathbf{F}_{\mathcal{U}}$ is positive definite with probability 1), a unique global minimum exists, and is given by

$$\tilde{\mathbf{\Lambda}}_{\mathcal{U}}^{(k)} = \mathbf{F}_{\mathcal{U}}^{-1} \mathbf{z}_{\mathcal{U}}^{(k)}. \tag{5.22}$$

However, the solution given by (5.22) need not lie in $\mathbb{A}_{\mathcal{U}}^{(k)}$. So, first round off the solution as

$$\widehat{\mathbf{\Lambda}}_{\mathcal{U}}^{(k)} = 2\left\lfloor 0.5\tilde{\mathbf{\Lambda}}_{\mathcal{U}}^{(k)} \right\rceil, \tag{5.23}$$

where the operation in (5.23) is done element-wise, since $\tilde{\mathbf{\Lambda}}_{\mathcal{U}}^{(k)}$ is a vector. Further, let $\widehat{\mathbf{\Lambda}}_{\mathcal{U}}^{(k)} \triangleq [\widehat{\lambda}_{i_1}^{(k)} \widehat{\lambda}_{i_2}^{(k)} \cdots \widehat{\lambda}_{i_K}^{(k)}]^T$. It is still possible that the solution $\widehat{\mathbf{\Lambda}}_{\mathcal{U}}^{(k)}$ in (5.23) need not lie in $\mathcal{A}_{\mathcal{U}}^{(k)}$. This would result in $x_{i_j}^{(k+1)} \notin \mathcal{A}_{i_j}$ for some j. For example, if $\mathcal{A}_{i_j}$ is M-PAM, then $x_{i_j}^{(k+1)} \notin \mathcal{A}_{i_j}$ if $x_{i_j}^{(k)} + \widehat{\lambda}_{i_j}^{(k)} > (M-1)$ or $x_{i_j}^{(k)} + \widehat{\lambda}_{i_j}^{(k)} < -(M-1)$. In such cases, the following adjustment to $\widehat{\lambda}_{i_j}^{(k)}$ for $j = 1, 2, \ldots, K$

can be used:

$$\widehat{\lambda}_{i_j}^{(k)} = \begin{cases} (M-1) - x_{i_j}^{(k)}, & \text{when } \widehat{\lambda}_{i_j}^{(k)} + x_{i_j}^{(k)} > (M-1), \\ -(M-1) - x_{i_j}^{(k)}, & \text{when } \widehat{\lambda}_{i_j}^{(k)} + x_{i_j}^{(k)} < -(M-1). \end{cases} \tag{5.24}$$

After these adjustments, it is guaranteed that $\widehat{\Lambda}_{\mathcal{U}}^{(k)} \in \mathbb{A}_{\mathcal{U}}^{(k)}$. Therefore, the new cost difference function value is given by $\Delta C_{\mathcal{U}}^{k+1}(\widehat{\lambda}_{i_1}^{(k)}, \widehat{\lambda}_{i_2}^{(k)}, \ldots, \widehat{\lambda}_{i_K}^{(k)})$. It is noted that the complexity of this approximate method does not depend on the size of the set $\mathcal{A}_{\mathcal{U}}^{(k)}$, i.e., it has constant complexity. Through simulations, it has been observed that this approximation results in a performance close to that of the brute-force method for $K = 2$ and 3. Defining the optimum $\mathcal{U}$ for the approximate method as $\widehat{\mathcal{U}}$,

$$\widehat{\mathcal{U}} \stackrel{\triangle}{=} (\hat{i}_1, \hat{i}_2, \ldots, \hat{i}_K) = \underset{\mathcal{U}}{\operatorname{argmin}} \ \Delta C_{\mathcal{U}}^{k+1}(\widehat{\lambda}_{i_1}^{(k)}, \widehat{\lambda}_{i_2}^{(k)}, \ldots, \widehat{\lambda}_{i_K}^{(k)}). \tag{5.25}$$

The K-symbol update is successful and the update is done only if

$$\Delta C_{\widehat{\mathcal{U}}}^{k+1}(\widehat{\lambda}_{\hat{i}_1}^{(k)}, \widehat{\lambda}_{\hat{i}_2}^{(k)}, \ldots, \widehat{\lambda}_{\hat{i}_K}^{(k)}) < 0.$$

The update rules for the $\mathbf{z}^{(k)}$ and $\mathbf{x}^{(k)}$ vectors are given by

$$\mathbf{z}^{(k+1)} = \mathbf{z}^{(k)} - \sum_{j=1}^{K} \widehat{\lambda}_{\hat{i}_j}^{(k)} \mathbf{g}_{\hat{i}_j}, \tag{5.26}$$

$$\mathbf{x}^{(k+1)} = \mathbf{x}^{(k)} + \sum_{j=1}^{K} \widehat{\lambda}_{\hat{i}_j}^{(k)} \mathbf{e}_{\hat{i}_j}. \tag{5.27}$$

5.1.3 Complexity

The complexity of the LAS algorithm comprises three main components, namely, (*i*) computation of the initial vector $\mathbf{x}^{(0)}$, (*ii*) computation of $\mathbf{H}^T\mathbf{H}$, and (*iii*) the search operation. For $n_t = n_r$, because of the matrix inversion involved, the complexity of computing the ZF or MMSE initial solution vector is $O(n_t^3)$, i.e., $O(n_t^2)$ per-symbol complexity. Likewise, $\mathbf{H}^T\mathbf{H}$ can be computed in $O(n_t^2)$ per-symbol complexity. From simulations, it has been found that the LAS search requires an average per-symbol complexity of $O(n_t)$. So the total complexity of the algorithm is dominated by the initial solution computation rather than the search operation. The overall average per-symbol complexity is $O(n_t^2)$ which scales well for large MIMO systems.

5.1.4 Generation of soft outputs

The output solutions from local search techniques are from discrete space. In the MIMO detection problem, the elements of the output vector from the local search are from the modulation alphabet which is discrete. The discrete-valued output symbols can be mapped back to their constituent bits using the modulation demapping function. This results in 'hard' bit decisions. It is preferred, however, to generate soft values of the individual bits so that the performance of channel decoding that follows detection in a coded system can improve. The following procedure is a way to generate soft bit values from discrete symbol values.

Let $\mathbf{d} = [\widehat{x}_1, \widehat{x}_2, \ldots, \widehat{x}_{2n_t}]$, $\widehat{x}_i \in \mathbb{A}_i$, denote the detected output symbol vector from the LAS algorithm. Let the symbol $\widehat{x}_i$ map to the bit vector $\mathbf{b}_i = [b_{i,1}, b_{i,2}, \ldots, b_{i,N_i}]^T$, where $N_i = \log_2 |\mathbb{A}_i|$, and $b_{i,j} \in \{+1, -1\}$, $i = 1, 2, \ldots, 2n_t$ and $j = 1, 2, \ldots, N_i$. Let $\tilde{b}_{i,j} \in \mathbb{R}$ denote the soft value for the jth bit of the ith symbol. Given $\mathbf{d}$, we need to find $\tilde{b}_{i,j}$, $\forall\,(i,j)$.

Note that the quantity $\|\mathbf{y} - \mathbf{Hd}\|^2$ is inversely related to the likelihood that $\mathbf{d}$ is indeed the transmitted symbol vector. Let the $\mathbf{d}$ vector with its jth bit of the ith symbol forced to $+1$ be denoted as vector $\mathbf{d}_i^{j+}$. Likewise, let $\mathbf{d}_i^{j-}$ be the vector $\mathbf{d}$ with its jth bit of the ith symbol forced to -1. Then the quantities $\|\mathbf{y} - \mathbf{Hd}_i^{j+}\|^2$ and $\|\mathbf{y} - \mathbf{Hd}_i^{j-}\|^2$ are inversely related to the likelihoods that the jth bit of the ith transmitted symbol is $+1$ and -1, respectively. So, if $\|\mathbf{y} - \mathbf{Hd}_i^{j-}\|^2 - \|\mathbf{y} - \mathbf{Hd}_i^{j+}\|^2$ is positive (or negative), it indicates that the jth bit of the ith transmitted symbol has a higher likelihood of being $+1$ (or -1). So, the quantity $\|\mathbf{y} - \mathbf{Hd}_i^{j-}\|^2 - \|\mathbf{y} - \mathbf{Hd}_i^{j+}\|^2$, appropriately normalized to avoid unbounded increase for increasing n_t, can be a good soft value for the jth bit of the ith symbol. With this motivation, a soft output value for the jth bit of the ith symbol can be generated as

$$\tilde{b}_{i,j} = \frac{\|\mathbf{y} - \mathbf{Hd}_i^{j-}\|^2 - \|\mathbf{y} - \mathbf{Hd}_i^{j+}\|^2}{\|\mathbf{h}_i\|^2}, \tag{5.28}$$

where the normalization by $\|\mathbf{h}_i\|^2$ is to contain unbounded increase of $\tilde{b}_{i,j}$ for increasing n_t. The right-hand side in the above can be efficiently computed in terms of $\mathbf{z}$ and $\mathbf{G}$ as follows. Since $\mathbf{d}_i^{j+}$ and $\mathbf{d}_i^{j-}$ differ only in the ith entry,

$$\mathbf{d}_i^{j-} = \mathbf{d}_i^{j+} + \lambda_{i,j}\mathbf{e}_i. \tag{5.29}$$

Since $\mathbf{d}_i^{j-}$ and $\mathbf{d}_i^{j+}$ are known, $\lambda_{i,j}$ is known from (5.29). Substituting (5.29) in (5.28),

$$\begin{aligned}
\tilde{b}_{i,j}\,\|\mathbf{h}_i\|^2 &= \|\mathbf{y} - \mathbf{Hd}_i^{j+} - \lambda_{i,j}\mathbf{h}_i\|^2 - \|\mathbf{y} - \mathbf{Hd}_i^{j+}\|^2 \\
&= \lambda_{i,j}^2\|\mathbf{h}_i\|^2 - 2\lambda_{i,j}\mathbf{h}_i^T(\mathbf{y} - \mathbf{Hd}_i^{j+}) \qquad (5.30)\\
&= -\lambda_{i,j}^2\|\mathbf{h}_i\|^2 - 2\lambda_{i,j}\mathbf{h}_i^T(\mathbf{y} - \mathbf{Hd}_i^{j-}). \qquad (5.31)
\end{aligned}$$

If $b_{i,j} = 1$, then $\mathbf{d}_i^{j+} = \mathbf{d}$, and substituting this in (5.30) and dividing by $\|\mathbf{h}_i\|^2$,

$$\tilde{b}_{i,j} = \lambda_{i,j}^2 - 2\lambda_{i,j}\frac{z_i}{(\mathbf{G})_{i,i}}. \tag{5.32}$$

If $b_{i,j} = -1$, then $\mathbf{d}_i^{j-} = \mathbf{d}$, and substituting this in (5.31) and dividing by $\|\mathbf{h}_i\|^2$,

$$\tilde{b}_{i,j} = -\lambda_{i,j}^2 - 2\lambda_{i,j}\frac{z_i}{(\mathbf{G})_{i,i}}. \tag{5.33}$$

It is noted that $\mathbf{z}$ and $\mathbf{G}$ are already available upon the termination of the K-LAS algorithm, and hence the complexity of computing $\tilde{b}_{i,j}$ in (5.32) and

(5.33) is constant. Hence, the overall complexity in computing the soft values for all the bits is $O(n_t \log_2 M)$ for M-PAM. It is seen from (5.32) and (5.33) that the magnitude of $\tilde{b}_{i,j}$ depends upon $\lambda_{i,j}$. For large-size signal sets, the possible values of $\lambda_{i,j}$ will also be large in magnitude. Therefore $\tilde{b}_{i,j}$ has to be normalized for the channel decoder to function properly. For turbo codes, it has been observed through simulations that normalizing $\tilde{b}_{i,j}$ by $\left(\lambda_{i,j}/2\right)^2$ results in good performance. In [22], it was shown that this soft decision output generation method, when used in large V-BLAST MIMO systems, offers about 1–1.5 dB improvement in coded BER performance compared to that achieved using hard decision outputs from the K-LAS algorithm.

5.1.5 Near-optimal performance in large dimensions

The BER performance of the 1-LAS algorithm in V-BLAST MIMO systems with $n_t = n_r$ and BPSK modulation is shown in Fig. 5.1. ZF detector output is used as the initial solution. Hence, the algorithm is referred to as ZF-1LAS in the figure. The low complexity attribute of the algorithm enables the simulation of 1-LAS detection performance for a large number of antennas (simulation results are shown for antennas up to $n_t = n_r = 400$ in Fig. 5.1). Since ML performance in hundreds of dimensions cannot be simulated because of its exponential complexity, its performance is plotted only for up to 32 antennas in Fig. 5.1(b). For more than 32 antennas, unfaded SISO AWGN performance is plotted as a lower bound on ML performance, which is tight at high SNRs in large dimensions. From Figs. 5.1(a) and (b), it is seen that the ZF-1LAS algorithm performs increasingly better for increasing values of $n_t = n_r$, which is an attribute of the ML detector. Such behavior does not happen with detectors like MF, ZF, and MMSE detectors. It is seen that the ZF-1LAS performance very closely approaches that of unfaded SISO AWGN for hundreds of antennas. For example, in Fig. 5.1(b), it is observed that the SNR required to achieve a target BER of 10^{-3} for $n_t = n_r = 400$ is about 7 dB, while the SNR required in the unfaded SISO AWGN channel to achieve the same BER is also about 7 dB. This illustrates the ability of a simple local search like 1-LAS to achieve near optimal performance in MIMO systems with hundreds of antennas.

While 1-LAS is very attractive in the hundreds of antennas regime, its performance in the small number of antennas regime is not competitive as can be seen in Fig. 5.1(b) for $n_t = n_r = 2$ to $n_t = n_r = 10$. In this small dimension regime, complexity is not a bottleneck, and hence the SD and several of its low complexity variants can be appropriate. It is of interest to achieve near-optimal performance in the tens of antennas regime using LAS, since the SD becomes prohibitive in this regime. This can be made possible by allowing some additional complexity compared to 1-LAS complexity (without increasing the order of complexity). 2- or 3-LAS and tabu search can be attempted to achieve this.

Figure 5.2(a) depicts how 3-LAS performance compares with 1-LAS performance. 3-LAS and 1-LAS performances in V-BLAST MIMO with $n_t = n_r$ and

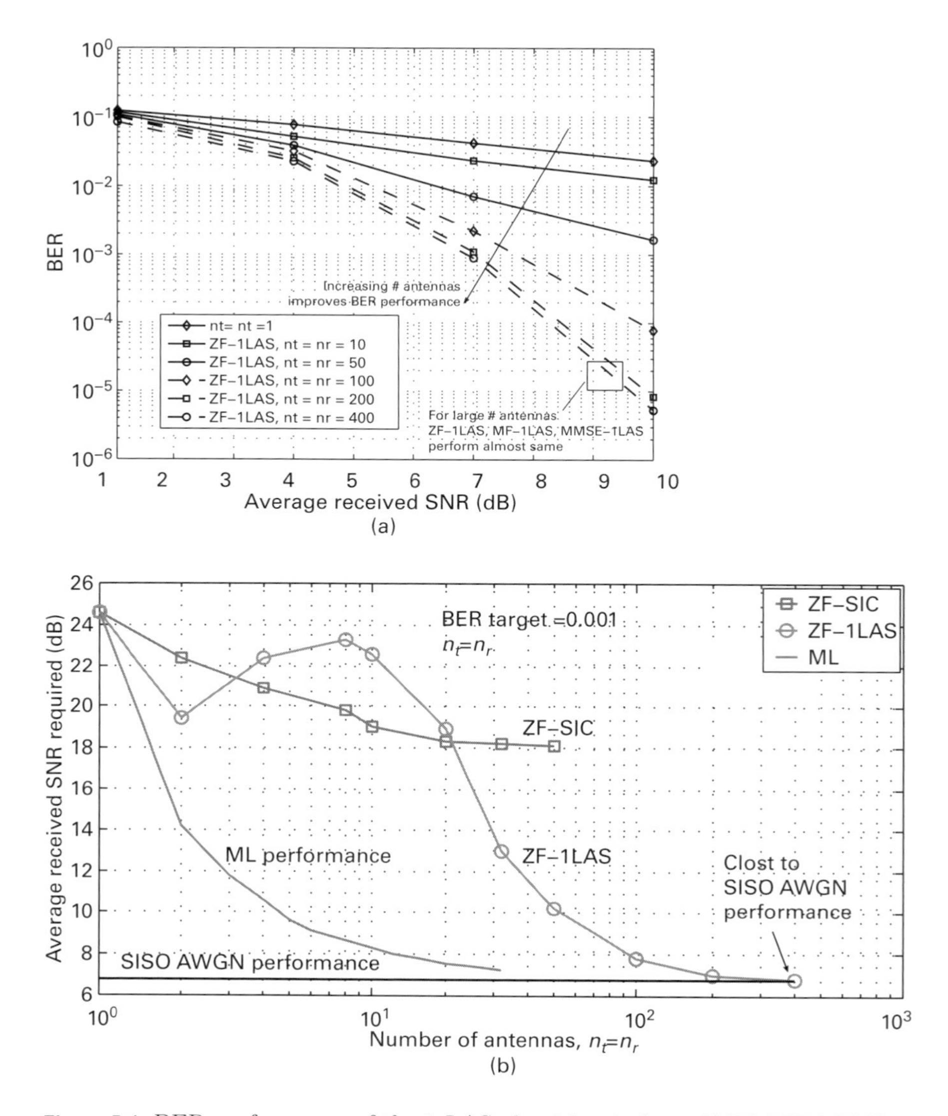

Figure 5.1 BER performance of the 1-LAS algorithm in large V-BLAST MIMO systems with $n_t = n_r$ and BPSK: (a) BER vs. average received SNR; (b) average received SNR required to achieve a target BER of 0.001 vs. number of antennas.

4-QAM are shown. MMSE detector output is used as the initial vector. Hence the algorithms are referred to as MMSE-1LAS and MMSE-3LAS in the figure. The unfaded SISO AWGN performance for 4-QAM is also shown as a lower bound. It is noted that the performance of the MMSE detector is quite poor for $n_t = n_r = 64$, whereas the performance of MMSE-3LAS much better. As expected, 3-LAS achieves a better performance than 1-LAS in Fig. 5.2(a); however,

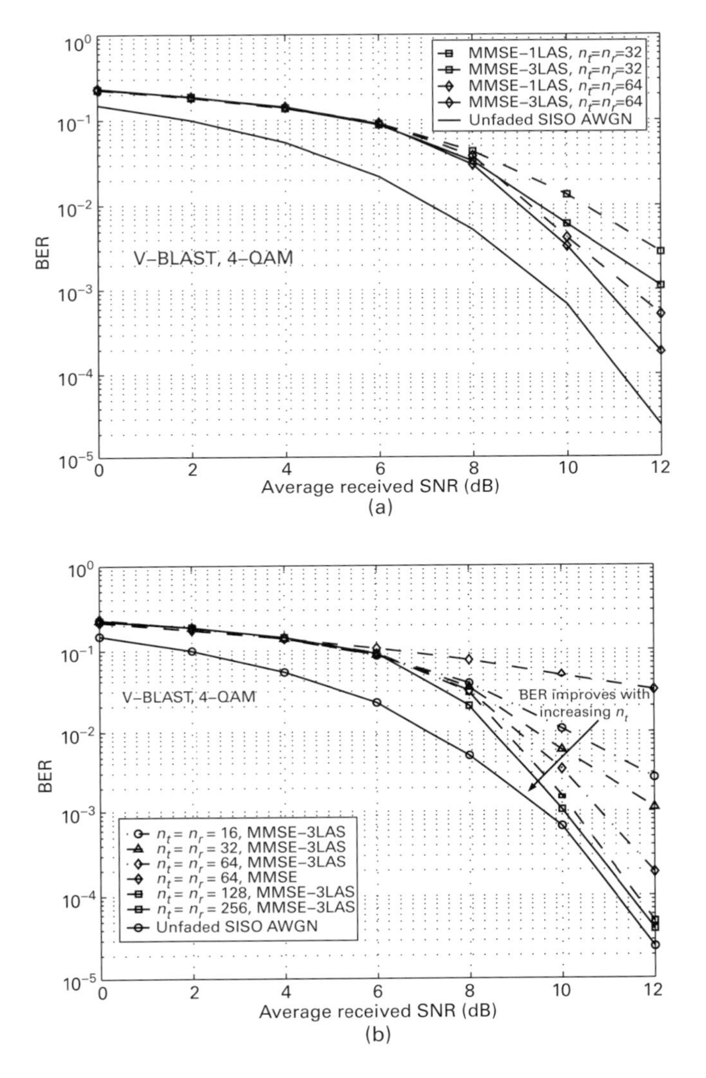

Figure 5.2 BER performance of 3-LAS algorithm in large V-BLAST MIMO systems with 4-QAM: (a) 3-LAS vs. 1-LAS performance for $n_t = n_r = 32, 64$; (b) 3-LAS performance for $n_t = n_r = 16, 32, 64, 128, 256$.

the improvement is not very significant. In Fig. 5.2(b), it is seen that 3-LAS approaches unfaded SISO AWGN performance for increasing values of $n_t = n_r$, and that 3-LAS, like 1-LAS, is attractive mainly in the hundreds of antennas regime (e.g., the 3-LAS performances for $n_t = n_r = 128$ and 256 are close to the unfaded SISO AWGN performance). Therefore, it is of interest to consider alternative ways to create the large dimension advantage of the LAS algorithm in the tens of antennas regime as well.

Large dimensions: the key

As seen above, the key to achieving near-optimal performance using the LAS algorithm is large dimensions. LAS can be effective with tens of antennas if the overall signaling is done in hundreds of dimensions. One way to create large dimensions using small spatial dimensions is to create additional dimensions in time. This must be done without reducing the rate (in symbols per channel use) compared to V-BLAST. Full rate NO-STBCs from CDA [27] can create hundreds of dimensions with tens of transmit antennas without reducing the rate. LAS can be used to near-optimally decode such large-dimension STBCs with tens of antennas.

5.1.6 Decoding of large NO-STBCs using LAS

Consider an STBC MIMO system with multiple transmit and receive antennas. An (n, p, k) STBC is represented by a matrix $\mathbf{X}_c \in \mathbb{C}^{n \times p}$, where $n = n_t$ and p denote the number of transmit antennas and the number of channel uses, respectively, and k denotes the number of complex data symbols sent in one STBC matrix. The (i, j)th entry in $\mathbf{X}_c$ represents the complex number transmitted from the ith transmit antenna in the jth channel use. The rate of an STBC is given by $r \triangleq k/p$.

NO-STBCs from CDA

A full-rate NO-STBC from CDA is an $n_t \times n_t$ matrix constructed using n_t^2 symbols. The matrix is sent using n_t transmit antennas in n_t channel uses so that the rate of the STBC is n_t symbols per channel use (the same as that of V-BLAST). In addition to full rate, this STBC gives the full transmit diversity order of n_t as well. The number of dimensions is n_t^2, which creates hundreds of dimensions with tens of antennas (e.g., the number of complex dimensions is 256 for a 16×16 NO-STBC from CDA). The construction of $n \times n$ NO-STBCs from CDA for arbitrary n (i.e., for an arbitrary number of transmit antennas) is given by the matrix [27]

$$\mathbf{X} = \begin{bmatrix} \sum_{i=0}^{n-1} x_{0,i}\, t^i & \delta \sum_{i=0}^{n-1} x_{n-1,i}\, \omega_n^i\, t^i & \cdots & \delta \sum_{i=0}^{n-1} x_{1,i}\, \omega_n^{(n-1)i}\, t^i \\ \sum_{i=0}^{n-1} x_{1,i}\, t^i & \sum_{i=0}^{n-1} x_{0,i}\, \omega_n^i\, t^i & \cdots & \delta \sum_{i=0}^{n-1} x_{2,i}\, \omega_n^{(n-1)i}\, t^i \\ \sum_{i=0}^{n-1} x_{2,i}\, t^i & \sum_{i=0}^{n-1} x_{1,i}\, \omega_n^i\, t^i & \cdots & \delta \sum_{i=0}^{n-1} x_{3,i}\, \omega_n^{(n-1)i}\, t^i \\ \vdots & \vdots & \vdots & \vdots \\ \sum_{i=0}^{n-1} x_{n-2,i}\, t^i & \sum_{i=0}^{n-1} x_{n-3,i}\, \omega_n^i\, t^i & \cdots & \delta \sum_{i=0}^{n-1} x_{n-1,i}\, \omega_n^{(n-1)i} t^i \\ \sum_{i=0}^{n-1} x_{n-1,i}\, t^i & \sum_{i=0}^{n-1} x_{n-2,i}\, \omega_n^i\, t^i & \cdots & \sum_{i=0}^{n-1} x_{0,i}\, \omega_n^{(n-1)i}\, t^i \end{bmatrix}, \tag{5.34}$$

where $\omega_n = e^{\mathbf{j}2\pi/n}$, $\mathbf{j} = \sqrt{-1}$, and $x_{u,v}$, $0 \leq u, v \leq n-1$ are the data symbols from a QAM alphabet. When $\delta = e^{\sqrt{5}\mathbf{j}}$ and $t = e^{\mathbf{j}}$, the STBC achieves full transmit diversity (under ML decoding) as well as information losslessness. When $\delta = t = 1$, the code ceases to be of full diversity, but continues to be information lossless.

Equivalent linear vector channel model

The received space-time signal matrix, $\mathbf{Y}_c \in \mathbb{C}^{n_r \times p}$, can be written as

$$\mathbf{Y}_c = \mathbf{H}_c \mathbf{X}_c + \mathbf{N}_c, \tag{5.35}$$

where $\mathbf{H}_c \in \mathbb{C}^{n_r \times n_t}$ is the channel gain matrix, which is assumed to be constant over one STBC matrix duration (quasi-static assumption). $\mathbf{N}_c \in \mathbb{C}^{n_r \times p}$ is the noise matrix at the receiver and its entries are modeled as iid $\mathcal{CN}(0, \sigma^2 = n_t E_s/\gamma)$, where E_s is the average energy of the transmitted symbols and γ is the average received SNR per receive antenna, and the (i,j)th entry in $\mathbf{Y}_c$ is the received signal at the ith receive antenna in the jth channel use. NO-STBCs from CDA are linear dispersion STBCs, where the matrix $\mathbf{X}_c$ can be decomposed into a linear combination of weight matrices corresponding to each data symbol as [28]

$$\mathbf{X}_c = \sum_{i=1}^{k} x_c^{(i)} \mathbf{A}_c^{(i)}, \tag{5.36}$$

where $x_c^{(i)}$ is the ith complex data symbol and $\mathbf{A}_c^{(i)} \in \mathbb{C}^{n_t \times p}$ is its weight matrix. From (5.35) and (5.36), and applying the vec (.) operation

$$\text{vec}\,(\mathbf{Y}_c) = \sum_{i=1}^{k} x_c^{(i)} \text{vec}\,(\mathbf{H}_c \mathbf{A}_c^{(i)}) + \text{vec}\,(\mathbf{N}_c). \tag{5.37}$$

If $\mathbf{U}$,$\mathbf{V}$,$\mathbf{W}$,$\mathbf{D}$ are matrices such that $\mathbf{D} = \mathbf{UWV}$, then it is true that $\text{vec}\,(\mathbf{D}) = (\mathbf{V}^T \otimes \mathbf{U})\,\text{vec}\,(\mathbf{W})$, where $\otimes$ denotes tensor product of matrices. Using this, (5.37) can be written as

$$\text{vec}\,(\mathbf{Y}_c) = \sum_{i=1}^{k} x_c^{(i)} (\mathbf{I}_p \otimes \mathbf{H}_c)\,\text{vec}\,(\mathbf{A}_c^{(i)}) + \text{vec}\,(\mathbf{N}_c), \tag{5.38}$$

where $\mathbf{I}_p$ is the $p \times p$ identity matrix. Further, define $\mathbf{y}_c \triangleq \text{vec}\,(\mathbf{Y}_c)$, $\widehat{\mathbf{H}}_c \triangleq (\mathbf{I}_p \otimes \mathbf{H}_c)$, $\mathbf{a}_c^{(i)} \triangleq \text{vec}\,(\mathbf{A}_c^{(i)})$, and $\mathbf{n}_c \triangleq \text{vec}\,(\mathbf{N}_c)$. From these definitions, it is clear that $\mathbf{y}_c \in \mathbb{C}^{n_r p \times 1}$, $\widehat{\mathbf{H}}_c \in \mathbb{C}^{n_r p \times n_t p}$, $\mathbf{a}_c^{(i)} \in \mathbb{C}^{n_t p \times 1}$, and $\mathbf{n}_c \in \mathbb{C}^{n_r p \times 1}$. Define a matrix $\widetilde{\mathbf{H}}_c \in \mathbb{C}^{n_r p \times k}$, whose ith column is $\widehat{\mathbf{H}}_c\,\mathbf{a}_c^{(i)}$, $i = 1, \ldots, k$. Let $\mathbf{x}_c \in \mathbb{C}^{k \times 1}$, whose ith entry is the data symbol $x_c^{(i)}$. With the above definitions, (5.38) can be written as

$$\mathbf{y}_c = \sum_{i=1}^{k} x_c^{(i)} \,(\widehat{\mathbf{H}}_c\,\mathbf{a}_c^{(i)}) + \mathbf{n}_c \;=\; \widetilde{\mathbf{H}}_c \mathbf{x}_c + \mathbf{n}_c. \tag{5.39}$$

Each element of $\mathbf{x}_c$ is an M-PAM or M-QAM symbol. Let $\mathbf{y}_c$, $\widetilde{\mathbf{H}}_c$, $\mathbf{x}_c$, and $\mathbf{n}_c$ be decomposed into real and imaginary parts as

$$\mathbf{y}_c = \mathbf{y}_I + j\mathbf{y}_Q, \quad \mathbf{x}_c = \mathbf{x}_I + j\mathbf{x}_Q, \quad \mathbf{n}_c = \mathbf{n}_I + j\mathbf{n}_Q, \quad \widetilde{\mathbf{H}}_c = \widetilde{\mathbf{H}}_I + j\widetilde{\mathbf{H}}_Q.$$

Further, define $\mathbf{x}_{eff} \in \mathbb{R}^{2k\times 1}$, $\mathbf{y}_{eff} \in \mathbb{R}^{2n_rp\times 1}$, $\mathbf{H}_{eff} \in \mathbb{R}^{2n_rp\times 2k}$, and $\mathbf{n}_{eff} \in \mathbb{R}^{2n_rp\times 1}$ as

$$\mathbf{x}_{eff} = [\mathbf{x}_I^T \ \ \mathbf{x}_Q^T]^T, \qquad \mathbf{y}_{eff} = [\mathbf{y}_I^T \ \ \mathbf{y}_Q^T]^T,$$

$$\mathbf{H}_{eff} = \begin{pmatrix} \widetilde{\mathbf{H}}_I & -\widetilde{\mathbf{H}}_Q \\ \widetilde{\mathbf{H}}_Q & \widetilde{\mathbf{H}}_I \end{pmatrix}, \qquad \mathbf{n}_{eff} = [\mathbf{n}_I^T \ \ \mathbf{n}_Q^T]^T.$$

Now, an equivalent real-valued linear vector channel model for the NO-STBC MIMO system can be written in the form

$$\mathbf{y}_{eff} = \mathbf{H}_{eff}\mathbf{x}_{eff} + \mathbf{n}_{eff}. \tag{5.40}$$

The LAS algorithm can be applied on this equivalent linear vector channel model to detect $\mathbf{x}_{eff}$, and hence the symbol vector $\mathbf{x}_c$.

Performance

The uncoded BER performance of 1-, 2-, and 3-LAS algorithms in decoding 4×4, 8×8, 16×16, 32×32 STBCs from CDA with $\delta = t = 1$ for $n_t = n_r = 4, 8, 16, 32$ and 4-QAM is illustrated in Fig. 5.3(a). The corresponding performance with the MMSE detector along with the unfaded SISO AWGN performance are also plotted for comparison. Note that 32×32 STBC has 2048 real dimensions – a problem size that can well exploit the large dimension advantage of LAS. The MMSE detector performance is found not to improve with increasing STBC size (i.e., increasing $n_t = n_r$), whereas, the performance of the MMSE-LAS algorithm (LAS with MMSE detector output as the initial solution) improves for increasing $n_t = n_r$. For example, decoding of 16×16 and 32×32 STBCs (with 512 and 2048 real dimensions, respectively) using LAS achieves a performance very close to that of unfaded SISO AWGN. With such large dimensions, 1-LAS itself is found to be adequate to achieve near optimal performance without the need for 2- or 3-LAS. In terms of coded BER performance, from Fig. 5.3(b) it is observed that 1-LAS with soft output followed by turbo decoding is able to achieve a very good coded performance which is close to within about 4 dB from the ergodic MIMO capacity.

In Fig. 5.4, it is observed that LAS when applied to V-BLAST MIMO with $n_t = n_r = 16$ and 4-QAM (32 real dimensions) achieves a performance which is far from the performance achieved by the SD in the same system. Whereas, for the same number of antennas and modulation in 16×16 NO-STBC (512 real dimensions), LAS achieves better performance even compared to the SD performance in 16×16 V-BLAST MIMO. This is because of the availability of transmit diversity in STBC and the lack of it in V-BLAST. The complexity

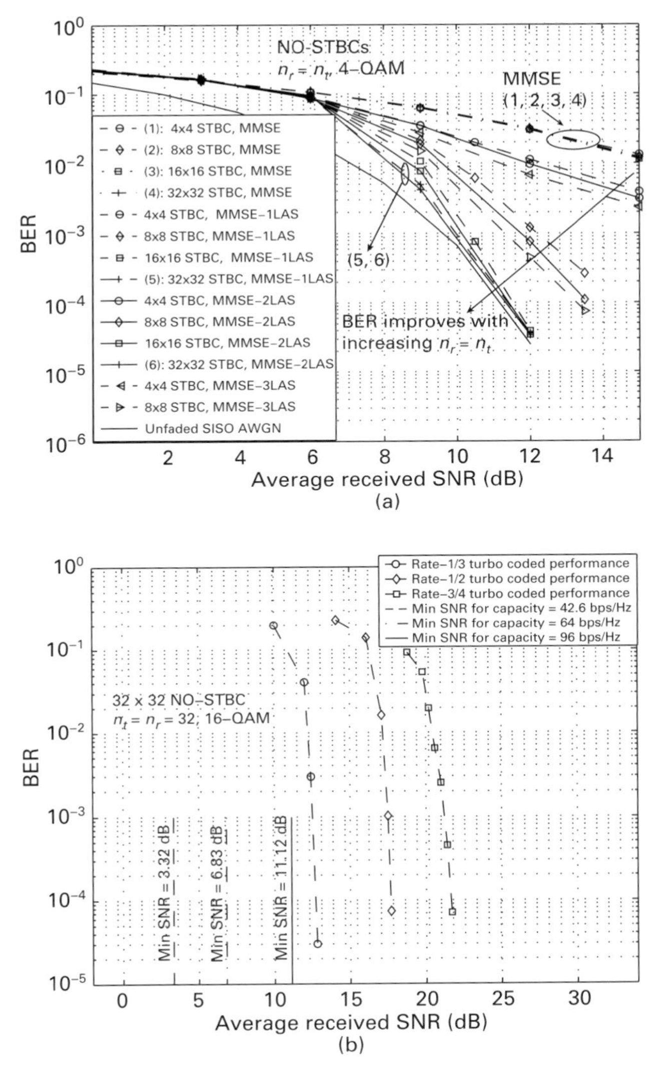

Figure 5.3 BER performance of 1-LAS algorithm in large NO-STBC MIMO systems: (a) uncoded BER performance; (b) coded BER performance.

involved in LAS decoding of 16×16 NO-STBC is less than that of sphere decoding of 16×16 V-BLAST at low to moderate SNRs [23].

Complexity

Two good properties of NO-STBCs from CDA are useful in achieving low orders of complexity for the computation of $\mathbf{x}_{eff}^{(0)}$ and $\mathbf{H}_{eff}^T\mathbf{H}_{eff}$. They are: (i) the weight matrices $\mathbf{A}_c^{(i)}$s are permutation type, and (ii) the $n_t^2 \times n_t^2$ matrix formed

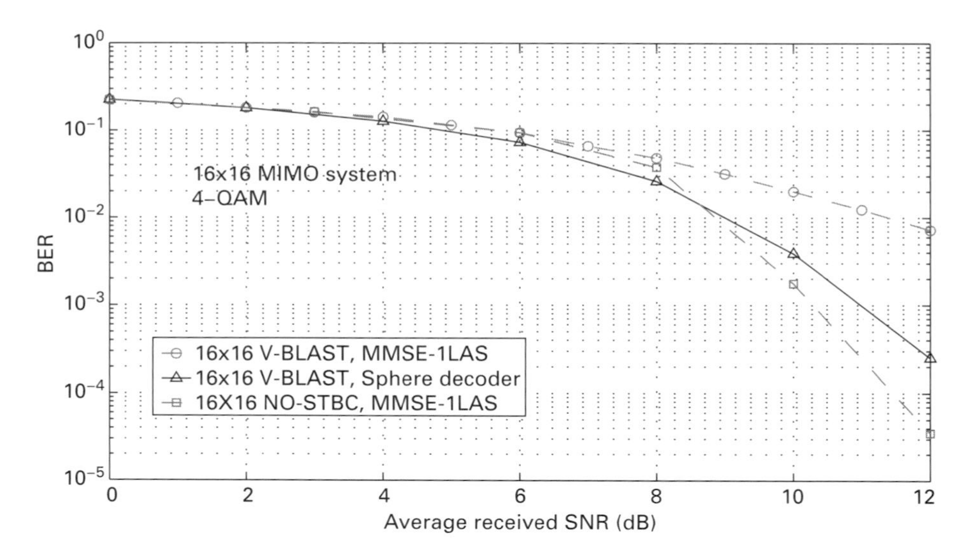

Figure 5.4 Comparison of 1-LAS performance in 16×16 V-BLAST versus NO-STBC MIMO systems.

with $(n_t^2 \times 1)$-sized $\mathbf{a}_c^{(i)}$ vectors as columns is a scaled unitary matrix. These properties allow the computation of the MMSE/ZF initial solution in $O(n_t^3 n_r)$ complexity, i.e., in $O(n_t n_r)$ per-symbol complexity (since there are n_t^2 symbols in one STBC matrix). Likewise, the computation of $\mathbf{H}_{eff}^T \mathbf{H}_{eff}$ can be done in $O(n_t^3)$ per-symbol complexity.

The average per-symbol complexities of the 1-LAS and 2-LAS search operations are $O(n_t^2)$ and $O(n_t^2 \log n_t)$, respectively, which is explained as follows. The average search complexity is the complexity of one search stage times the mean number of search stages till the algorithm terminates. For 1-LAS, the number of search stages is always 1. There are multiple iterations in the search, and in each iteration all possible $\binom{2n_t^2}{1}$ one-symbol updates are considered. So, the per-iteration complexity in 1-LAS is $O(n_t^2)$, i.e., $O(1)$ complexity per symbol. Further, the mean number of iterations before the algorithm terminates in 1-LAS was found to be $O(n_t^2)$ through simulations. So, the overall per-symbol complexity of 1-LAS is $O(n_t^2)$. In 2-LAS, the complexity of the two-symbol update dominates the one-symbol update. Since there are $\binom{2n_t^2}{2}$ possible two-symbol updates, the complexity of one search stage is $O(n_t^2)$, i.e., $O(n_t^2)$ complexity per symbol. The mean number of stages till the algorithm terminates in 2-LAS was found to be $O(\log n_t)$ through simulations [27]. Therefore, the overall per-symbol complexity of 2-LAS is $O(n_t^2 \log n_t)$.

For the special case of information-lossless-only STBCs (i.e., STBCs with $\delta = t = 1$), the complexity involved in computing $\mathbf{x}_{eff}^{(0)}$ and $\mathbf{H}_{eff}^T \mathbf{H}_{eff}$ can be reduced further. This becomes possible due to the following property of information-lossless-only STBCs. Let $\mathbf{V}_a$ be the complex $n_t^2 \times n_t^2$ matrix with $\mathbf{a}_c^{(i)}$ as its ith column. The computation of $\mathbf{x}_{eff}^{(0)}$ (or $\mathbf{H}_{eff}^T\mathbf{H}_{eff}$) involves multiplication of $\mathbf{V}_a^H$ with another vector (or matrix). The columns of $\mathbf{V}_a^H$ can be permuted in such a way that the permuted matrix is block-diagonal, where each block is an $n_t \times n_t$ discrete Fourier transform (DFT) matrix for $\delta = t = 1$. So, the multiplication of $\mathbf{V}_a^H$ by any vector becomes equivalent to a n_t-point DFT operation, which can be efficiently computed using a fast Fourier transform (FFT) in $O(n_t \log n_t)$ complexity. Using this simplification, the per-symbol complexity of computing $\mathbf{H}_{eff}^T\mathbf{H}_{eff}$ is reduced from $O(n_t^3)$ to $O(n_t^2 \log n_t)$. Computing $\mathbf{x}_{eff}^{(0)}$ using the MMSE filter involves the computation of $(1/n_t)\mathbf{V}_a^H(\mathbf{I} \otimes ((\mathbf{H}_c^H\mathbf{H}_c + (1/\gamma n_t)\,\mathbf{I})^{-1}\mathbf{H}_c^H))\mathbf{y}_c$. The complexity of computing the vector $(\mathbf{I} \otimes ((\mathbf{H}_c^H\mathbf{H}_c + (1/\gamma n_t)\,\mathbf{I})^{-1}\mathbf{H}_c^H))\mathbf{y}_c$ is $O(n_t^2 n_r)$, and the complexity of computing $\mathbf{V}_a^H(\mathbf{I} \otimes ((\mathbf{H}_c^H\mathbf{H}_c + (1/\gamma n_t)\,\mathbf{I})^{-1}\mathbf{H}_c^H))\mathbf{y}_c$ is $O(n_t^3 n_r)$. In the case of information-lossless-only STBC, because of the above-mentioned property, the complexity of computing $\mathbf{V}_a^H(\mathbf{I}\otimes((\mathbf{H}_c^H\mathbf{H}_c+(1/\gamma n_t)\,\mathbf{I})^{-1}\mathbf{H}_c^H))\mathbf{y}_c$ is reduced to $O(n_t^2 \log n_t)$ from $O(n_t^3 n_r)$. So, the total complexity for computing $\mathbf{x}_{eff}^{(0)}$ in information-lossless-only STBC is $O(n_t^2 n_r) + O(n_t^2 \log n_t)$, which gives a per-symbol complexity of $O(n_r) + O(\log n_t)$. So, the overall per-symbol complexity for 1-LAS detection of information-lossless-only STBCs is $O(n_t^2 \log n_t)$.

5.2 Randomized search (RS)

Local search algorithms can be designed based on random selection methods for choosing the set of vectors to be tested in a local neighborhood. The RS algorithm [29] presented in this section is one such algorithm. The algorithm also keeps track of the symbol positions changed in the previous iterations to improve the search efficiency. Used along with multiple restarts, the RS algorithm achieves near-optimal performance in large MIMO systems at a complexity of $O(n_t^{1.4})$.

5.2.1 RS algorithm

Consider the V-BLAST MIMO system model defined in Section 5.1.1. Given $\mathbf{y}$ and $\mathbf{H}$, the RS algorithm starts with an initial solution vector $\mathbf{x}^{(0)}$, a fixed index set $S = \{1, 2, \ldots, 2n_t\}$, and two dynamic index sets C and D which are initialized to be empty. The algorithm is iterative where each iteration results in a solution vector, which, in turn, is used as the input in the next iteration. The set C is updated only once per iteration, whereas the set D may be updated multiple times or may not be updated within each iteration. The set C contains the set of those indices (i.e., symbol positions) where a symbol change in those positions relative to the solution vector of the previous iteration led to an ML cost improvement. In other words, in iteration t, the set C adds the index of the element in $\mathbf{x}^{(t)}$ which when changed, improved the ML cost. The set D contains the set of indices where a symbol change in those positions within an iteration did not result in an ML cost improvement.

Define the neighborhood set of $\mathbf{x}^{(t)}$, denoted by $\mathcal{N}(\mathbf{x}^{(t)})$, as

$$\mathcal{N}(\mathbf{x}^{(t)}) = \Big\{\mathbf{p} \in \mathbb{A}^{2n_t} : \sum_{i=1}^{2n_t} I_{(x_i^{(t)} \neq p_i)} = 1 \text{ and } \\ j \notin \{C\} \; \forall \; x_j^{(t)} \neq p_j, \; j = 1, \ldots, 2n_t\Big\}, \tag{5.41}$$

where $x_i^{(t)}$ and p_i represent the ith components of $\mathbf{x}^{(t)}$ and $\mathbf{p}$, respectively, I is an indicator function ($= 1$ if $x_i^{(t)} \neq p_i$ and 0 otherwise), and $\mathcal{N}(\mathbf{x}^{(t)})$ represents all feasible vectors of $\mathbb{A}^{2n_t}$ which are one symbol away from $\mathbf{x}^{(t)}$, and the index corresponding to the symbol in which they differ is not in set C.

Step 1
Given an initial solution vector $\mathbf{x}^{(t=0)}$, find its neighborhood set, $\mathcal{N}(\mathbf{x}^{(t=0)})$.

Step 2
Randomly select an element m from the index set $\{S - C - D\}$. Choose a subset of vectors from $\mathcal{N}(\mathbf{x}^{(t)})$, denoted by $\{\mathbf{d}(j), \, j = 1, 2, \ldots, |\mathbb{A}| - 1\}$, such that the $\mathbf{d}(j)$s differ from $\mathbf{x}^{(t)}$ in the mth position, $m \in \{S - C - D\}$. It is noted that $j \in \{1, \ldots, |\mathbb{A}| - 1\}$, since, for each symbol in a given position, there are $|\mathbb{A}| - 1$ possible other symbols. Let $g(\mathbf{x}^{(t)} \to \mathbf{d}(j))$ denote the difference in the ML cost between $\mathbf{x}^{(t)}$ and $\mathbf{d}(j)$, i.e.,

$$\begin{aligned} g(\mathbf{x}^{(t)} \to \mathbf{d}(j)) &= f(\mathbf{x}^{(t)}) - f(\mathbf{d}(j)) \\ &= \|\mathbf{y} - \mathbf{H}\mathbf{x}^{(t)}\|^2 - \|\mathbf{y} - \mathbf{H}\mathbf{d}(j)\|^2 \\ &= \mathbf{x}^{(t)T}\mathbf{H}^T\mathbf{H}\mathbf{x}^{(t)} - \mathbf{d}(j)^T\mathbf{H}^T\mathbf{H}\mathbf{d}(j) - 2\mathbf{y}^T\mathbf{H}(\mathbf{x}^{(t)} - \mathbf{d}(j)). \end{aligned} \tag{5.42}$$

Let $\mathbf{G} \triangleq \mathbf{H}^T\mathbf{H}$, $\mathbf{z} \triangleq \mathbf{H}^T\mathbf{y}$, and $\beta(j) \triangleq g(\mathbf{x}^{(t)} \to \mathbf{d}(j))$. By definition, $\mathbf{d}(j)$ can be rewritten as

$$\mathbf{d}(j) = \mathbf{x}^{(t)} + \lambda_m \mathbf{e}_m, \tag{5.43}$$

where $\mathbf{e}_m$ denotes the vector with its mth entry only as 1 and all other entries as zeros, and λ_m belongs to a set of integers such that $\mathbf{d} \in \mathbb{A}^{2n_t}$. For example, if $\mathbb{A} = \{-3, -1, +1, +3\}$, then the possible integer values that λ_m can take are $\{-6, -4, -2, 0, 2, 4, 6\}$. Now, (5.42) can be simplified as

$$\beta(j) = 2\lambda_m z_m - 2\lambda_m \mathbf{e}_m^T \mathbf{G}\mathbf{x}^{(t)} - \lambda_m^2 \mathbf{G}_{m,m}, \tag{5.44}$$

where z_m denotes the mth element of $\mathbf{z}$, and $\mathbf{G}_{i,j}$ denotes the element in the ith row and jth column of $\mathbf{G}$.

Step 3
Compute

$$\beta_{max} = \max_j \, \{\beta(j)\}_{j=1}^{|\mathbb{A}|-1} \tag{5.45}$$

$$max\_idx = \operatorname*{argmax}_j \, \{\beta(j)\}_{j=1}^{|\mathbb{A}|-1}. \tag{5.46}$$

Two cases, namely, $\beta_{max} \geq 0$ and $\beta_{max} < 0$, are possible.

(a) If $\beta_{max} \geq 0$, then
make $t = t + 1$, $\mathbf{x}^{(t)} = \mathbf{d}(max\_idx)$, add m to C, find the new neighborhood set $\mathcal{N}(\mathbf{x}^{(t)})$, and go to Step 2 if $C \neq S$; else output $\mathbf{x}^{(t)}$ as the final solution and stop.

(b) If $\beta_{max} < 0$, then
include m in D, and go to Step 2 if $D \neq \{S - C\}$; else output $\mathbf{x}^{(t)}$ as the final solution and stop.

It is noted that the maximum number of iterations possible is $2n_t$, and the size of the neighborhood set $\mathcal{N}(\mathbf{x}^{(t)})$ decreases by $|\mathbb{A}| - 1$ in each iteration.

Multistart RS

Running the above RS algorithm with a random initial vector allows only "some" parts of the solution space to be explored. Exploring other parts of the solution space can yield better solution vectors. This can be achieved by running the RS algorithm several times (parameterized by $L - 1$, referred to as the *number of restarts*) such that each time a "different" part of the solution space is likely to be explored without increasing the order of complexity. This can be realized through starting the RS algorithm with a different random initial vector each time; this works as follows:

- Choose a random initial vector.
- Run the RS algorithm.
- Repeat the above two steps $L - 1$ times.
- Output the solution vector having the least ML cost among the L solution vectors as the final solution vector and stop.

5.2.2 Performance and complexity

The BER performance of the RS algorithm with no restarts (i.e., $L - 1 = 0$) is found to be far from the SD performance. However, with an increasing number of restarts, the RS performance closely approaches the SD performance. Figure 5.5(a) shows the BER performance of the RS algorithm as a function of number of restarts, $L - 1$, in a 16×16 V-BLAST MIMO system with 4-QAM at SNR $= 10$ dB and 12 dB. The SD performance is also shown for comparison. It is observed that, for both the SNRs, about 60 restarts are required for the BER performance of RS to closely approach SD performance. In general, the number of restarts required to achieve a good performance is found to be a function of the operating SNR, the dimension of the MIMO system, and the modulation alphabet.

Figure 5.5(b) shows the BER as a function of SNR for 8×8, 16×16, 32×32, and 64×64 V-BLAST MIMO systems with 4-QAM. The number of restarts for each point was chosen through simulation such that any further increase in

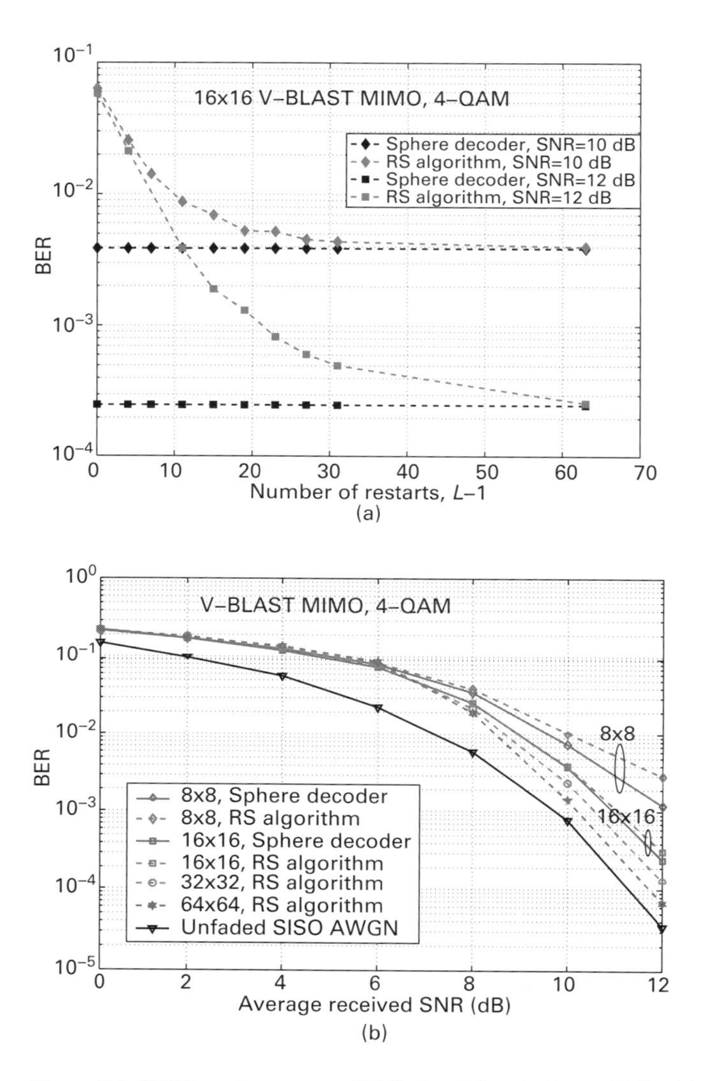

Figure 5.5 BER performance of RS algorithm in large V-BLAST MIMO systems with 4-QAM: (a) BER vs number of restarts; (b) BER vs average received SNR.

number of restarts does not yield significant BER improvement. For example, at 8 dB SNR, the chosen values for L are: $L = 16$ for 8×8, $L = 24$ for 16×16, $L = 32$ for 32×32, and $L = 32$ for 64×64. From Fig. 5.5(b), it is observed that the RS algorithm achieves almost the same performance as the SD for 16×16 MIMO. The RS performance in 32×32 and 64×64 is also close to unfaded SISO AWGN performance which is a lower bound on the ML performance. Through simulations, it has been found that the complexity of the RS algorithm scales as $O(n_t^{1.4})$ [29], which is quite attractive for large MIMO systems.

The performances of the LAS and RS algorithms are close to the optimum in large MIMO systems for 4-QAM modulation. But their performance for higher-order modulation is far from the optimum performance. The issue of improving the higher-order modulation performance is addressed by variants of tabu search in the following section.

5.3 Reactive tabu search (RTS)

Tabu search, attributed to F. W. Glover [5],[6], is a mathematical optimization method that can be used to solve combinatorial optimization problems. It is a heuristic method which is found to be very effective when the problem size becomes large to the extent that the computational burden of finding the exact solution becomes prohibitive given its combinatorial complexity. Tabu search methods have yielded impressive successes in a wide range of application domains. Some of the vastly diverse application domains where tabu search has yielded good results include: the constraint satisfaction problem (with practical applications related to resource assignments, planning, and time tabling), the graph coloring problem, DNA sequencing, internet traffic engineering, vehicle routing problems, multiperiod forest harvesting, the spare parts supply chain, the design of electromagnetic devices, power distribution systems, real-time video tracking, image watermarking, cluster building in wireless sensor networks, high-level synthesis in electronic chip design, to name a few. In many of these applications, determination of the global optimum is often not as critical as finding a set of near-optimal solutions quickly, and the ability of tabu search to efficiently locate near-optimal solutions has made it quite appealing in such applications. In the context of application to communication problems, tabu search has been employed in MUD in CDMA and MIMO detection [12],[16],[17],[20].

Tabu search uses a local search procedure that iteratively moves from one solution to another solution in the neighborhood until some stopping criterion is satisfied. It is a search procedure that explores the solution space beyond local optimality. A key ingredient in the search is that revisits to already visited solutions are prohibited for some time. Already visited solutions are marked as "tabu" for a certain number of iterations (termed as "tabu period") and visits to those solutions marked as tabu are not permitted. This prohibition enhances the search efficiency since it can avoid repeated journeys through the same search path (cycles). The tabu mechanism is implemented by making use of certain memory structures (often referred to as the "tabu list") that keep track of the search dynamics, including the tabu solutions. Allowing revisits to solutions once their tabu periods are over is permitted. This can be helpful because by the time a tabu period ends, the tabu list (and hence the non-tabu neighborhood) of the solution might have changed, thereby the search can move to unexplored regions in the solution space.

Enhanced search efficiencies are achieved in a tabu search through the use of "adaptive memory" and "responsive exploration" of the search space. For example, the tabu period can be kept fixed or changed dynamically. The search is referred to as a fixed tabu search if the tabu period is kept constant throughout. But a good fixed tabu period that gives good performance can turn out to be tedious to obtain. Alternatively, the tabu period can be adaptively changed based on the evolution of the search dynamics. For example, the tabu period can be changed based on the number of repetitions of solutions observed in the search path; increase the tabu period if more repetitions are observed and reduce otherwise. Such a strategy that adapts the tabu period is referred to as a "reactive tabu search." Because of its adaptive nature, RTS is more robust than fixed tabu search. Responsive exploration of the search space refers to intensification of the search in good regions, diversification of the search to promising new regions, and strategically oscillating between intensification and diversification that yields non-monotonic and efficient search patterns.

Tabu search

The basis for the tabu search can be explained as follows. Given a function $\phi(\mathbf{x})$ to be optimized over a set $\mathcal{S}$, a tabu search begins the same way as an ordinary local search, proceeding iteratively from one solution to another until a specified stopping criterion is satisfied. Each $\mathbf{x} \in \mathcal{S}$ has an associated *neighborhood* $\mathcal{N}(\mathbf{x}) \subset \mathcal{S}$. A tabu search differs from an ordinary local search in that it employs the strategy of modifying $\mathcal{N}(\mathbf{x})$ as the search progresses, effectively replacing it by another neighborhood $\mathcal{N}'(\mathbf{x})$. This allows the search to avoid the local minima traps encountered in an ordinary local search. The use of special memory structures (e.g., the tabu list) serves to determine $\mathcal{N}'(\mathbf{x})$, and hence organizes the way in which the solution space is explored. The tabu mechanism described above is one way to determine the solutions admitted to $\mathcal{N}'(\mathbf{x})$.

In the following sections, tabu search algorithms and their variants that have been adopted for MIMO detection and have been shown to achieve near-optimal performance in large MIMO systems are discussed.

System model

Consider a V-BLAST MIMO system with n_t transmit and n_r receive antennas. The transmitted symbols take values from a modulation alphabet $\mathbb{A}$ (e.g., M-QAM/M-PSK). Let $\mathbf{x} \in \mathbb{A}^{n_t}$ denote the transmitted vector. Let $\mathbf{H} \in \mathbb{C}^{n_r \times n_t}$ denote the channel gain matrix, whose entries are assumed to be iid Gaussian with zero mean and unit variance. The received vector $\mathbf{y}$ is given by

$$\mathbf{y} = \mathbf{H}\mathbf{x} + \mathbf{n}, \tag{5.47}$$

where $\mathbf{n}$ is the noise vector whose entries are modeled as iid $\mathbb{C}\mathcal{N}(0, \sigma^2)$. The ML detection rule is given by

$$\widehat{\mathbf{x}}_{ML} = \underset{\mathbf{x} \in \mathbb{A}^{n_t}}{\operatorname{argmin}} \; \|\mathbf{y} - \mathbf{H}\mathbf{x}\|^2 \; = \; \underset{\mathbf{x} \in \mathbb{A}^{n_t}}{\operatorname{argmin}} \; \phi(\mathbf{x}), \tag{5.48}$$

where

$$\phi(\mathbf{x}) \stackrel{\triangle}{=} \mathbf{x}^H \mathbf{H}^H \mathbf{H}\mathbf{x} - 2\Re(\mathbf{y}^H \mathbf{H}\mathbf{x}) \tag{5.49}$$

is the ML cost function. The computational complexity in (5.48) is M^{n_t}. This exponential complexity in n_t is prohibitive for large n_t. A tabu search can be an attractive low complexity approach to obtaining near-optimal solutions to the optimization problem in (5.48).

5.3.1 RTS algorithm

The RTS algorithm for MIMO detection starts with an initial solution vector, defines a neighborhood around it (i.e., defines a set of neighboring vectors based on a neighborhood criterion), and moves to the best vector among the neighboring vectors (even if the best neighboring vector is worse, in terms of ML cost, than the current solution vector; this allows the algorithm to escape from local minima). This process is continued for a certain number of iterations, after which the algorithm is terminated and the best among the solution vectors in all the iterations is declared as the final solution vector.

In defining the neighborhood of the solution vector in a given iteration, the algorithm attempts to avoid cycling by marking the moves to solution vectors of the past few iterations as "tabu" (i.e., prohibits these moves) to ensure an efficient search of the solution space. The number of these past iterations is parameterized as the "tabu period," which is dynamically changed depending on the number of repetitions of the solution vectors that are observed in the search path.

Neighborhood definition

Symbol neighborhood Let M denote the cardinality of the modulation alphabet $\mathbb{A} = \{a_1, a_2, \ldots, a_M\}$. Define a set $\mathcal{N}(a_q)$, $q \in \{1, \ldots, M\}$, as a fixed subset of $\mathbb{A} \backslash a_q$, which is referred to as the *symbol-neighborhood* of a_q. Choose the cardinality of this set to be the same for all a_q, $q = 1, \ldots, M$; i.e., take $|\mathcal{N}(a_q)| = N$, $\forall q$. Note that the maximum and minimum values of N are $M - 1$ and 1, respectively. Let the symbol-neighborhood definition be based on *Euclidean distance*, i.e., for a given symbol, those N symbols which are the nearest will form its neighborhood; the nearest symbol will be the first neighbor, the next nearest symbol will be the second neighbor, and so on. An example of a symbol-neighborhood with $N = 2$ for the alphabet shown in Fig. 5.6 is $\mathcal{N}(a_1) = \{a_2, a_3\}$, $\mathcal{N}(a_2) = \{a_1, a_3\}$, $\mathcal{N}(a_3) = \{a_2, a_4\}$, $\mathcal{N}(a_4) = \{a_3, a_2\}$. Likewise, for $N = 3$, the symbol-neighborhood is $\mathcal{N}(a_1) = \{a_2, a_3, a_4\}$, $\mathcal{N}(a_2) = \{a_1, a_3, a_4\}$, $\mathcal{N}(a_3) = \{a_2, a_4, a_1\}$, $\mathcal{N}(a_4) = \{a_3, a_2, a_1\}$.

Figure 5.6 Illustration of a symbol-neighborhood.

In the case of 4-PAM modulation, $\mathbb{A} = \{-3, -1, 1, 3\}$. For $N = 2$, the symbol neighborhood is $\mathcal{N}(-3) = \{-1, 1\}$, $\mathcal{N}(-1) = \{-3, 1\}$, $\mathcal{N}(1) = \{-1, 3\}$, $\mathcal{N}(3) =$

$\{1, -1\}$. Let $w_v(a_q)$, $v = 1, \ldots, N$ denote the vth element in $\mathcal{N}(a_q)$. Call $w_v(a_q)$ the vth symbol-neighbor of a_q.

Vector neighborhood Let $\mathbf{x}^{(m)} = [x_1^{(m)}\ x_2^{(m)} \cdots x_{n_t}^{(m)}]$ denote the data vector belonging to the solution space in the mth iteration, where $x_i^{(m)} \in \mathbb{A}$. The vector

$$\mathbf{z}^{(m)}(u, v) = \left[z_1^{(m)}(u, v)\ \ z_2^{(m)}(u, v)\ \cdots\ z_{n_t}^{(m)}(u, v)\right], \tag{5.50}$$

is referred to as the (u, v)th *vector-neighbor* (or simply the (u, v)th neighbor) of $\mathbf{x}^{(m)}$, $u = 1, \ldots, n_t$, $v = 1, \ldots, N$, if (i) $\mathbf{x}^{(m)}$ differs from $\mathbf{z}^{(m)}(u, v)$ in the uth coordinate only, and (ii) the uth element of $\mathbf{z}^{(m)}(u, v)$ is the vth symbol-neighbor of $x_u^{(m)}$. That is,

$$z_i^{(m)}(u, v) = \begin{cases} x_i^{(m)} & \text{for } i \neq u, \\ w_v(x_u^{(m)}) & \text{for } i = u. \end{cases} \tag{5.51}$$

There will be $n_t N$ vectors which differ from a given vector in the solution space in only one coordinate. These $n_t N$ vectors form the neighborhood of the given vector. As an example, for the symbol-neighborhood definition with $N = 2$ in Fig. 5.6, the vector-neighbors of a three-element vector $\mathbf{x} = [a_3\ a_2\ a_4]^T$ are shown in Fig. 5.7.

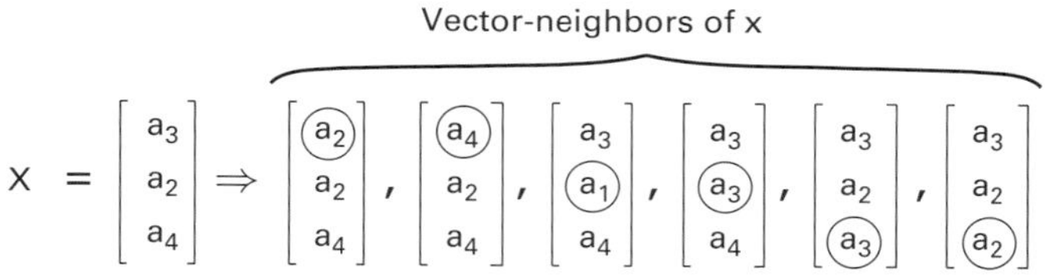

$$\overbrace{\hspace{1em}}^{\text{Vector-neighbors of x}}$$

$$\mathrm{x} = \begin{bmatrix} a_3 \\ a_2 \\ a_4 \end{bmatrix} \Rightarrow \begin{bmatrix} \circled{a_2} \\ a_2 \\ a_4 \end{bmatrix}, \begin{bmatrix} \circled{a_4} \\ a_2 \\ a_4 \end{bmatrix}, \begin{bmatrix} a_3 \\ \circled{a_1} \\ a_4 \end{bmatrix}, \begin{bmatrix} a_3 \\ \circled{a_3} \\ a_4 \end{bmatrix}, \begin{bmatrix} a_3 \\ a_2 \\ \circled{a_3} \end{bmatrix}, \begin{bmatrix} a_3 \\ a_2 \\ \circled{a_2} \end{bmatrix}$$

Figure 5.7 Illustration of a vector-neighborhood.

An operation on $\mathbf{x}^{(m)}$ which gives $\mathbf{x}^{(m+1)}$ belonging to the vector-neighborhood of $\mathbf{x}^{(m)}$ is called a *move*. The algorithm is said to execute a move (u, v) if $\mathbf{x}^{(m+1)} = \mathbf{z}^{(m)}(u, v)$. It is noted that the number of candidates to be considered for a move in any one iteration is $n_t N$. Also, the overall number of "distinct" moves possible is $n_t M N$, which is the cardinality of the union of all moves from all M^{n_t} possible solution vectors. The tabu value of a move, which is a non-negative integer, means that the move cannot be considered for that many number of subsequent iterations.

Tabu matrix

A tabu matrix $\mathbf{T}$ of size $n_t M \times N$ is the matrix whose entries, $t_{r,s}$, $r = 1, \ldots, n_t M$ and $s = 1, \ldots, N$, denote the tabu values of moves. For each coordinate of the solution vector (there are n_t coordinates), there are M rows in $\mathbf{T}$, where each row corresponds to one symbol in the modulation alphabet $\mathbb{A}$; the indices of the rows corresponding to the uth coordinate are from $(u - 1)M + 1$ to uM, $u \in \{1, \ldots, n_t\}$ (see Fig. 5.8). The N columns of the $\mathbf{T}$ matrix correspond to the N symbol-neighbors of the symbol corresponding to each row. In other words, the

(r,s)th entry of the tabu matrix, $r = 1, \ldots, n_t M$, $s = 1, \ldots, N$, corresponds to the move (u, v) from $\mathbf{x}^{(m)}$ when $u = \lfloor (r-1)/M \rfloor + 1$, $v = s$ and $x_u^{(m)} = a_q$, where $q = \text{mod}(r-1, M) + 1$. The entries of the tabu matrix, which are non-negative integers, are updated in each iteration, and they are used to decide the direction in which the search proceeds (as described in the algorithm description below).

$$
\begin{array}{ccc}
 & & \begin{array}{ccccc} \text{1st} & \text{2nd} & & N\text{th } \textit{neighbor} \end{array} \\
\begin{array}{c} x_1 \in \mathbb{A} \rightarrow \\ \\ \\ \\ x_2 \in \mathbb{A} \rightarrow \\ \\ \\ \\ \vdots \\ x_{n_t} \in \mathbb{A} \rightarrow \\ \\ \\ \\ \end{array} &
\begin{array}{c} a_1 \\ a_2 \\ \vdots \\ a_M \\ a_1 \\ a_2 \\ \vdots \\ a_M \\ \vdots \\ a_1 \\ a_2 \\ \vdots \\ a_M \end{array} &
\left(\begin{array}{cccc}
t_{1,1} & t_{1,2} & \cdots & t_{1,N} \\
t_{2,1} & t_{2,2} & \cdots & t_{2,N} \\
\vdots & \vdots & \vdots & \vdots \\
t_{M,1} & t_{M,2} & \cdots & t_{M,N} \\
t_{M+1,1} & t_{M+1,2} & \cdots & t_{M+1,N} \\
t_{M+2,1} & t_{M+2,2} & \cdots & t_{M+2,N} \\
\vdots & \vdots & \vdots & \vdots \\
t_{2M,1} & t_{2M,2} & \cdots & t_{2M,N} \\
\vdots & \vdots & \vdots & \vdots \\
t_{(n_t-1)M+1,1} & t_{(n_t-1)M+1,2} & \cdots & t_{(n_t-1)M+1,N} \\
t_{(n_t-1)M+2,1} & t_{(n_t-1)M+2,2} & \cdots & t_{(n_t-1)M+2,N} \\
\vdots & \vdots & \vdots & \vdots \\
t_{n_t M,1} & t_{n_t M,2} & \cdots & t_{n_t M,N}
\end{array} \right)
\end{array}
$$

Figure 5.8 Tabu matrix.

Search algorithm

Let $\mathbf{g}^{(m)}$ be the vector which has the least ML cost found till the mth iteration of the algorithm. Let l_{rep} be the average length (in number of iterations) between two successive occurrences of a solution vector (repetitions). The tabu period, P, a dynamic non-negative integer parameter, is defined as follows: if a move is marked as tabu in an iteration, it will remain as tabu for P subsequent iterations unless the move results in a better solution. A binary flag, $lflag \in \{0, 1\}$, is used to indicate whether the algorithm has reached a local minimum in a given iteration or not; this flag is used in the evaluation of the stopping criterion of the algorithm. The algorithm starts with an initial solution vector $\mathbf{x}^{(0)}$. Set $\mathbf{g}^{(0)} = \mathbf{x}^{(0)}$, $l_{rep} = 0$, and $P = P_0$. All entries of the tabu matrix are set to zero. The following steps (1)–(3) are performed in each iteration. Consider the mth iteration in the algorithm, $m \geq 0$.

Step (1)

Initialize $lflag = 0$. The ML costs of the $n_t N$ neighbors of $\mathbf{x}^{(m)}$, $\phi(\mathbf{z}^{(m)}(u,v))$, $u = 1, \ldots, n_t$, $v = 1, \ldots, N$, are computed. Let

$$(u_1, v_1) = \underset{u,v}{\text{argmin}} \; \phi(\mathbf{z}^{(m)}(u,v)). \tag{5.52}$$

The move (u_1, v_1) is accepted if any one of the following two conditions is satisfied:

$$\phi(\mathbf{z}^{(m)}(u_1, v_1)) \;<\; \phi(\mathbf{g}^{(m)}), \tag{5.53}$$
$$\mathbf{T}((u_1 - 1)M + q, v_1) \;=\; 0, \tag{5.54}$$

where q is such that $a_q = x_{u_1}^{(m)}, a_q \in \mathbb{A}$. If move (u_1, v_1) is not accepted (i.e., neither of the conditions (5.53) and (5.54) is satisfied), find (u_2, v_2) such that

$$(u_2, v_2) = \underset{u,v:u \neq u_1, v \neq v_1}{\operatorname{argmin}} \quad \phi(\mathbf{z}^{(m)}(u, v)), \tag{5.55}$$

and check for acceptance of the move (u_2, v_2). If this also cannot be accepted, repeat the procedure for (u_3, v_3), and so on. If all the $n_t N$ moves are tabu, then all the tabu matrix entries are decremented by the minimum value in the tabu matrix; this goes on till one of the moves becomes acceptable. Let (u', v') be the index of the neighbor with the minimum cost for which the move is permitted. Make

$$\mathbf{x}^{(m+1)} = \mathbf{z}^{(m)}\,(u', v')\,. \tag{5.56}$$

The variables q', q'', v'' are implicitly defined by $a_{q'} = x_{u'}^{(m)} = w_{v''}(x_{u'}^{(m+1)})$, and $a_{q''} = x_{u'}^{(m+1)}$, where $a_{q'}, a_{q''} \in \mathbb{A}$. It is noted that in this Step (1) of the algorithm, essentially the best permissible vector-neighbor is chosen as the solution vector for the next iteration.

Step (2)
The new solution vector obtained from Step (1) is checked for repetition. For the linear vector channel model in (5.47), repetition can be checked by comparing the ML costs of the solutions in the previous iterations. If there is a repetition, the length of the repetition from the previous occurrence is found, the average length, l_{rep}, is updated, and the tabu period P is modified as $P = P + 1$. If the number of iterations elapsed since the last change of the value of P exceeds βl_{rep}, for a fixed $\beta > 0$, make $P = \max(1, P - 1)$. After a move (u', v') is accepted, if $\phi(\mathbf{x}^{(m+1)}) < \phi(\mathbf{g}^{(m)})$, make

$$\mathbf{T}((u' - 1)M + q', v') = \mathbf{T}((u' - 1)M + q'', v'') = 0,$$
$$\mathbf{g}^{(m+1)} = \mathbf{x}^{(m+1)},$$

else

$$\mathbf{T}((u' - 1)M + q', v') = \mathbf{T}((u' - 1)M + q'', v'') = P + 1,$$
$$lflag = 1, \quad \mathbf{g}^{(m+1)} = \mathbf{g}^{(m)}.$$

It is noted that this Step of the algorithm implements the “reactive” part in the search, by dynamically changing P.

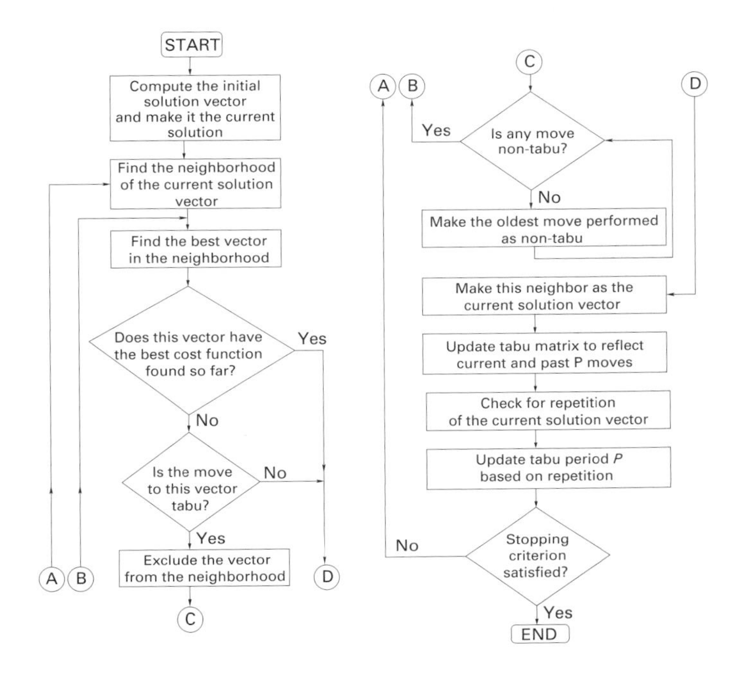

Figure 5.9 RTS algorithm flowchart.

Step (3)
Update the entries of the tabu matrix as

$$\mathbf{T}(r, s) = \max\{\mathbf{T}(r, s) - 1, 0\}, \tag{5.57}$$

for $r = 1, \ldots, n_t M$, $s = 1, \ldots, N$. The algorithm terminates in Step (3) if the following stopping criterion is satisfied, else it goes back to Step (1).

Stopping criterion
The search algorithm described above is stopped if the maximum number of iterations max_iter is reached. If the current solution is a local minimum ($lflag = 1$) and the total number of repetitions of solutions is greater than max_rep, the algorithm is also stopped. The solution of the algorithm is then the vector with the least ML cost which has been found before the algorithm was stopped.

This completes the description of the RTS algorithm. A flowchart of the algorithm described above is shown in Fig. 5.9. The algorithm is parameterized by the parameters listed in Table 5.1.

5.3.2 RTS algorithm versus LAS algorithm

The LAS algorithm presented in Section 5.1 is also a local neighborhood search based algorithm, where the basic definition of neighborhood is the same as in RTS. However, LAS differs from RTS in the following aspects: (i) while the

Table 5.1. *RTS algorithm parameters*

Parameters	Function
max_iter	Maximum number of iterations allowed. Used in the stopping criterion.
max_rep	Maximum number of repetitions allowed. Used in the stopping criterion.
P_0	Initial value of the tabu period.
β	A positive constant. Used in adaptation of the tabu period.

definition of neighborhood is static in LAS for all iterations, in RTS, in addition to the basic neighborhood definition, there is also a dynamic aspect to the neighborhood definition by way of prohibiting certain vectors from being included in the neighbor list (implemented through repetition checks/tabu period), and (*ii*) while LAS is trapped in the local minimum that it first encounters and declares this minimum to be the final solution vector, RTS can potentially find better minima because of the escape strategy embedded in the algorithm (by allowing moves to the best neighbor even if that neighbor has a smaller likelihood than the current solution vector). The multistage LAS algorithm executes a different escape mechanism when it encounters a local minimum, by changing the neighborhood definition: it considers vectors which differ in two or more coordinates (as opposed to only one coordinate in the basic neighborhood definition) as neighbors. On escaping from a local minimum, the algorithm reverts back to the basic neighborhood definition till the next local minimum is encountered and stops when no escape from a local minimum is possible. While the performance gain in multistage LAS compared to LAS is found to be small, the performance gain in RTS compared to LAS is significant.

5.3.3 Performance and complexity of RTS

Performance Figure 5.10 shows the simulated uncoded BER performance of the RTS algorithm as a function of maximum number of iterations, max_iter, in 8×8, 16×16, 32×32, and 64×64 V-BLAST MIMO systems with 4-QAM modulation at an average SNR of 10 dB under the assumption of perfect CSIR. The RTS parameters used in the simulations were: MMSE initial vector, $P_0 = 2, \beta = 0.1, max\_rep = 75$. Two key observations can be made from Fig. 5.10: (*i*) for the system parameters considered, the BER converges (i.e., the change in BER between successive iterations becomes very small) for max_iter greater than 300, and (*ii*) the converged BER of RTS exhibits large-dimension behavior (i.e., the converged BER improves with increasing $n_t = n_r$); e.g., the converged BER improves from 8.3×10^{-3} for 8×8 V-BLAST MIMO to 1.3×10^{-3} for 64×64 V-BLAST MIMO. This is significant considering that the BER in the unfaded SISO AWGN channel itself is 7.8×10^{-4} for 4-QAM.

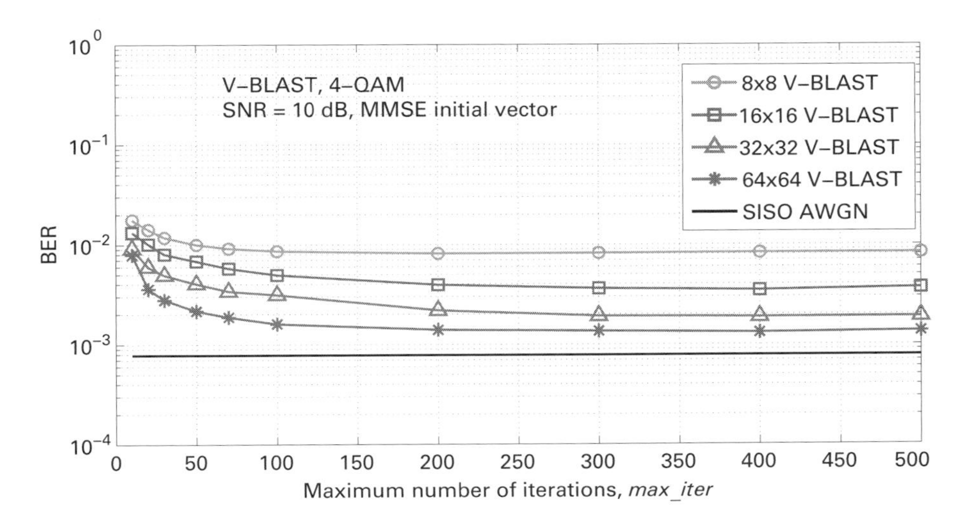

Figure 5.10 Convergence behavior of the RTS algorithm in large MIMO systems.

Figure 5.11 shows the BER performance of the RTS algorithm in comparison with that of the LAS algorithm in 16×16, 32×32, and 64×64 V-BLAST MIMO systems with 4-QAM. It is seen that for the number of dimensions (i.e., n_t) considered, RTS performs better than LAS; e.g., LAS requires 128 real dimensions (i.e., 64×64 V-BLAST with 4-QAM) to achieve a performance close to within 1.8 dB of the unfaded SISO AWGN performance at 10^{-3} BER, whereas the performance of RTS is closer to that of SISO AWGN with just 32 real dimensions (i.e., 16×16 V-BLAST with 4-QAM). Also, in 64×64 V-BLAST MIMO, RTS achieves 10^{-3} BER at an SNR that is just 0.4 dB away from SISO AWGN performance. RTS is able to achieve this better performance because, while the basic neighborhood definitions are similar in both RTS and LAS, the inherent escape strategy in RTS allows it to move out of local minima and move towards better solutions. Because of the escape strategy, RTS incurs some extra complexity compared to LAS.

Complexity

The total complexity of the RTS algorithm comprises three components, namely, (i) computation the initial solution vector $\mathbf{x}^{(0)}$, (ii) computation of $\mathbf{H}^H\mathbf{H}$, and (iii) the RTS operation. The MMSE initial solution vector can be computed in $O(n_t^2 n_r)$ complexity, i.e., in $O(n_t n_r)$ per-symbol complexity since there are n_t symbols per channel use. Likewise, the computations of $\mathbf{H}^H\mathbf{H}$ can be done in $O(n_t n_r)$ per-symbol complexity. Since computations of $\mathbf{x}^{(0)}$ and $\mathbf{H}^H\mathbf{H}$ are needed in both RTS and LAS, complexity components (i) and (ii) will be the same for both these algorithms. Further, while complexity components (i) and (ii) are deterministic, component (iii), which is due to the search part alone, is random and its average complexity is obtained from simulations. Figure 5.12 shows the

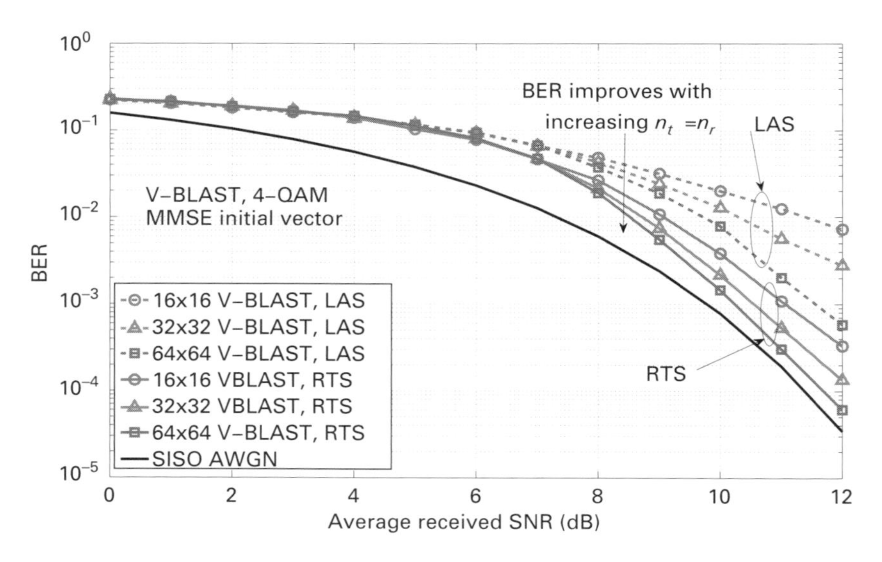

Figure 5.11 BER performance of RTS and LAS algorithms in 16×16, 32×32, and 64×64 V-BLAST MIMO with 4-QAM.

complexity plots for the search part alone (i.e., component (*iii*)) as well as the overall complexity plots of the RTS and LAS algorithms for V-BLAST MIMO with $n_t = n_r$ and 4-QAM at a BER of 10^{-2}. Figure 5.12 shows that the RTS search part has a higher complexity than the LAS search part. This is expected, because the RTS can escape from a local minimum and look for better solutions, whereas LAS settles in the first local minimum itself. However, since the overall complexity is dominated by the computation of $\mathbf{H}^H\mathbf{H}$ and $\mathbf{x}^{(0)}$, the difference in overall complexity between RTS and LAS is not high. This low complexity attribute of the RTS algorithm is attractive for large MIMO signal detection.

RTS performance in higher-order QAM

Though the RTS performance is close to that of SISO AWGN for 4-QAM (as illustrated Fig. 5.11), it is far from SISO AWGN performance for higher-order QAM. This is illustrated in Fig. 5.13, where the RTS performance plots for 4-QAM, 16-QAM, and 64-QAM in 32×32 V-BLAST MIMO are shown. The RTS parameters used in the simulations were: MMSE initial vector, $P_0 = 2$, $\beta = 0.01$; $max\_rep = 250$, $max\_iter = 1000$ for 16-QAM; $max\_rep = 1000$, $max\_iter = 3000$ for 64-QAM. While the RTS performance is just about 0.5 dB away from that of SISO AWGN at 10^{-3} BER in the case of 4-QAM, the gap between RTS performance and SISO AWGN performance at 10^{-3} BER widens for 16-QAM and 64-QAM; the gap is 7.5 dB for 16-QAM and 16.5 dB for 64-QAM. This gap can be viewed as a potential indicator of how much further the RTS detection of higher-order QAM signals can be improved. The RTS algorithm in conjunction with layering and random restart approaches can offer this possible improvement

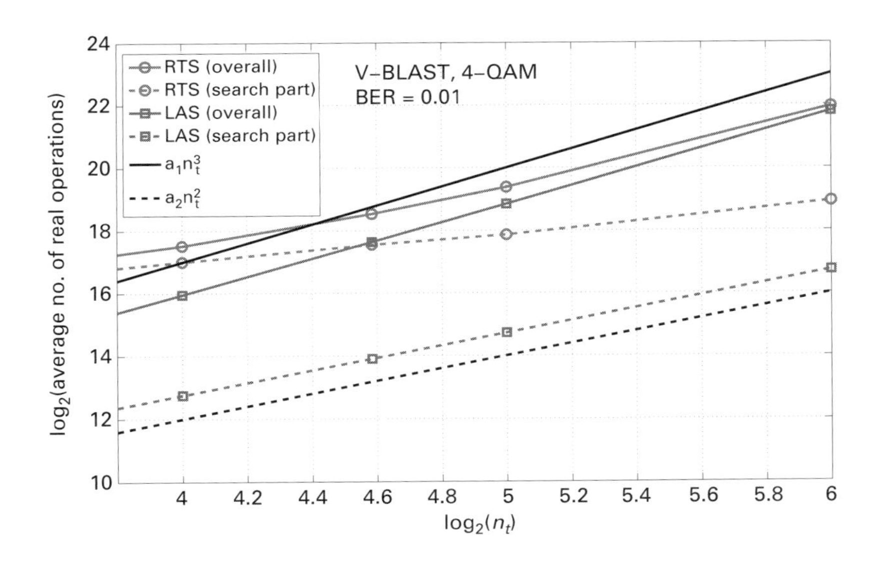

Figure 5.12 Complexity comparison of the RTS and LAS algorithms.

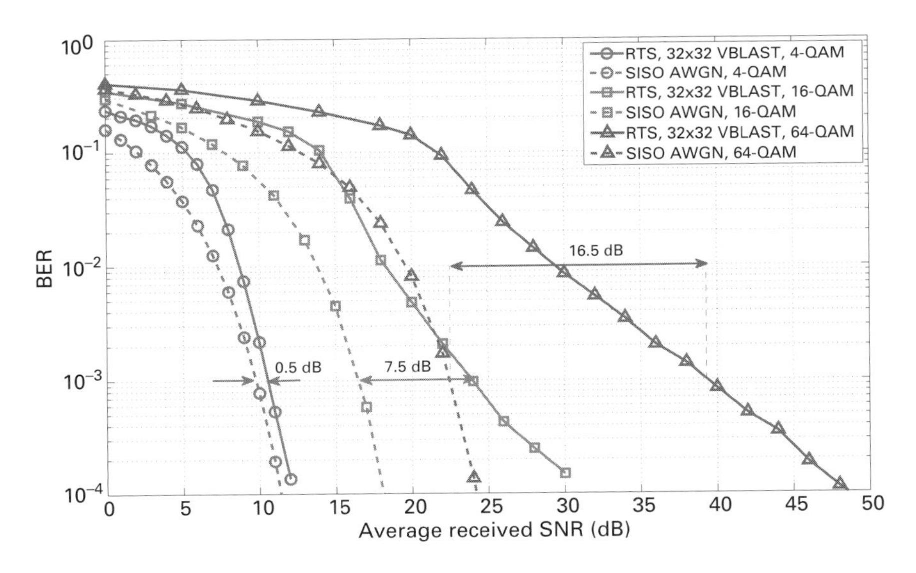

Figure 5.13 BER performance of RTS in 32×32 V-BLAST MIMO with higher-order QAM.

in large MIMO systems with higher-order QAM. Layered RTS (LTS) [25] and random-restart RTS (R3TS) [26] algorithms which use the RTS algorithm as the basic core are treated in the next two sections.

5.3.4 LTS

The LTS algorithm adopts a strategy of detecting symbols in a layered manner, where the RTS algorithm (presented in the previous section) is applied in each layer. In each layer, RTS is used to detect a subvector of the transmitted symbol vector. The subvector size is increased from one layer to the next layer. In addition, the detected subvector in a given layer is used to form the initializing solution for the search in the next layer. The layered structure was inspired by previously suggested approaches based on successive cancelation (or decision feedback) systems, along with the use of QR decomposition for detection and detection ordering. However, unlike cancelation, the LTS approach can update the solution vector for all symbols under consideration within the specific layer.

Let $\mathbf{U}$ denote the upper triangular matrix obtained from the QR decomposition of $\mathbf{H}$. Then, the objective equivalent to (5.48) will be to find the transmitted vector $\mathbf{x}$ which minimizes $\|\mathbf{U}(\mathbf{x} - \bar{\mathbf{x}})\|^2$, where

$$\bar{\mathbf{x}} = \mathbf{H}^{\dagger}\mathbf{y}, \tag{5.58}$$

and $\mathbf{H}^{\dagger}$ is the Moore–Penrose pseudo-inverse of $\mathbf{H}$. Let u_{ij} denote the element in the ith row and jth column of the $\mathbf{U}$ matrix, and x_i denote the ith element of the vector $\mathbf{x}$.

The algorithm processes one layer at a time. It starts with the n_tth layer first. In the kth layer, $k = n_t, (n_t - 1), (n_t - 2), \ldots, 1$, the algorithm detects the $(n_t - k + 1)$-sized subvector $[x_k, x_{k+1}, \ldots, x_{n_t}]$. The symbols of this subvector are detected jointly because they interfere with each other due to the structure of the $\mathbf{U}$ matrix. For example, since $\mathbf{U}$ is upper triangular, there will be no interference to the symbol x_{n_t} in the n_tth layer. In the $(n_t - 1)$th layer, there will be one interferer, x_{n_t}. In the $(n_t - 2)$th layer there will be two interferers, x_{n_t-1} and x_{n_t}, and so on in the subsequent layers. The joint detection method employed in each layer is the RTS algorithm described in the previous section. The complexity can be reduced by skipping the joint detection search in a layer if a simple cancelation of interference due to the already detected symbols in the previous layer results in a good quality output. The LTS algorithm based on the above principles is described below.

Let $\check{\mathbf{x}}$ be the quantized version of $\bar{\mathbf{x}}$, i.e., each element in $\bar{\mathbf{x}}$ is rounded off to its nearest symbol in the alphabet to get $\check{\mathbf{x}}$, so that $\check{\mathbf{x}} \in \mathbb{A}^{n_t}$. Let d_{min} be the minimum Euclidean distance between any two symbols in the alphabet $\mathbb{A}$. The steps performed in the kth layer, $k = n_t, (n_t - 1), \ldots, 1$, are as follows:

Step (1)
Calculate

$$r_k = \bar{x}_k - \sum_{l=k+1}^{n_t} \frac{u_{kl}}{u_{kk}} (\hat{x}_l - \bar{x}_l), \tag{5.59}$$

which is a cancelation operation that removes the interference due to the symbols detected in the previous layer (i.e., $\hat{x}_l$s). Note that for $k = n_t$ (i.e., for the n_tth layer, which is processed first), there will be no second term on the right-hand side in (5.59).

Step (2)
Find the symbol in the alphabet $\mathbb{A}$ which is closest to r_k in Euclidean distance. Let this symbol be a_q.

(*i*) If $|r_k - a_q| < \delta d_{min}$, $0 < \delta \leq 0.5$, then $\hat{x}_k = a_q$ ($\hat{x}_k$ is the detected symbol corresponding to x_k). Make $k = k - 1$ and return to Step (1). Execution of this part of the step essentially skips the joint detection using RTS. The nearness of r_k to an element in $\mathbb{A}$ to within δd_{min}, $0 < \delta \leq 0.5$ is used as the criterion to decide to carry out or skip RTS in layer k. Figure 5.14 shows the BER performance and complexity (in average number of real operations per symbol) of the LTS algorithm as a function of δ for 16×16 V-BLAST MIMO with 16-QAM at an SNR of 19 dB. It can be seen that the BER improves as δ is decreased from 0.5 towards 0. This is because a smaller δ means an increased chance of carrying out joint detection using RTS in Step (2)(*ii*), which results in improved performance while incurring increase in complexity.

(*ii*) If $|r_k - a_q| \geq \delta d_{min}$, then set $\hat{x}_k = \check{x}_k$. Run the RTS algorithm in Section 5.3.1, by replacing $\mathbf{x}^{(0)}$ with $\tilde{\mathbf{x}}^{(0)}$, $\mathbf{H}$ with $\tilde{\mathbf{H}}$, $\mathbf{y}$ with $\tilde{\mathbf{y}}$, where $\tilde{\mathbf{x}}^{(0)}$, $\tilde{\mathbf{H}}$, $\tilde{\mathbf{y}}$ for the kth layer are taken as

$$\tilde{\mathbf{x}}^{(0)} = [\hat{x}_k, \hat{x}_{k+1}, \ldots, \hat{x}_{n_t}], \tag{5.60}$$

$$\tilde{\mathbf{H}} = \begin{bmatrix} u_{kk} & u_{k(k+1)} & \cdots & u_{kn_t} \\ 0 & u_{(k-1)(k-1)} & \cdots & u_{(k-1)n_t} \\ \vdots & \vdots & \vdots & \vdots \\ 0 & 0 & \cdots & u_{n_t n_t} \end{bmatrix}, \tag{5.61}$$

$$\tilde{\mathbf{y}} = \tilde{\mathbf{H}}\,[\bar{x}_k \;\; \bar{x}_{k+1} \;\; \cdots \;\; \bar{x}_{n_t}]^T. \tag{5.62}$$

The output vector from the RTS algorithm is the updated $[\hat{x}_k, \hat{x}_{k+1}, \cdots, \hat{x}_{n_t}]$ subvector. Make $k = k - 1$ and return to Step (1).

After processing all the n_t layers, the vector $\hat{\mathbf{x}} = [\hat{x}_1, \hat{x}_2, \ldots, \hat{x}_{n_t}]$ is declared to be the final detected data vector. Note that in the RTS algorithm, RTS is carried out once on the full $n_t \times n_r$ system model. Whereas, in the LTS algorithm RTS is performed multiple times, once on each layer (depending on the effectiveness of the interference cancelation performed in that layer as per (5.59)). The dimension of the problem increases by 1 from one layer to the next.

Detection with ordering

A way to further improve the performance in any layered scheme is to follow an optimum order while detecting the symbols. One needs to find an optimum order

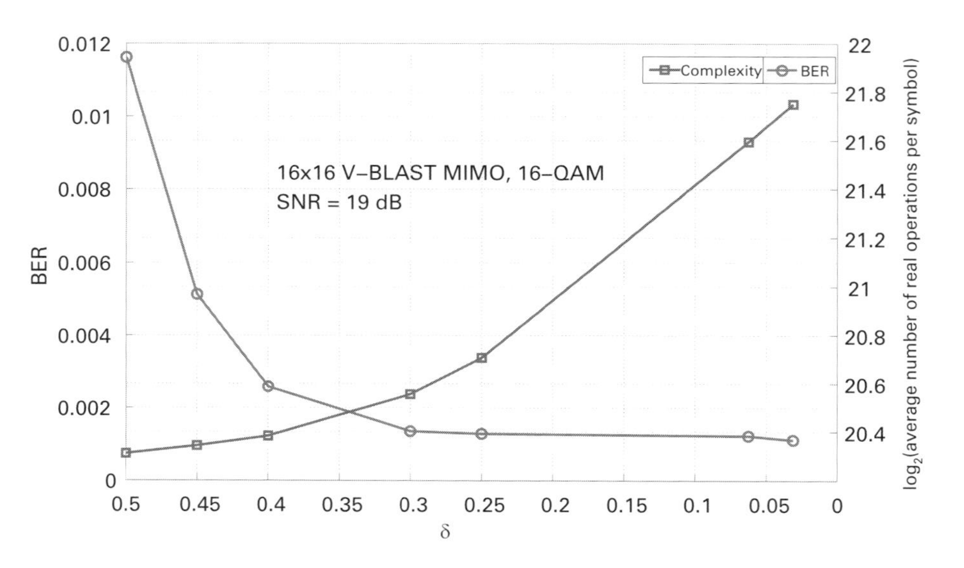

Figure 5.14 BER performance and complexity of LTS algorithm for 16×16 V-BLAST MIMO with 16-QAM.

$(p_1, p_2, \ldots, p_{n_t})$ which is a permutation of $(1, 2, \ldots, n_t)$. One way to obtain the optimum ordering is based on the post-detection SNR of the symbols as follows.

Perform the following steps for $i = n_t, \ldots, 1$ with $\mathbf{H}_{n_t} = \mathbf{H}$. (i) Find $\mathbf{H}_i^{\dagger}$, the Moore–Penrose pseudo-inverse of $\mathbf{H}_i$, where $\mathbf{H}_i$ is obtained by zeroing $(p_{i+1}, p_{i+2}, \ldots, p_{n_t})$ columns of $\mathbf{H}$. (ii) Find p_i, the index that corresponds to the row with the least norm among all rows of $\mathbf{H}_i^{\dagger}$. Detection is then carried out in the following order: p_{n_t}, p_{n_t-1}, $p_{n_{t-2}}$, and so on.

Performance and complexity of LTS

Figure 5.15 shows the BER performance of the LTS algorithm with ordering in $n_t \times n_r$ V-BLAST MIMO systems with $n_t = n_r = 4, 8, 32$, and 16-QAM. The parameters used in all LTS simulations in each layer were: $max\_rep = 10$, $max\_iter = \beta = 20$ for 4-QAM, $max\_rep = 10$, $max\_iter = \beta = 100$ for 16-QAM, $max\_rep = 20$, $max\_iter = \beta = 200$ for 64-QAM, $P_0 = 1$, and $\delta = 1/4$. Figure 5.15 shows that the LTS algorithm exhibits large-dimension advantage, where the achieved BER performance improves and approaches the unfaded SISO AWGN performance with increasing $n_t = n_r$. ZF-SIC and MMSE-SIC detectors with ordering perform less well. Figure 5.15 also shows the BER performance of LTS in 32×32 MIMO with 8-PSK, where the following parameters were used: $max\_rep = 10$, $max\_iter = 50$, $\beta = 50$, $P_0 = 1$, $\delta = 1/8$.

Figure 5.16 shows a comparison between the BER performances of the LTS algorithm without and with ordering, and the RTS algorithm in a 32×32 V-BLAST MIMO system with 16-QAM and 64-QAM. The LTS approach significantly improves the BER performance compared to RTS. For example, RTS needs an

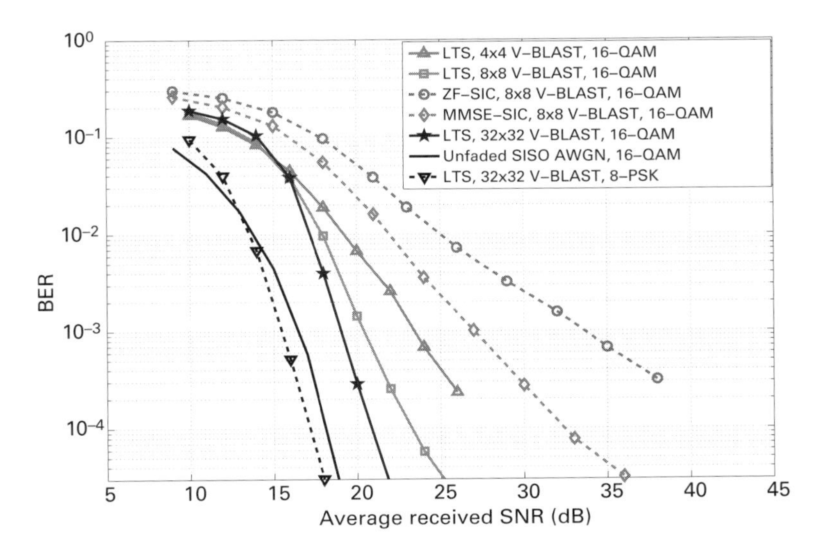

Figure 5.15 BER performance of LTS algorithm in 32×32 V-BLAST MIMO with 16-QAM and 8-PSK.

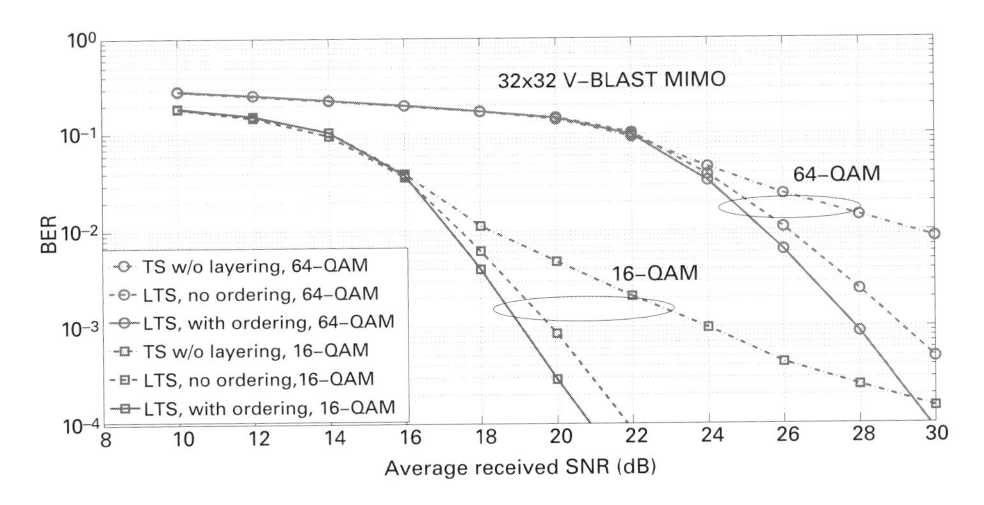

Figure 5.16 BER performance of LTS and RTS algorithms in 32×32 V-BLAST MIMO with 16-QAM and 64-QAM.

SNR of 24 dB to achieve 10^{-3} BER for 16-QAM, whereas the LTS algorithm with ordering achieves the same BER at 19 dB, which amounts to an SNR gain of 5 dB. For 64-QAM, this SNR gain is even higher.

Figure 5.17 shows a complexity comparison between the LTS and RTS algorithms, where the average number of real operations as a function $n_t = n_r$ for 16-QAM at 10^{-2} BER is plotted. Though the order of complexity for RTS is

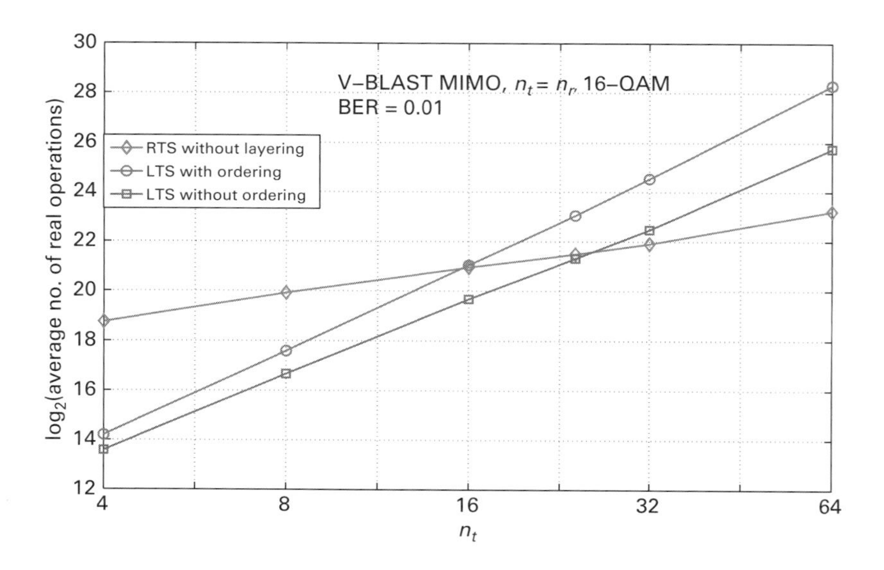

Figure 5.17 Complexity comparison between LTS and RTS algorithms.

less, the constant is high; at $n_t = n_r = 16$ the LTS with ordering has a complexity similar to that of RTS. Also, LTS without ordering has about the same complexity as RTS for $n_t = n_r = 32$; LTS without ordering, however, achieves a better performance than that of RTS.

5.3.5 R3TS

Using multiple random restarts is an efficient technique for achieving improved search performance. The idea is to run the basic search algorithm multiple times, each time with a random initial vector and choose the best among the resulting solution vectors. By doing so, opportunities to search different parts of the solution space are created leading to good solutions. A good strategy to limit the number of restarts is essential to limit the complexity.

In R3TS, the RTS algorithm is used as the basic search algorithm. Three parameters (MAX, Θ, p) are defined for the purpose of limiting the number of searches. The R3TS algorithm works as follows.

- ***Step (1)*** Choose a random initial vector. Run the RTS algorithm using this initial vector and obtain the corresponding solution vector.
- ***Step (2)*** Check if MAX number of RTS searches have been done. If yes, go to Step (5); else go to Step (3).
- ***Step (3)*** If the ML cost of the solution vector from Step (1) is less than Θ, then output the solution vector from Step (1) as the final solution vector and stop; else go to Step (4).

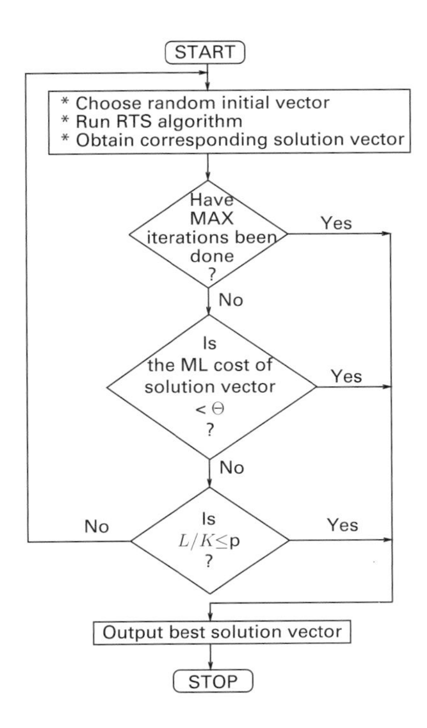

Figure 5.18 Flowchart of R3TS algorithm.

- ***Step (4)*** Let K denote the number of searches done so far. Let L denote the number of distinct solution vectors from Step (1) so far. If $L/K \leq p$, go to Step (5); else go to Step (1).
- ***Step (5)*** Output the best (in terms of ML cost) among the solution vectors obtained so far and stop.

The choice of the value of Θ is made as follows. If the solution vector is the same as the transmitted vector, then the ML cost is $\|\mathbf{n}\|^2$, which has a non-central chi-square distribution with mean $n_r\sigma^2$ and variance $n_r\sigma^4$. The Θ value can be taken empirically to be $n_r\sigma^2 + 2\sqrt{n_r\sigma^4}$, i.e., the Θ value is taken to be the mean plus twice the standard deviation of the ML cost variable corresponding to error-free detection. The threshold comparison in Step (3) reduces the number of searches and hence the complexity. Also, the motivation to do Step (4) is to reduce complexity in realizations where $\|\mathbf{n}\|^2$ happens to be greater than Θ. The parameter values $p = 0.2$ and MAX $= 50$ are found to result in good performance. A flowchart of the R3TS algorithm is shown in Fig. 5.18.

Performance and complexity of R3TS

Figures 5.19, 5.20, 5.21 show the simulated BER performance of the R3TS algorithm with 4-, 16-, and 64-QAM for 16×16, 32×32, and 64×64 V-BLAST MIMO systems, respectively. The following RTS parameters were used

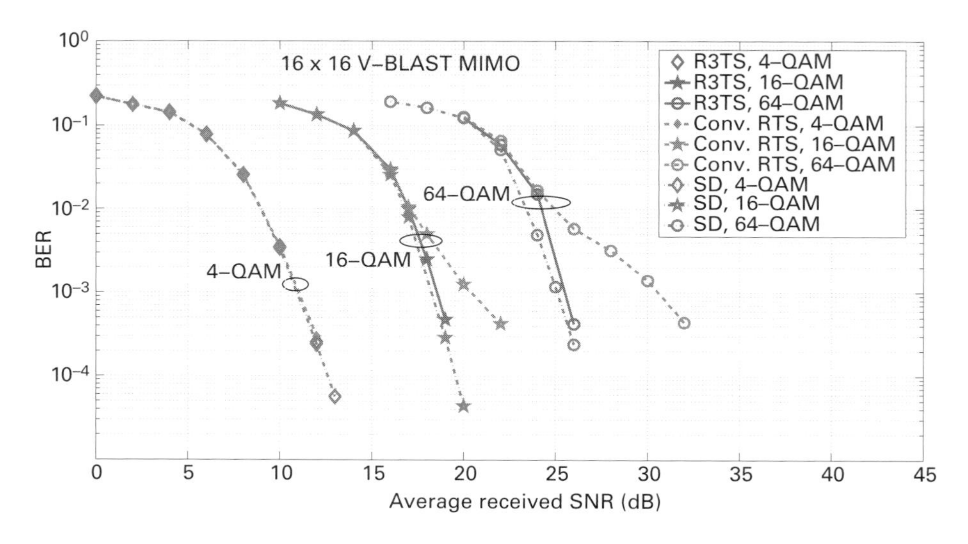

Figure 5.19 BER performance of the R3TS algorithm in 16×16 V-BLAST MIMO with 4-, 16-, 64-QAM.

in the simulations: $max\_rep = 75$, $max\_iter = 300$, $\beta = 0.1$, $P_0 = 2$ for 4-QAM, $max\_rep = 250$, $max\_iter = 1000$, $\beta = 0.01$, $P_0 = 2$ for 16-QAM, and $max\_rep = 1000$, $max\_iter = 3000$, $\beta = 0.01$, $P_0 = 2$ for 64-QAM. MMSE initial vector is used in RTS.

In Fig. 5.19, the R3TS performance is compared with the performances of RTS and the SD. Comparison of R3TS is made with the SD only for 16×16 MIMO, and not for 32×32 and 64×64 MIMO, because of the prohibitively high complexity of the SD in such large dimensions. So, for 32×32 and 64×64 MIMO, comparison is made between R3TS performance and that of the unfaded SISO AWGN, which is a lower bound on the ML performance.

Figure 5.19 shows that the performance of the R3TS algorithm almost matches that of SD in 16×16 MIMO for all the modulations considered (4-, 16-, 64-QAM). It achieves this excellent performance (i.e., almost SD performance) at a much lower complexity than that of SD. Also, the performance improvement achieved by R3TS compared to RTS is quite significant for 16-QAM and 64-QAM (e.g., for 64-QAM in 16×16 MIMO, R3TS outperforms conventional RTS performance by about 5 dB at 10^{-3} BER). Figures 5.20 and 5.21 show that R3TS performs very well in 32×32 and 64×64 MIMO as well.

Table 5.2 presents a performance (in terms of SNR required to achieve 10^{-2} BER) and complexity (in terms of number of real operations at 10^{-2} BER) comparison between R3TS and RTS 32×32 and 64×64 MIMO with 16- and 64-QAM. At 10^{-2} BER, R3TS outperforms RTS by about 5.3 dB in 32×32 MIMO with 64-QAM, and by about 6.6 dB in 64×64 MIMO with 64-QAM, at additional complexities incurred due to multiple restarts.

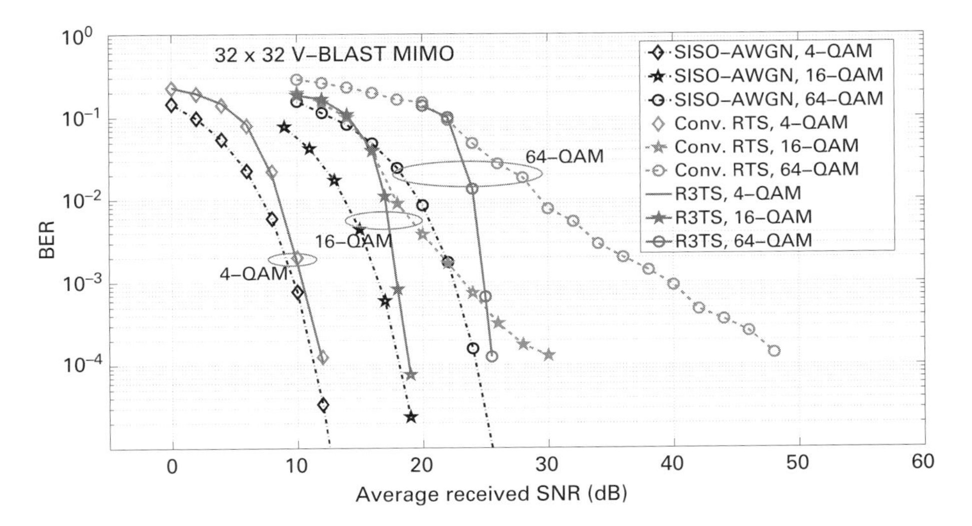

Figure 5.20 BER performance of the R3TS algorithm in 32×32 V-BLAST MIMO with 4-QAM, 16-QAM, 64-QAM.

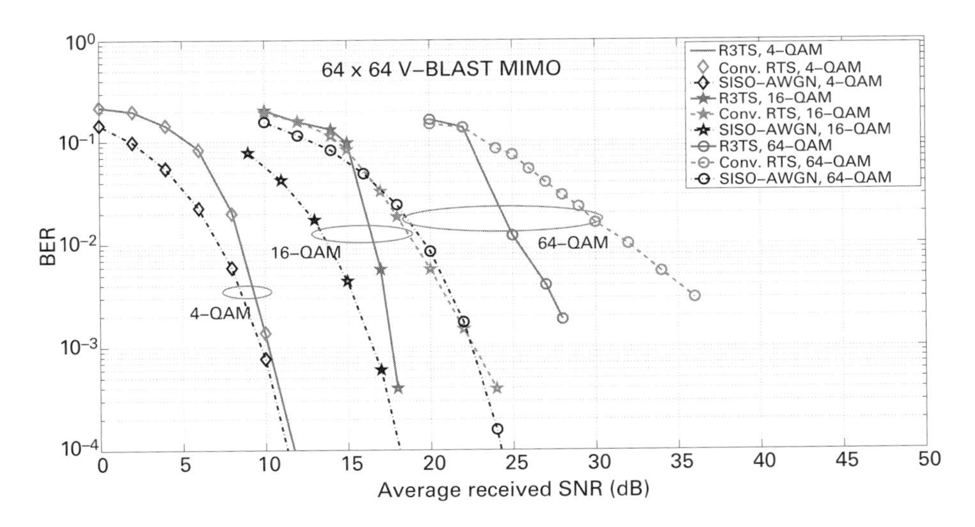

Figure 5.21 BER performance of the R3TS algorithm in 64×64 V-BLAST MIMO with 4-, 16-, 64-QAM.

5.3.6 Lower bounds on ML performance using RTS

Assessing the closeness of the bit error performance of various detection algorithms with respect to the true ML performance is of interest. This is challenging in large MIMO systems because it is prohibitively complex to predict the true ML performance for large n_t (e.g., for 32×32, 64×64 V-BLAST MIMO), either

Table 5.2. *Complexity and performance comparison between R3TS and RTS algorithms in* 16×16, 32×32, 64×64 *V-BLAST MIMO with 16-QAM and 64-QAM*

Modulation	Algorithm	Complexity in av. no. of real operations $\times 10^6$ and SNR required to achieve 10^{-2} BER			
		32×32 MIMO		64×64 MIMO	
		Complexity	SNR	Complexity	SNR
16-QAM	RTS	6.014432	17.9 dB	12.539648	19 dB
	R3TS	7.40464	17 dB	37.750656	16.6 dB
64-QAM	RTS	27.635104	29.4 dB	32.863872	32 dB
	R3TS	77.08784	24.1 dB	467.373248	25.4 dB

through a brute-force search or by using sphere decoding. Although unfaded SISO AWGN performance serves as a lower bound on ML performance and is easy to compute, there is a need for other better bounds which are tight as well as computable at low complexities for large MIMO systems. Upper bounds on the ML performance based on union bounding are known. But the complexity of computing these bounds is M^{n_t}, which is prohibitive for large n_t and M (M is the modulation alphabet size). Also, the tightness of such upper bounds for large n_t and M for a given SNR/BER is difficult to predict because of the lack of knowledge of the true ML performance for those n_t, M, and SNR/BER values. This brings the need for good low complexity lower bounds on the ML performance for large n_t, so that the closeness to the ML performance can be predicted for large MIMO detection algorithms.

The local neighborhood search algorithms presented in this chapter can be used to obtain lower bounds on ML performance. Since these algorithms scale well for large n_t, they are attractive for obtaining ML bounds for large MIMO systems. The use of the RTS algorithm to obtain ML lower bounds is described below.

The actually transmitted vector $\mathbf{x}$ will be used as the initial vector in the RTS algorithm to obtain lower bounds. Define the n-symbol neighborhood of a certain vector to be the set of all vectors which differ from that vector in i coordinates, $1 \leq i \leq n$. With the transmitted vector $\mathbf{x}$ as the initial vector, the RTS algorithm is run and the output solution vector is obtained. Let $\mathbf{x}_{TS}$ denote the output solution vector obtained from the RTS algorithm, and let e_{TS} denote the number of symbol errors in $\mathbf{x}_{TS}$ compared to $\mathbf{x}$. For each realization in the simulations, $\mathbf{x}$, $\mathbf{x}_{TS}$, and hence e_{TS} are known. Let $\mathcal{N}_{\mathbf{x}}$ denote the n-symbol neighborhood of $\mathbf{x}$. Also, let $\mathbf{x}_{ML}$ denote the true ML vector, and e_{ML} denote the number of symbol errors in $\mathbf{x}_{ML}$ (which we do not know, and on which we seek a lower bound). Note that the solution vector $\mathbf{x}_{TS}$ may or may not lie in the n-symbol-neighborhood of $\mathbf{x}$, $\mathcal{N}_{\mathbf{x}}$. Since the RTS algorithm chooses $\mathbf{x}_{TS}$ to be the vector with the least ML cost among all the tested vectors, if $\mathbf{x}_{TS} \notin \mathcal{N}_{\mathbf{x}}$, then $\mathbf{x}_{ML} \notin \mathcal{N}_{\mathbf{x}}$. Also, by the definition of $\mathcal{N}_{\mathbf{x}}$, the number of errors in $\mathbf{x}_{TS}$ and

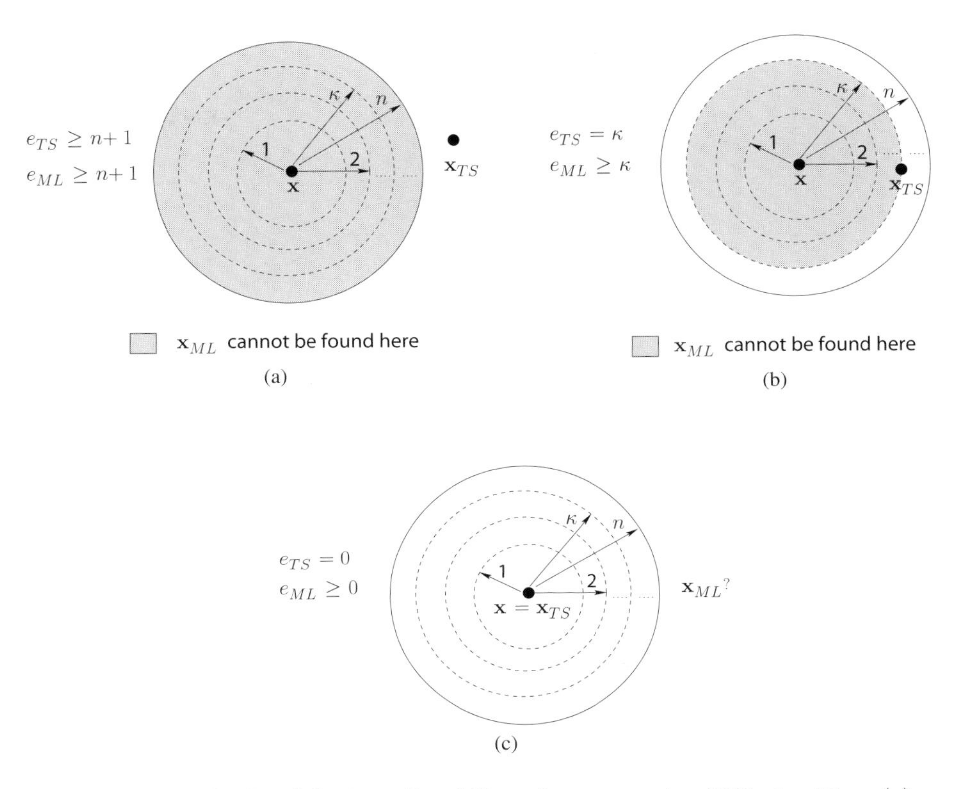

Figure 5.22 Neighborhood for bounding ML performance using RTS algorithm: (a) $e_{TS} \geq n+1$; (b) $e_{TS} = \kappa$, $\kappa \leq n$, (c) $e_{TS} = 0$.

$\mathbf{x}_{ML}$ is lower bounded by $n+1$, i.e., $e_{TS}, e_{ML} \geq n+1$ (Fig. 5.22(a)). So, in the simulations, if $e_{TS} \geq n+1$ in a given realization, then take e_{ML} as $n+1$ as a lower bound on the number of symbol errors in the ML vector. On the other hand, if $\mathbf{x}_{TS} \in \mathcal{N}_{\mathbf{x}}$ which implies that $e_{TS} = \kappa$, $1 \leq \kappa \leq n$, then two cases are possible: (1) $\mathbf{x}_{TS}$ is the ML vector, and (2) $\mathbf{x}_{TS}$ is not the ML vector. In case (1) $e_{TS} = e_{ML} = \kappa$, and in case (2) $e_{TS} = \kappa$ and $\mathbf{x}_{ML}$ being outside $\mathcal{N}_{\mathbf{x}}$, $e_{ML} \geq n+1$ (Fig. 5.22(b)). So, in the simulations, if $e_{TS} = \kappa$, $\kappa \leq n$, then take e_{ML} as κ as a lower bound. Lastly, if $e_{TS} = 0$, then $\mathbf{x}_{TS} = \mathbf{x}$ which may or may not be the ML vector (Fig. 5.22(c)); in such a realization, take $e_{ML} = 0$ as a lower bound. In summary, in the simulations,

- if $e_{TS} = \kappa$, $\kappa \leq n$, then take e_{ML} as κ, and
- if $e_{TS} \geq n+1$, then take e_{ML} as n+1,

which result in a lower bound on the ML symbol error performance. Since the number of symbol errors is a lower bound on the number of bit errors, it is a bit error bound as well.

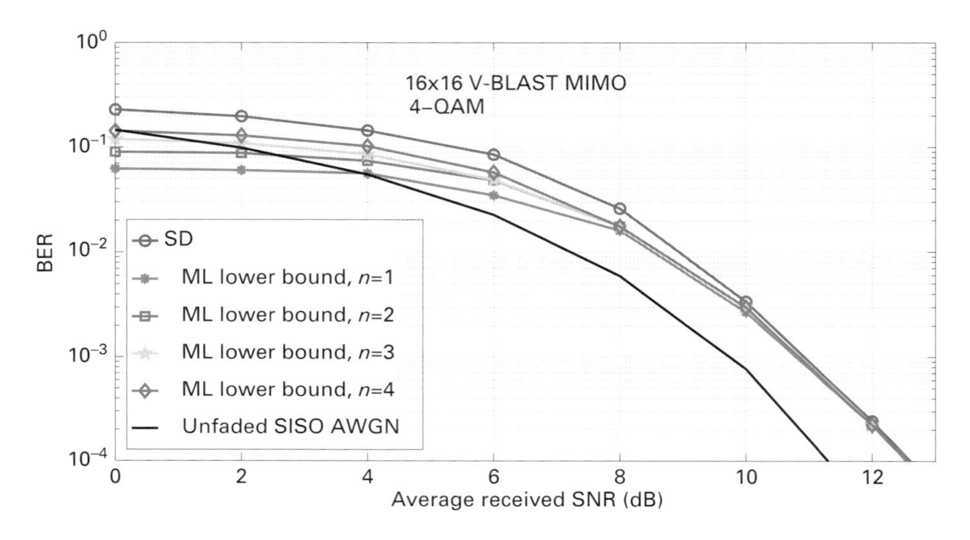

Figure 5.23 ML lower bounds using the RTS algorithm for 16×16 V-BLAST MIMO with 4-QAM.

Results and discussions on the lower bound

The lower bounds obtained as per the above simulation procedure for a 16×16 V-BLAST MIMO system with 4-QAM, 16-QAM, and 64-QAM are plotted in Fig. 5.23. The parameters used in the RTS algorithm simulations were: $max\_rep = 75$, $max\_iter = 300$, $\beta = 0.1$, $P_0 = 2$ for 4-QAM; $max\_rep = 250$, $max\_iter = 1000$, $\beta = 0.01$, $P_0 = 2$ for 16-QAM; and $max\_rep = 1000$, $max\_iter = 3000$, $\beta = 0.01$, $P_0 = 2$ for 64-QAM. In Fig. 5.23, the ML lower bounds for $n = 1, 2, 3, 4$ obtained using RTS simulations are compared with the actual ML performance obtained by sphere decoding. In Fig. 5.23, the lower bound is quite tight (within just 0.5 dB) for BERs less than 10^{-2}, and becomes increasingly tighter for smaller BERs. Even at low SNRs the bound tightens with increasing n. The bounds are found to be tighter than the unfaded SISO AWGN performance. Because of the low complexity of the RTS algorithm, these bounds are easily computed for large n_t.

An approximate prediction of ML performance

The improved tightness of the lower bounds for increasing n is observed to be quite significant at low SNRs in Fig. 5.23. However, a large n means increased complexity. As a low complexity alternative, an approximation to the true ML error performance is given by the error performance of the RTS solution when the transmitted vector $\mathbf{x}$ is used as the initial vector, i.e., when it is assumed that $e_{ML} = e_{TS}$. From the previous discussion on the lower bound, it is noted that e_{TS} indeed corresponds to an upper bound to the ML lower bound. But this upper bound need not be a lower or upper bound to true ML performance. So the performance obtained by equating e_{TS} to e_{ML} is an "approximate ML

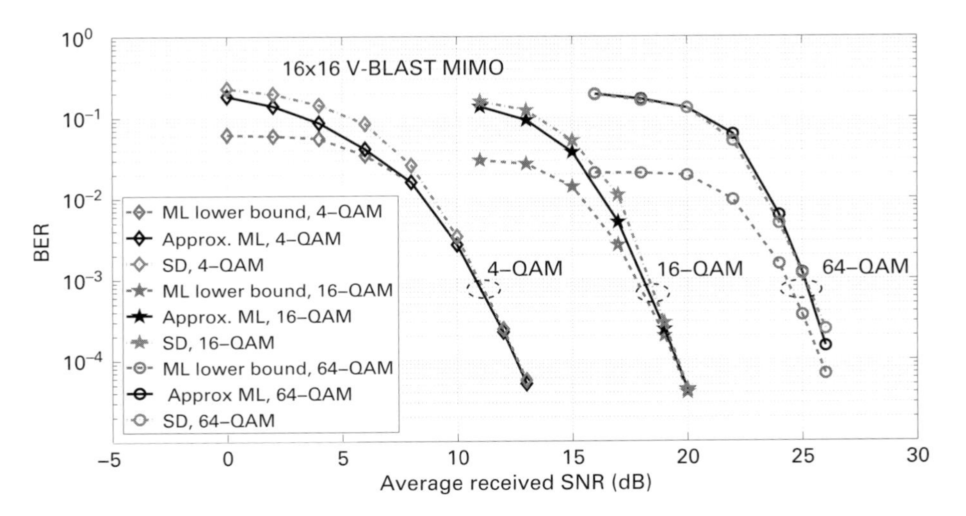

Figure 5.24 ML lower bound, approximate ML performance, and sphere decoding performance for 16×16 V-BLAST MIMO with 4-QAM, 16-QAM, 64-QAM.

performance." It can be noted that, complexity-wise, like the lower bound, the approximate ML performance is also easily obtained for large n_t. In Fig. 5.24, the ML lower bound for $n = 1$, the approximate ML performance, and the SD performance for 16×16 V-BLAST MIMO with 4-, 16-, and 64-QAM are compared. The approximate ML performance is quite close to the actual ML performance obtained by sphere decoding even at low SNRs.

References

[1] E. Aarts and J. K. Lenstra, Eds., *Local Search in Combinatorial Optimization.* Princeton, NJ: Princeton University Press, New Age International Ltd, 2007.

[2] P. Merz and B. Freisleben, "Greedy and local search heuristics for unconstrained binary quadratic programming," *J. Heuristics*, vol. 8, pp. 197–213, 2002.

[3] B. Kernighan and S. Lin, "An efficient heuristic procedure for partitioning graphs," *Bell Syst. Tech. J.*, vol. 49, pp. 291–307, 1972.

[4] S. Lin and B. Kernighan, "An effective heuristic algorithm for the traveling salesman problem," *Oper. Res.*, vol. 21, pp. 498–516, 1973.

[5] F. Glover, "Tabu search – Part I," *ORSA J. Computing*, vol. 1, no. 3, pp. 190–206, Summer 1989.

[6] ——, "Tabu search – Part II," *ORSA J. Computing*, vol. 2, no. 1, pp. 4–32, Winter 1990.

[7] S. Verdu, *Multiuser Detection.* Cambridge, UK: Cambridge University Press, 1998.

[8] Y. Sun, "Search algorithms based on elimination-highest-error and fastest-metric-descent criteria for bit-synchronous CDMA multiuser detection," in *IEEE ICC'1998*, Atlanta, GA, Jun. 1998, pp. 390–394.

[9] J. Luo, G. Levchuk, K. Pattipati, and P. Willett, "A class of coordinate descent methods for multiuser detection," in *IEEE ICASSP2000*, Istanbul, Jun. 2000, pp. 2853–2856.

[10] J. Hu and R. S. Blum, "A gradient guided search algorithm for multiuser detection," *IEEE Commun. Lett.*, vol. 4, no. 11, pp. 340–342, Nov. 2000.

[11] A. AlRustamani and B. R. Vojcic, "A new approach to greedy multiuser detection," *IEEE Trans. Commun.*, vol. 50, no. 8, pp. 1326–1336, Aug. 2002.

[12] H. S. Lim and B. Venkatesh, "An efficient local search heuristics for asynchronous multiuser detection and heuristic search methods," *IEEE Commun. Lett.*, vol. 7, no. 7, pp. 299–301, Jul. 2003.

[13] Z. Qin, K. Cai, and X. Zou, "Turbo multiuser detection based on local search algorithms," in *IEEE ICC'2007*, Glasgow, Jun. 2007, pp. 5987–5992.

[14] Y. Sun, "A family of likelihood ascent search multiuser detectors: an upper bound of bit error rate and a lower bound of asymptotic multiuser efficiency," *IEEE Trans. Commun.*, vol. 57, no. 6, pp. 1743–1752, Jun. 2009.

[15] ——, "A family of likelihood ascent search multiuser detectors: approaching optimum performance via random multicodes with linear complexity," *IEEE Trans. Commun.*, vol. 57, no. 8, pp. 2215–2220, Aug. 2009.

[16] P. H. Tan and L. K. Rasmussen, "A reactive tabu search heuristic for multiuser detection in CDMA," in *IEEE ISIT'2002*, Lausanne, Jun.–Jul. 2002, p. 472.

[17] ——, "Multiuser detection in CDMA – a comparison of relaxations, exact, and heuristic search methods," *IEEE Trans. Wireless Commun.*, vol. 3, no. 5, pp. 1802–1809, Sep. 2004.

[18] Z. Qin and K. C. Teh, "Reduced-complexity turbo equalization for coded intersymbol interference channels based on local search algorithms," *IEEE Trans. Veh. Tech.*, vol. 57, no. 1, pp. 630–635, Jan. 2008.

[19] J. H.-Y. Fan, R. D. Murch, and W. H. Mow, "Near maximum likelihood detection schemes for wireless MIMO systems," *IEEE Trans. Wireless Commun.*, vol. 3, no. 5, pp. 1427–1430, Sep. 2004.

[20] H. Zhao, H. Long, and W. Wang, "Tabu search detection for MIMO systems," in *IEEE PIMRC'2007*, Athens, Sep. 2007, pp. 1–5.

[21] K. V. Vardhan, S. K. Mohammed, A. Chockalingam, and B. S. Rajan, "A low-complexity detector for large MIMO systems and multicarrier CDMA systems," *IEEE J. Sel. Areas Commun.*, vol. 26, no. 3, pp. 473–485, Apr. 2008.

[22] S. K. Mohammed, A. Chockalingam, and B. S. Rajan, "A low-complexity near-ml performance achieving algorithm for large MIMO detection," in *IEEE ISIT'2008*, Toronto, Jul. 2008, pp. 2012–2016.

[23] S. K. Mohammed, A. Zaki, A. Chockalingam, and B. S. Rajan, "High-rate space-time coded large-MIMO systems: low-complexity detection and channel estimation," *IEEE J. Sel. Topics Sig. Proc.*, vol. 3, no. 6, pp. 958–974, Dec. 2009.

[24] P. Li and R. D. Murch, "Multiple output selection-las algorithm in large mimo systems," *IEEE Commun. Lett.*, vol. 14, no. 5, pp. 399–401, May 2010.

[25] N. Srinidhi, T. Datta, A. Chockalingam, and B. S. Rajan, "Layered tabu search algorithm for large-MIMO detection and a lower bound on ML performance," *IEEE Trans. Commun.*, vol. 59, no. 11, pp. 2955–2963, Nov. 2011.

[26] T. Datta, N. Srinidhi, A. Chockalingam, and B. S. Rajan, "Random-restart reactive tabu search algorithm for detection in large-MIMO systems," *IEEE Commun. Lett.*, vol. 14, no. 12, pp. 1107–1109, Dec. 2010.

[27] B. A. Sethuraman, B. S. Rajan, and V. Shashidhar, "Full-diversity high-rate space-time block codes from division algebras," *IEEE Trans. Inform. Theory*, vol. 49, no. 10, pp. 2596–2616, Oct. 2003.

[28] H. Jafarkhani, *Space-Time Coding: Theory and Practice.* Cambridge, UK: Cambridge University Press, 2005.

[29] A. Kumar, S. Chandrasekaran, A. Chockalingam, and B. S. Rajan, "Near-optimal large-MIMO detection using randomized MCMC and randomized search algorithms," in *IEEE ICC'2011*, Kyoto, Jun. 2011.

6 Detection based on probabilistic data association (PDA)

Algorithms based on PDA can be employed for low complexity MIMO detection. PDA was originally developed for target tracking in remote sensing applications like radar, sonar, etc. [1]–[3]. In these applications, tracking a target (potentially a non-cooperative hostile target) or multiple targets is of interest. Signals from such targets of interest can be weak. To detect such weak signals, the detection threshold has to be lowered, which may lead to the detection of other background signals (e.g., signals from other unwanted targets) and sensor noise, yielding spurious measurements which are clutter or false-alarm originated. Target tracking loss may result when such spurious measurements are used in a tracking filter (e.g., a Kalman filter). The selection of which measurements to use to update the estimate of the target state for tracking is known as data association. Data association uncertainty arises from ambiguity as to which measurements are appropriate for use in the tracking filter. Tracking of targets of interest in the presence of data association uncertainty has been studied extensively. The goal is to obtain accurate estimates of the target state and the associated uncertainty.

Optimal estimation of the target state involves recursive computation of the conditional probability density function (pdf) of the state, given all the information available, namely, (*i*) the prior information about the initial state, (*ii*) the intervening known inputs, and (*iii*) the set of measurements up to that time. The conditional pdf of the state in optimal estimation has a number of mixture terms that increases exponentially in time. So several practical suboptimal target state estimation algorithms have been proposed. PDA is one such algorithm. Other algorithms include particle filter, multiple hypothesis tracker, etc.

A key feature in the PDA approach is that it approximates the conditional pdf of the target state at every stage as a single Gaussian with moments matched to the mixture. This repeated conversion of a multimodal Gaussian mixture probability structure to a single Gaussian with matched mean and covariance is the source of PDA's suboptimality. However, this "Gaussian forcing" has been shown to be attractive in reducing complexity and effective in achieving good performance in several successful real-world systems that use PDA.

6.1 PDA in communication problems

Because of its broad appeal in several application domains, PDA has been put to good use in solving several communication problems as well. For example, the PDA approach has been successfully adopted to carry out signal detection, equalization, and channel estimation/tracking in a variety of communication systems. PDA is a reduced complexity alternative to the a posteriori probability (APP) detector/equalizer.

The early and prominent adoption of the PDA approach in communication problems happened in MUD in CDMA [4]–[6]. In CDMA, the basic premise of PDA adoption is the approximation of the inter-user interference (IUI) as Gaussian with an appropriately modified covariance matrix, and the iterative update of the probability associated with each user signal. While Gaussian forcing occurs once per scan without any revisit in tracking applications, in CDMA Gaussian forcing is done for each user and there is iteration. PDA uses a soft IUI cancelation by increasing the covariance of the effective noise (IUI + noise) based on the uncertainty in the other user signals. Demonstrated advantages of PDA include: (*i*) near-optimal performance, and (*ii*) $O(K^3)$ complexity, where K is the number of users. Another advantage is that, since the PDA detector generates APPs directly, it can be applied in coded CDMA for iterative multiuser decoding with only minor modifications [7]. Other related works in MUD that used the PDA framework have focused on CDMA with QAM [8], asynchronous CDMA [9],[10], PIC schemes based on PDA [11], reducing complexity by using soft-decision values from zero-forced output to generate the initial probability values for PDA [12], connection between PDA and soft interference cancelation [13], and large system analysis [14].

The PDA approach is increasingly being adopted in MIMO signal processing. One popular application is in MIMO detection. Symbol based PDA detection of V-BLAST MIMO signals has been reported using a real-valued [15] as well as a complex-valued [16],[17] formulation of the detection problem. PDA based detection has been shown to be equivalent to MMSE based iterative soft-decision interference cancelation (MMSE-ISDIC) [9],[18]. A PDA-aided sphere decoding approach has been shown to offer better performance–complexity tradeoff compared to PDA and the SD [19]. The idea is that the dimensionality reduction achieved through a single-stage PDA preprocessing can provide significant computational relief to sphere decoding at a small performance cost. Equalization in MIMO inter-symbol interference (MIMO-ISI) channels is another popular area where PDA is increasingly employed [20]–[22]. PDA has also been used as a frequency domain equalizer to remove inter-symbol interference (ISI) and inter-carrier interference (ICI) in OFDM without a cyclic prefix [23]. Iterative/joint detection and decoding (turbo equalization) [24]–[26], channel estimation/tracking [27],[28], decoding of space-frequency block code (SFBC) signals [29], and decoding of large dimension NO-STBC signals [30] are other receiver functions where PDA has been successfully employed. PDA has also been used in emerging areas like MIMO wireless relaying [31] and BS cooperation [32].

It has been observed that PDA performs increasingly well with increasing problem size [30],[33],[34]. This is because the quality of the Gaussian approximation used in PDA improves as the number of dimensions (e.g., number of antennas in MIMO systems) increases. This attribute of PDA makes it well suited for signal detection in large MIMO systems. The development of the PDA approach for MIMO detection and its performance in large MIMO systems are discussed next.

6.2 PDA based MIMO detection

Consider a MIMO channel with n_t transmit and n_r receive antennas. The $n_t n_r$ channel gains are modeled as iid $\mathcal{CN}(0,1)$, which are assumed to be known at the receiver but not at the transmitter. Let us consider two types of MIMO transmit architectures, namely, V-BLAST MIMO and NO-STBC MIMO. The following equivalent real-valued system model is applicable for both architectures:

$$\mathbf{y} = \mathbf{H}'\mathbf{x} + \mathbf{n}, \tag{6.1}$$

where $\mathbf{H}' \in \mathbb{R}^{2n_r p \times 2k}$ is the equivalent channel matrix, $\mathbf{y} \in \mathbb{R}^{2n_r p}$ is the equivalent received vector, $\mathbf{x} \in \mathbb{A}^{2k}$ is the vector of transmitted symbols, and $\mathbf{n} \in \mathbb{R}^{2n_r p}$ is the noise vector. For a V-BLAST MIMO system with n_r receive and n_t transmit antennas, $p = 1$ and $k = n_t$ in the above model. For a MIMO system using NO-STBCs from CDA, $p = n_t$ and $k = n_t^2$ [35].

The entries of the noise vector $\mathbf{n}$ are modeled as iid $\mathcal{N}\left(0, \sigma^2 = n_t E_s/2\gamma\right)$, where E_s is the average energy of the transmitted complex symbols, and γ is the average received SNR per receive antenna. Considering square M-QAM without loss of generality, each element of $\mathbf{x}$ is a $\sqrt{M}$-PAM symbol. $\sqrt{M}$-PAM symbols take discrete values from

$$\mathbb{A} \stackrel{\triangle}{=} \{a_q, q = 1, \ldots, \sqrt{M}\},$$

where

$$a_q = \sqrt{\frac{3E_s}{2(M-1)}}(2q - 1 - \sqrt{M}).$$

6.2.1 Real-valued bit-wise system model

In the real-valued system model in (6.1), each entry of $\mathbf{x}$ belongs to a $\sqrt{M}$-PAM constellation, where M is the size of the original square QAM constellation of the transmitted complex symbols. Let $b_i^{(0)}, b_i^{(1)}, \ldots, b_i^{(q-1)}$ denote the $q = \log_2(\sqrt{M})$ constituent bits of the ith entry x_i of $\mathbf{x}$. We can write the value of each entry of $\mathbf{x}$ as a linear combination of its constituent bits as

$$x_i = \sum_{j=0}^{q-1} 2^j \, b_i^{(j)}, \qquad i = 0, 1, \ldots, 2k - 1. \tag{6.2}$$

Let $\mathbf{b} \in \{\pm 1\}^{2qk}$, defined as

$$\mathbf{b} \triangleq \left[b_0^{(0)} \cdots b_0^{(q-1)} b_1^{(0)} \cdots b_1^{(q-1)} \cdots b_{2k-1}^{(0)} \cdots b_{2k-1}^{(q-1)}\right]^T, \tag{6.3}$$

denote the transmitted bit vector. Defining $\mathbf{c} \triangleq [2^0\, 2^1 \cdots 2^{q-1}]$, we can write $\mathbf{x}$ as

$$\mathbf{x} = (\mathbf{I}_{2k} \otimes \mathbf{c})\mathbf{b}, \tag{6.4}$$

using which we can rewrite (6.1) as

$$\mathbf{y} = \underbrace{\mathbf{H}'(\mathbf{I}_{2k} \otimes \mathbf{c})}_{\triangleq \mathbf{H}}\, \mathbf{b} + \mathbf{n}, \tag{6.5}$$

where $\mathbf{H} \in \mathbb{R}^{2n_r p \times 2qk}$ is the effective channel matrix. The MAP estimate of bit $b_i^{(j)}$ is then given by

$$\hat{b}_i^{(j)} = \underset{a \in \{\pm 1\}}{\operatorname{argmax}}\; p\big(b_i^{(j)} = a \,|\, \mathbf{y}, \mathbf{H}\big), \tag{6.6}$$

whose computational complexity is exponential in k. The goal is to obtain $\widehat{\mathbf{b}}$, an estimate of $\mathbf{b}$, using PDA. For this, iteratively update the statistics of each bit in $\mathbf{b}$ (as described in the following section) for a certain number of iterations, and make hard decisions on the final statistics to get $\widehat{\mathbf{b}}$.

6.2.2 Iterative procedure

The PDA algorithm for MIMO detection is iterative in nature, where $2qk$ statistic updates, one for each of the constituent bits, are performed in each iteration. The algorithm starts by initializing the a priori probabilities as $P(b_i^{(j)} = +1) = P(b_i^{(j)} = -1) = 0.5$, $\forall\, i = 0, \ldots, 2k-1$ and $j = 0, \ldots, q-1$. In an iteration, the statistics of the bits are updated sequentially, i.e., the ordered sequence of updates in an iteration is $\{b_0^{(0)}, \ldots, b_0^{(q-1)}, \ldots, b_{2k-1}^{(0)}, \ldots, b_{2k-1}^{(q-1)}\}$. The steps involved in each iteration of the algorithm are as follows.

The likelihood ratio of bit $b_i^{(j)}$ in an iteration, denoted by $\Lambda_i^{(j)}$, is given by

$$\Lambda_i^{(j)} \triangleq \frac{P(b_i^{(j)} = +1|\mathbf{y})}{P(b_i^{(j)} = -1|\mathbf{y})} = \underbrace{\frac{P\big(\mathbf{y}|b_i^{(j)} = +1\big)}{P\big(\mathbf{y}|b_i^{(j)} = -1\big)}}_{\triangleq \beta_i^{(j)}} \underbrace{\frac{P\big(b_i^{(j)} = +1\big)}{P\big(b_i^{(j)} = -1\big)}}_{\triangleq \alpha_i^{(j)}}. \tag{6.7}$$

Denoting the tth column of $\mathbf{H}$ by $\mathbf{h}_t$, (6.5) can be written as

$$\mathbf{y} = \mathbf{h}_{qi+j}\, b_i^{(j)} + \underbrace{\sum_{l=0}^{2k-1} \sum_{\substack{m=0 \\ m \neq q(i-l)+j}}^{q-1} \mathbf{h}_{ql+m}\, b_l^{(m)} + \mathbf{n}}_{\triangleq \tilde{\mathbf{n}}}, \tag{6.8}$$

where $\widetilde{\mathbf{n}} \in \mathbb{R}^{2n_r p}$ is the interference-plus-noise vector (i.e., the effective noise vector). To calculate $\beta_i^{(j)}$, the distribution of the vector $\widetilde{\mathbf{n}}$ is approximated to be Gaussian. Because of this approximation, $\mathbf{y}$ is Gaussian conditioned on $b_i^{(j)}$s. Since there are $2qk-1$ terms in the double summation in (6.8), this vector Gaussian approximation becomes increasingly accurate as k increases, and, hence, n_t increases (note that k is proportional to n_t). Since a Gaussian distribution is fully characterized by its mean and covariance, it suffices to evaluate the mean and covariance of $\mathbf{y}$ given $b_i^{(j)} = +1$ and $b_i^{(j)} = -1$. For notational simplicity, define $p_i^{j+} \triangleq P(b_i^{(j)} = +1)$ and $p_i^{j-} \triangleq P(b_i^{(j)} = -1)$, where $p_i^{j+} + p_i^{j-} = 1$. Let $\boldsymbol{\mu}_i^{j+} \triangleq \mathbb{E}(\mathbf{y}|b_i^{(j)} = +1)$ and $\boldsymbol{\mu}_i^{j-} \triangleq \mathbb{E}(\mathbf{y}|b_i^{(j)} = -1)$. From (6.8), $\boldsymbol{\mu}_i^{j+}$ can be written as

$$\boldsymbol{\mu}_i^{j+} = \mathbf{h}_{qi+j} + \sum_{l=0}^{2k-1} \sum_{\substack{m=0 \\ m \neq q(i-l)+j}}^{q-1} \mathbf{h}_{ql+m}(2p_l^{m+} - 1). \tag{6.9}$$

Similarly, $\boldsymbol{\mu}_i^{j-}$ can be written as

$$\begin{aligned} \boldsymbol{\mu}_i^{j-} &= -\mathbf{h}_{qi+j} + \sum_{l=0}^{2k-1} \sum_{\substack{m=0 \\ m \neq q(i-l)+j}}^{q-1} \mathbf{h}_{ql+m}(2p_l^{m+} - 1) \\ &= \boldsymbol{\mu}_i^{j+} - 2\mathbf{h}_{qi+j}. \end{aligned} \tag{6.10}$$

The $2n_r p \times 2n_r p$ covariance matrix, $\mathbf{C}_i^j$, of $\mathbf{y}$ given b_i^j is

$$\begin{aligned} \mathbf{C}_i^j = \mathbb{E}\Bigg\{ &\Bigg[\mathbf{n} + \sum_{l=0}^{2k-1} \sum_{\substack{m=0 \\ m \neq q(i-l)+j}}^{q-1} \mathbf{h}_{ql+m}(b_l^{(m)} - 2p_l^{m+} + 1)\Bigg] \\ &\cdot \Bigg[\mathbf{n} + \sum_{l=0}^{2k-1} \sum_{\substack{m=0 \\ m \neq q(i-l)+j}}^{q-1} \mathbf{h}_{ql+m}(b_l^{(m)} - 2p_l^{m+} + 1)\Bigg]^T \Bigg\}. \end{aligned} \tag{6.11}$$

Assuming independence among the constituent bits, $\mathbf{C}_i^j$ in (6.11) can be simplified as

$$\mathbf{C}_i^j = \sigma^2 \mathbf{I}_{2n_r p} + \sum_{l=0}^{2k-1} \sum_{\substack{m=0 \\ m \neq q(i-l)+j}}^{q-1} \mathbf{h}_{ql+m}\, \mathbf{h}_{ql+m}^T\, 4p_l^{m+}(1 - p_l^{m+}). \tag{6.12}$$

Using the above mean and covariance, the distribution of $\mathbf{y}$ given $b_i^{(j)} = \pm 1$ can be written as

$$P(\mathbf{y}|b_i^{(j)} = \pm 1) = \frac{e^{-(\mathbf{y}-\boldsymbol{\mu}_i^{j\pm})^T (\mathbf{C}_i^j)^{-1} (\mathbf{y}-\boldsymbol{\mu}_i^{j\pm})}}{(2\pi)^{n_r p} |\mathbf{C}_i^j|^{\frac{1}{2}}}. \tag{6.13}$$

Using (6.13), β_i^j can be written as

$$\beta_i^j = e^{-\left((\mathbf{y}-\boldsymbol{\mu}_i^{j+})^T (\mathbf{C}_i^j)^{-1} (\mathbf{y}-\boldsymbol{\mu}_i^{j+}) - (\mathbf{y}-\boldsymbol{\mu}_i^{j-})^T (\mathbf{C}_i^j)^{-1} (\mathbf{y}-\boldsymbol{\mu}_i^{j-})\right)}. \tag{6.14}$$

Using $\alpha_i^{(j)}$ and $\beta_i^{(j)}$, $\Lambda_i^{(j)}$ is computed using (6.7). Using the value of $\Lambda_i^{(j)}$ and $P(b_i^{(j)} = +1|\mathbf{y}) + P(b_i^{(j)} = -1|\mathbf{y}) = 1$, the statistics of $b_i^{(j)}$ are updated as

$$P(b_i^{(j)} = +1|\mathbf{y}) = \frac{\Lambda_i^{(j)}}{1+\Lambda_i^{(j)}}, \qquad P(b_i^{(j)} = -1|\mathbf{y}) = \frac{1}{1+\Lambda_i^{(j)}}. \tag{6.15}$$

This completes one iteration of the algorithm; i.e., each iteration involves the computation of $\alpha_i^{(j)}$ and (6.9), (6.10), (6.12), (6.14), (6.7), and (6.15) for all i, j. The updated values of $P(b_i^{(j)} = +1|\mathbf{y})$ and $P(b_i^{(j)} = -1|\mathbf{y})$ in (6.15) for all i, j are fed back as a priori probabilities to the next iteration. The algorithm terminates after a certain number of such iterations. At the end of the last iteration, a decision is made on the final statistics to obtain the bit estimate $\widehat{b}_i^{(j)}$ as $+1$ if $\Lambda_i^{(j)} \geq 1$, and -1 otherwise. In coded systems, the $\Lambda_i^{(j)}$s are fed as soft inputs to the decoder.

6.2.3 Complexity reduction

The most computationally expensive operation in computing $\beta_i^{(j)}$ is the evaluation of the inverse of the covariance matrix, $\mathbf{C}_i^j$, of size $2n_rp \times 2n_rp$ which requires $O(n_r^3p^3)$ complexity; this complexity can be reduced as follows. Define matrix $\mathbf{D}$ as

$$\mathbf{D} \triangleq \sigma^2\mathbf{I}_{2n_rp} + \sum_{l=0}^{2k-1}\sum_{m=0}^{q-1} \mathbf{h}_{ql+m}\mathbf{h}_{ql+m}^T 4p_l^{m+}(1-p_l^{m+}). \tag{6.16}$$

At the start of the algorithm, with $p_i^{j+} = p_i^{j-} = 0.5$, $\forall i, j$, $\mathbf{D}$ becomes $\sigma^2\mathbf{I}_{2n_rp} + \mathbf{H}\mathbf{H}^T$.

Computation of $\mathbf{D}^{-1}$

When the statistics of $b_i^{(j)}$ are updated using (6.15), the $\mathbf{D}$ matrix in (6.16) also changes. Inversion of this updated $\mathbf{D}$ would require $O(n_r^3p^3)$ complexity. However, $\mathbf{D}^{-1}$ can be obtained from the previously available $\mathbf{D}^{-1}$ in $O(n_r^2p^2)$ complexity as follows. Since the statistics of only $b_i^{(j)}$ are updated, the new $\mathbf{D}$ is just a rank 1 update of the old $\mathbf{D}$. So, using the matrix inversion lemma, the new $\mathbf{D}^{-1}$ can be obtained from the old $\mathbf{D}^{-1}$ as

$$\mathbf{D}^{-1} \leftarrow \mathbf{D}^{-1} - \frac{\mathbf{D}^{-1}\mathbf{h}_{qi+j}\mathbf{h}_{qi+j}^T\mathbf{D}^{-1}}{\mathbf{h}_{qi+j}^T\mathbf{D}^{-1}\mathbf{h}_{qi+j} + \frac{1}{\eta}}, \tag{6.17}$$

where $\eta = 4p_i^{j+}(1-p_i^{j+}) - 4p_{i,old}^{j+}(1-p_{i,old}^{j+})$, and p_i^{j+} and $p_{i,old}^{j+}$ are the new (i.e., after the update in (6.15)) and old (before the update) values, respectively. Both the numerator and denominator in the second term on the right-hand side of (6.17) can be computed in $O(n_r^2p^2)$ complexity. So, the computation of the new $\mathbf{D}^{-1}$ using the old $\mathbf{D}^{-1}$ can be done in $O(n_r^2p^2)$ complexity.

Computation of $(\mathbf{C}_i^j)^{-1}$

Using (6.16) and (6.12), $\mathbf{C}_i^j$ can be written in terms of $\mathbf{D}$ as

$$\mathbf{C}_i^j = \mathbf{D} - 4p_i^{j+}(1-p_i^{j+})\,\mathbf{h}_{qi+j}\,\mathbf{h}_{qi+j}^T. \tag{6.18}$$

$(\mathbf{C}_i^j)^{-1}$ can be computed from $\mathbf{D}^{-1}$ at reduced complexity using the matrix inversion lemma, as

$$(\mathbf{C}_i^j)^{-1} = \mathbf{D}^{-1} - \frac{\mathbf{D}^{-1}\,\mathbf{h}_{qi+j}\,\mathbf{h}_{qi+j}^T\,\mathbf{D}^{-1}}{\mathbf{h}_{qi+j}^T\,\mathbf{D}^{-1}\,\mathbf{h}_{qi+j} - \dfrac{1}{4p_i^{j+}(1-p_i^{j+})}}, \tag{6.19}$$

which can be computed in $O(n_r^2 p^2)$ complexity.

Computation of $\boldsymbol{\mu}_i^{j+}$ *and* $\boldsymbol{\mu}_i^{j-}$ Computation of $\beta_i^{(j)}$ involves the computation of $\boldsymbol{\mu}_i^{j+}$ and $\boldsymbol{\mu}_i^{j-}$ also. From (6.10), it is clear that $\boldsymbol{\mu}_i^{j-}$ can be computed from $\boldsymbol{\mu}_i^{j+}$ with a computational overhead of only $O(n_r p)$. From (6.9), it is seen that computing $\boldsymbol{\mu}_i^{j+}$ would require $O(qn_r pk)$ complexity. However, this complexity can be reduced as follows. Define vector $\mathbf{u}$ as

$$\mathbf{u} \triangleq \sum_{l=0}^{2k-1}\sum_{m=0}^{q-1} \mathbf{h}_{ql+m}\big(2p_l^{m+} - 1\big). \tag{6.20}$$

Using (6.9) and (6.20), $\boldsymbol{\mu}_i^{j+}$ can be written as

$$\boldsymbol{\mu}_i^{j+} = \mathbf{u} + 2\big(1 - p_i^{j+}\big)\mathbf{h}_{qi+j}. \tag{6.21}$$

$\mathbf{u}$ can be computed iteratively at $O(n_r p)$ complexity as follows. When the statistics of $b_i^{(j)}$ are updated, the new $\mathbf{u}$ can be obtained from the old $\mathbf{u}$ as

$$\mathbf{u} \leftarrow \mathbf{u} + 2\big(p_i^{j+} - p_{i,old}^{j+}\big)\mathbf{h}_{ni+j}, \tag{6.22}$$

whose complexity is $O(n_r p)$. So, the computation of $\boldsymbol{\mu}_i^{j+}$ and $\boldsymbol{\mu}_i^{j-}$ needs $O(n_r p)$ complexity. The full listing of the PDA algorithm is presented in Table 6.1.

Overall complexity

$\mathbf{HH}^T$ needs to be computed at the start of the algorithm. This requires $O(qkn_r^2p^2)$ complexity. So the computation of the initial $\mathbf{D}^{-1}$ requires $O(qkn_r^2p^2) + O(n_r^3p^3)$. Based on the complexity reduction described above, the complexity in updating the statistics of one constituent bit is $O(n_r^2p^2)$. Hence, the complexity for the update of all the $2qk$ constituent bits in an iteration is $O(qkn_r^2p^2)$. Since the number of iterations is fixed, the overall complexity of the algorithm is $O(qkn_r^2p^2) + O(n_r^3p^3)$. Note that for a $n_t = n_r$ V-BLAST MIMO system since $p = 1$ the per-bit complexity is $O(n_t^2)$, which is the same as that of the LAS algorithm presented in the Chapter 5. However, in the case of NO-STBC since $p = n_t$ PDA has a per-bit complexity of $O(n_t^4)$, which is an order higher than that of the LAS algorithm.

6.3 Performance results

The BER performance of PDA detection in large V-BLAST MIMO systems and large NO-STBC MIMO systems is illustrated in the following subsections. A comparison between the performance of the PDA and LAS algorithms is also illustrated.

Table 6.1. *PDA algorithm for MIMO detection*

Initialization

1. $p_i^{j+} = p_i^{j-} = 0.5, \quad \Lambda_i^{(j)} = 1, \quad \forall i = 0, 1, \ldots, 2k-1, j = 0, 1, \ldots, q-1$
2. $\mathbf{u} = \mathbf{0}, \ \mathbf{D}^{-1} = \left(\mathbf{H}\mathbf{H}^T + \sigma^2 \mathbf{I}\right)^{-1}$
3. num_iter: number of iterations
4. $\kappa = 1$; $\qquad \kappa$ is the iteration number

Statistics update in the κth iteration

5. for $i = 0$ to $2k - 1$
6. for $j = 0$ to $q - 1$

Update of statistics of bit $b_i^{(j)}$

7. $\boldsymbol{\mu}_i^{j+} = \mathbf{u} + 2(1 - p_i^{j+})\mathbf{h}_{qi+j}$
8. $\boldsymbol{\mu}_i^{j-} = \boldsymbol{\mu}_i^{j+} - 2\mathbf{h}_{qi+j}$
9. $(\mathbf{C}_i^j)^{-1} = \mathbf{D}^{-1} - \dfrac{\mathbf{D}^{-1}\,\mathbf{h}_{qi+j}\,\mathbf{h}_{qi+j}^T\,\mathbf{D}^{-1}}{\mathbf{h}_{qi+j}^T\,\mathbf{D}^{-1}\,\mathbf{h}_{qi+j} - \frac{1}{4p_i^{j+}(1-p_i^{j+})}}$
10. $\beta_i^j = e^{-\left((\mathbf{y}-\boldsymbol{\mu}_i^{j+})^T(\mathbf{C}_i^j)^{-1}(\mathbf{y}-\boldsymbol{\mu}_i^{j+}) - (\mathbf{y}-\boldsymbol{\mu}_i^{j-})^T(\mathbf{C}_i^j)^{-1}(\mathbf{y}-\boldsymbol{\mu}_i^{j-})\right)}$
11. $p_{i,old}^{j+} = p_i^{j+}, \qquad p_{i,old}^{j-} = p_i^{j-}$
12. $\alpha_i^{(j)} = \dfrac{p_{i,old}^{j+}}{p_{i,old}^{j-}}$
13. $\Lambda_i^{(j)} = \beta_i^{(j)}\alpha_i^{(j)}$
14. $p_i^{j+} = \dfrac{\Lambda_i^{(j)}}{1 + \Lambda_i^{(j)}}, \qquad p_i^{j-} = \dfrac{1}{1 + \Lambda_i^{(j)}}$

Update of $\mathbf{u}$ *and* $\mathbf{D}^{-1}$

15. $\mathbf{u} \leftarrow \mathbf{u} + 2(p_i^{j+} - p_{i,old}^{j+})\mathbf{h}_{qi+j}$
16. $\eta = 4p_i^{j+}(1 - p_i^{j+}) - 4p_{i,old}^{j+}(1 - p_{i,old}^{j+})$
17. $\mathbf{D}^{-1} \leftarrow \mathbf{D}^{-1} - \dfrac{\mathbf{D}^{-1}\mathbf{h}_{qi+j}\mathbf{h}_{qi+j}^T\mathbf{D}^{-1}}{\mathbf{h}_{qi+j}^T\mathbf{D}^{-1}\mathbf{h}_{qi+j} + \frac{1}{\eta}}$
18. end; $\qquad$ End of for loop starting at line 5
19. if ($\kappa = num\_iter$) go to line 21
20. $\kappa = \kappa + 1$, go to line 5
21. $\widehat{b}_i^{(j)} = \mathrm{sgn}\left(\log(\Lambda_i^{(j)})\right), \quad \forall i = 0, 1, \ldots, 2k-1, \ j = 0, 1, \ldots, q-1$
22. $\widehat{x}_i = \sum_{j=0}^{q-1} 2^j\, \widehat{b}_i^{(j)}, \quad \forall i = 0, 1, \ldots, 2k-1$
23. Terminate

6.3.1 Performance in large V-BLAST MIMO

Figure 6.1 shows the BER performance of PDA detection in an $n_r = n_t = 64$ V-BLAST MIMO system with 4-QAM, for an increasing number of PDA iterations ($m = 1, 2, 4, 8$). It is observed that the error performance improves as the number of iterations increases. The performance improvement after more than four iterations is marginal. Figure 6.2 shows the BER performance of PDA detection in V-BLAST MIMO for an increasing number of transmit and receive antennas ($n_t = n_r = 8, 16, 32, 64, 96$) with 4-QAM and $m = 5$ iterations. The error performance improves with increasing $n_t = n_r$, which is due to the

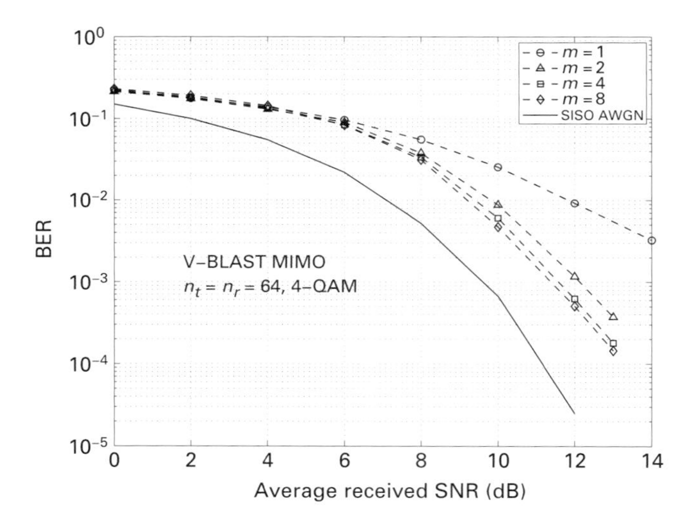

Figure 6.1 BER performance of PDA detection in a V-BLAST MIMO system with $n_t = n_r = 64$ and 4-QAM for different number of PDA iterations ($m = 1, 2, 4, 8$).

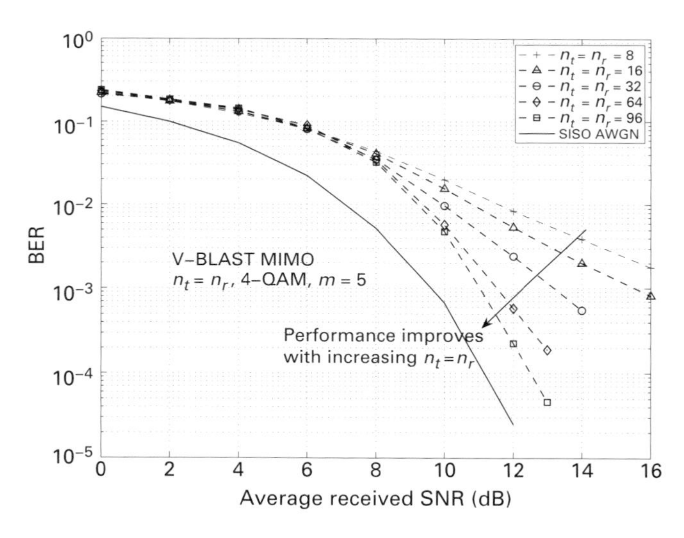

Figure 6.2 BER performance of PDA detection in V-BLAST MIMO for $n_t = n_r = 8, 16, 32, 64, 96$, 4-QAM, and $m = 5$ iterations.

Gaussian approximation becoming increasingly better for an increasing number of antennas.

6.3.2 PDA versus LAS performance in NO-STBC MIMO

In the evaluation of the performance in NO-STBC MIMO the quasi-static assumption (i.e., the assumption that fade remains constant for the duration of

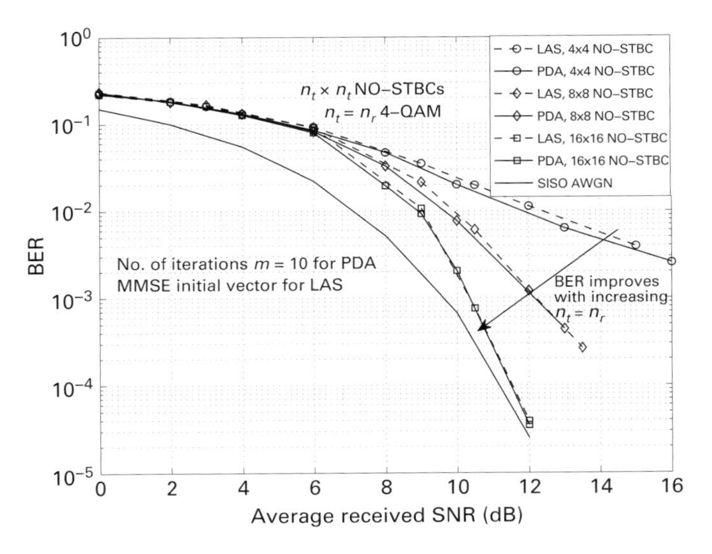

Figure 6.3 Comparison between the BER performances of the PDA and LAS algorithms in decoding 4×4, 8×8, 16×16 NO-STBCs with $n_t = n_r$ and 4-QAM.

one NO-STBC block of n_t channel uses, and iid from one NO-STBC block to another) is made.

4-QAM performance

Figure 6.3 shows the BER performance of the PDA algorithm in decoding 4×4, 8×8, and 16×16 NO-STBCs from CDA with $n_t = n_r$ and 4-QAM. Note that the numbers of real dimensions of the transmitted vectors for 4×4, 8×8, and 16×16 NO-STBCs with QAM are 32, 128, and 512, respectively. For the same settings, the performance of the 1-LAS algorithm with an MMSE initial vector is also plotted for comparison. From Fig. 6.3, it can be seen that (*i*) as in V-BLAST MIMO, the BER performance in NO-STBC improves as the number of dimensions is increased, and (*ii*) with 4-QAM, PDA and LAS algorithms achieve almost similar performances. A performance close to within about 1 dB from non-faded SISO AWGN performance is achieved at 10^{-3} BER in decoding 16×16 NO-STBC from CDA having 512 real dimensions. This illustrates the ability of the PDA to achieve good performance at low complexities in large dimensions.

16-QAM performance

Figure 6.4 presents a BER comparison between the PDA and LAS algorithms in decoding NO-STBCs from CDA with 16-QAM. It is seen that PDA performs better at low SNRs than LAS. For example, with 8×8 and 16×16 NO-STBCs, at low SNRs (e.g., < 25 dB for 16×16 NO-STBC), PDA performs better by about 2 dB compared to LAS at 10^{-2} BER. The PDA performance is still far from

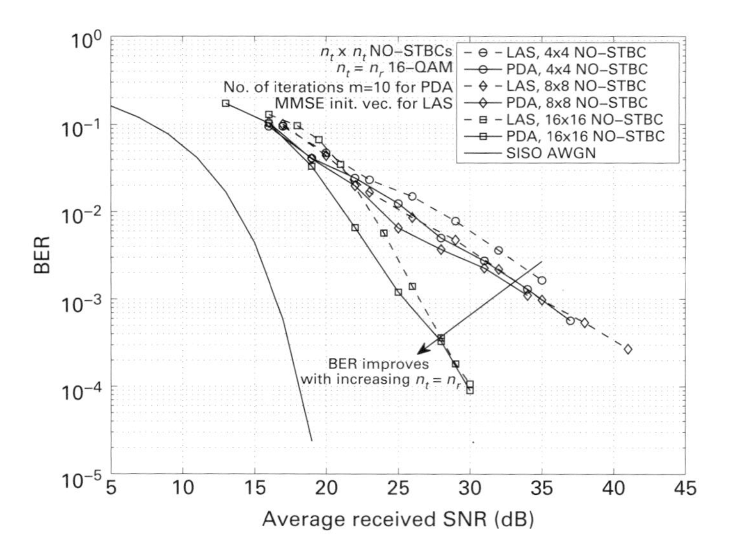

Figure 6.4 Comparison between the BER performances of the PDA and LAS algorithms in decoding 4×4, 8×8, 16×16 NO-STBCs. $n_t = n_r$, 16-QAM.

optimal for 16-QAM, and there appears to be room to improve PDA performance for higher-order QAM in large dimensions.

References

[1] Y. Bar-Shalom and T. E. Fortmann, *Tracking and Data Association.* San Diego, CA: Academic, 1988.

[2] Y. Bar-Shalom and X. R. Li, *Estimation and Tracking: Principles, Techniques and Software.* Dedham, MA: Artech House, 1993.

[3] Y. Bar-Shalom, F. Daum, and J. Huang, "The probabilistic data association filter," *IEEE Cont. Syst. Mag.*, vol. 29, no. 6, pp. 82–100, Dec. 2009.

[4] R. R. Muller and J. B. Huber, "Iterated soft-decision interference cancelation for CDMA," *Broadband Wireless Communications*, pp. 110–115, London: Springer, 1998.

[5] A. Lampe and J. B. Huber, "On improved multiuser detection with iterated soft decision interference cancelation," in *IEEE ICC'1999*, Vancouver Jun. 1999, pp. 172–176.

[6] J. Luo, K. R. Pattipati, P. K. Willett, and F. Hasegawa, "Near-optimal multiuser detection in synchronous CDMA using probabilistic data association," *IEEE Commun. Lett.*, vol. 5, no. 9, pp. 361–363, Sep. 2001.

[7] P. H. Tan, L. K. Rasmussen, and J. Luo, "Iterative multiuser decoding based on probabilistic data association," in *IEEE ISIT'2003*, Ykohama, Jun.–Jul. 2003, p. 301.

[8] J. F. RoBler and J. B. Huber, "Iterative soft decision interference cancelation receivers for DS-CDMA downlink employing 4-QAM and 16-QAM interference

cancelation," in *Asilomar Conf. Signals, Systems & Computers*, Pacific Grove, CA, Nov. 2002, pp. 1488–1494.

[9] D. Pham, J. Luo, K. Pattipati, and P. Willett, "A PDA-Kalman approach to multiuser detection in asynchronous CDMA," *IEEE Commun. Lett.*, vol. 6, no. 11, pp. 475–477, Nov. 2002.

[10] J. Luo, K. R. Pattipati, and P. K. Willett, "A sliding window PDA for asynchronous CDMA, and a proposal for deliberate asynchronicity," *IEEE Trans. Commun.*, vol. 51, no. 12, pp. 1970–1974, Dec. 2003.

[11] F. Vanhaverbeke and M. Moeneclaey, "Evaluation of the critical load of parallel interference cancelation based on the PDA approach," in *IEEE PIMRC'2003*, Beijing, Sep. 2003, pp. 1551–1554.

[12] C.-H. Li and J.-S. Leu, "Performance improvement of multiuser detector using probabilistic data association in CDMA systems," in *IEEE ICSMC'2007*, Montreal, Oct. 2006, pp. 1951–1955.

[13] Y. Huang and J. Zhang, "Generalized probabilistic data association multiuser detector," in *IEEE ISIT'2004*, Chicago IL, Jun.–Jul. 2004.

[14] P. H. Tan and L. K. Rasmussen, "Asymptotically optimal nonlinear MMSE multiuser detection based on multivariate Gaussian approximation," *IEEE Trans. Commun.*, vol. 54, no. 8, pp. 1427–1438, Aug. 2006.

[15] D. Pham, K. R. Pattipati, P. K. Willet, and J. Luo, "A generalized probabilistic data association detector for multiantenna systems," *IEEE Commun. Lett.*, vol. 8, no. 4, pp. 205–207, Apr. 2004.

[16] Y. Jia, C. Andrieu, R. J. Piechocki, and M. Sandell, "Gaussian approximation based mixture reduction for near optimum detection in MIMO systems," *IEEE Commun. Lett.*, vol. 9, no. 11, pp. 997–999, Nov. 2005.

[17] Y. Jia, C. M. Vithanage, C. Andrieu, and R. J. Piechocki, "Probabilistic data association for symbol detection in MIMO systems," *Electron. Lett.*, vol. 42, no. 1, pp. 38–40, Jan. 2006.

[18] F. Cao, J. Li, and J. Yang, "On the relation between PDA and MMSE-ISDIC," *IEEE Signal Process. Lett.*, vol. 14, no. 9, pp. 597–600, Sep. 2007.

[19] G. Latsoudas and N. D. Sidiropoulos, "A hybrid probabilistic data association-sphere decoding for multiple-input-multiple-output systems," *IEEE Signal Process. Lett.*, vol. 12, no. 4, pp. 309–312, Apr. 2005.

[20] S. Liu and Z. Tian, "Near-optimum soft decision equalization for frequency selective MIMO channels," *IEEE Trans. Signal Process.*, vol. 52, no. 3, pp. 721–733, Mar. 2004.

[21] ——, "A Kalman-PDA approach to soft-decision equalization for frequency-selective MIMO channels," *IEEE Trans. Signal Process.*, vol. 53, no. 10, pp. 3819–3830, Oct. 2005.

[22] C. Vithanage, C. Andrieu, R. Piechocki, and M. S. Yee, "Reduced complexity equalization of MIMO systems with a fixed-lag smoothed M-BCJR algorithm," in *IEEE SPAWC'2005*, New York, NY, Jun. 2005, pp. 136–140.

[23] S. Parsaeefard and H. Amindavar, "Frequency domain probabilistic data association equalizer for OFDM systems without cyclic-prefix," in *IEEE TENCON'2006*, Hong Kong, Nov. 2006.

[24] Y. Yin, Y. Huang, and J. Zhang, "Turbo equalization using probabilistic data association," in *IEEE GLOBECOM'2004*, Dallas, TX, vol. 4, Nov.-Dec. 2004.

[25] S. Liu, M. Zhao, Z. Luo, F. Li, and Y. Liu, "V-BLAST architecture employing joint iterative GPDA detection and decoding," in *IEEE ICC'2006*, Istanbul, Jun. 2006, pp. 4219–4224.

[26] L. Rugini, P. Banelli, K. Fang, and G. Leus, "Enhanced turbo MMSE equalization for MIMO-OFDM over rapidly time-varying frequency-selective channels," in *IEEE ICASSP'2009*, Taipei, Apr. 2009, pp. 36–40.

[27] Z. J. Wang, Z. Han, and K. J. R. Liu, "A MIMO-OFDM channel estimation approach using time of arrivals," *IEEE Trans. Wireless Commun.*, vol. 4, no. 3, pp. 1207–1213, May 2005.

[28] Y. Jia, C. Andrieu, R. J. Piechocki, and M. Sandell, "PDA multiple model approach for joint channel tracking and symbol detection in MIMO systems," *IEE Proc. Commun.*, vol. 153, no. 4, pp. 501–507, Aug. 2006.

[29] B. Yang, P. Gong, S. Feng, *et al.* "Monte Carlo probabilistic data association detector for SFBC-VBLAST-OFDM system," in *IEEE WCNC'2007*, Hong Kong, Mar. 2007, pp. 1502–1505.

[30] S. K. Mohammed, A. Chockalingam, and B. S. Rajan, "Low-complexity near-MAP decoding of large non-orthogonal STBCs using PDA," in *IEEE ISIT'2009*, Seoul, Jun.–Jul. 2009.

[31] Z. Mei, Z. Yang, X. Li, and L. Wu, "Probabilistic data association detectors for multi-input multi-output relaying system," *IET Commun.*, vol. 5, no. 4, pp. 534–541, Mar. 2011.

[32] S. Yang, T. Lv, R. G. Maunder, and L. Hanzo, "Distributed probabilistic-data-association-based soft reception employing base station cooperation in MIMO-aided multiuser multicell systems," *IEEE Trans. Veh. Tech.*, vol. 60, no. 7, pp. 3532–3538, Sep. 2011.

[33] J. Fricke, M. Sandell, J. Mietzner, and P. Hoeher, "Impact of the Gaussian approximation on the performance of the probabilistic data association MIMO decoder," *EURASIP J. Wireless Commun. Netw.*, vol. 5, no. 5, pp. 796–800, Oct. 2005.

[34] S. Yang, T. Lv, R. G. Maunder, and L. Hanzo, "Unified bit-based probabilistic data association aided MIMO detection for high-order QAM constellations," *IEEE Trans. Veh. Tech.*, vol. 60, no. 3, pp. 981–991, Mar. 2011.

[35] B. A. Sethuraman, B. S. Rajan, and V. Shashidhar, "Full-diversity high-rate space-time block codes from division algebras," *IEEE Trans. Inform. Theory*, vol. 49, no. 10, pp. 2596–2616, Oct. 2003.

7 Detection/decoding based on message passing on graphical models

Probability theory and graph theory are two branches of mathematics that are widely applicable in many different domains. Graphical models combine concepts from both these branches to provide a structured framework that supports representation, inference, and learning for a broad spectrum of problems [1]. Graphical models are graphs that indicate inter-dependencies between random variables [2],[3]. Distributions that exhibit some structure can generally be represented naturally and compactly using a graphical model, even when the explicit representation of the joint distribution is very large. The structure often allows the distribution to be used effectively for inference, i.e., answering certain queries of interest using the distribution. The framework also facilitates construction of these models by learning from data.

In this chapter we consider the use of graphical models in: (1) the representation of distributions of interest in MIMO systems, (2) formulation of the MIMO detection problem as an inference problem on such models (e.g., computation of posterior probability of variables of interest), and (3) efficient algorithms for inference (e.g., low complexity algorithms for computing the posterior probabilities). Three basic graphical models that are widely used to represent distributions include Bayesian belief networks [4], Markov random fields (MRFs) [5], and factor graphs [6]. Message passing algorithms like the BP algorithm [7] are efficient tools for inference on graphical models. In this chapter, a brief survey of various graphical models and BP techniques is presented. Application of BP to equalization in MIMO-ISI channels and large MIMO signal detection is also covered.

7.1 Graphical models

7.1.1 Bayesian belief networks

A Bayesian belief network is a directed acyclic graph [2],[7]. Every node in the graph denotes a discrete random variable. The edges are directed to reflect the statistical dependence between the variables. In particular, the variable pertaining to the node at the head of an arrow of an edge is statistically dependent on the variable pertaining to the node at the tail of the arrow. Therefore, the edge joining node x_1 to node x_2 is associated with the conditional probability $\mathrm{p}(x_1|x_2)$.

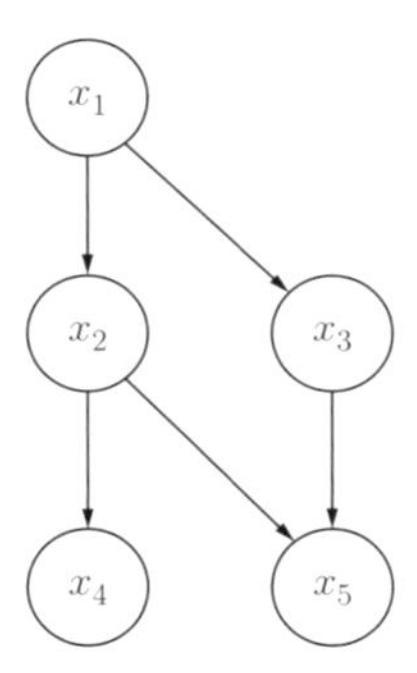

Figure 7.1 An example of a Bayesian belief network.

An example Bayesian belief network is depicted in Fig. 7.1. In the example shown in Fig. 7.1, the joint probability of all the variables is given by

$$\mathrm{p}(x_1, x_2, x_3, x_4, x_5) = \mathrm{p}(x_1)\,\mathrm{p}(x_2|x_1)\,\mathrm{p}(x_3|x_1)\,\mathrm{p}(x_4|x_2)\,\mathrm{p}(x_5|x_2, x_3). \quad (7.1)$$

In general, in a problem with variables $x_1, x_2, \ldots, x_N$, the joint probability is given by

$$\mathrm{p}(x_1, x_2, \ldots, x_N) = \prod_{i=1}^{N} \mathrm{p}\left[x_i|\mathcal{P}(x_i)\right], \quad (7.2)$$

where $\mathcal{P}(x_i)$ denotes the *parents* of the node pertaining to the variable x_i. The probability of any variable $x_k, k \in \{1, 2, \ldots, N\}$ can then be obtained by marginalizing the joint probability function over all the other variables, i.e.,

$$\mathrm{p}(x_k) = \sum_{x_1} \cdots \sum_{x_{k-1}} \sum_{x_{k+1}} \cdots \sum_{x_N} \mathrm{p}(x_1, x_2, \ldots, x_N). \quad (7.3)$$

7.1.2 Markov random fields

An MRF is an undirected graph whose vertices are random variables [2],[7]. The variables are such that any variable is independent of all the other variables, given its neighbors. That is,

$$\mathrm{p}(x_k|x_1, \ldots, x_{k-1}, x_{k+1}, \ldots, x_N) = \mathrm{p}\left[x_k|\mathcal{N}(x_k)\right], \quad (7.4)$$

where $\mathcal{N}(x_k)$ represents the set of all nodes neighboring the node pertaining to the variable x_k.

Usually, the variables in an MRF are constrained by a *compatibility function*, also known as a *clique potential* in the literature. A *clique* of an MRF is a fully connected subgraph. A *maximal clique* is a clique which does not remain fully connected if any additional vertex of the MRF is included in it. Often, the term clique is used to refer to a maximal clique. Let there be N_C cliques in the MRF, and let $\mathbf{x}_j$ be the variables in clique j. Let $\psi_j(\mathbf{x}_j)$ be the clique potential of clique

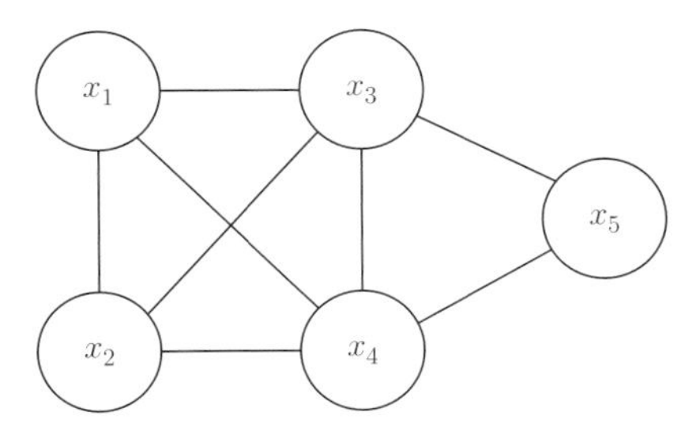

Figure 7.2 An example of an MRF.

j. Then the joint distribution of the variables is given by

$$\mathrm{p}(\mathbf{x}) \propto \prod_{j=1}^{N_C} \psi_j(\mathbf{x}_j). \tag{7.5}$$

For example, consider the MRF shown in Fig. 7.2. There are two cliques in the MRF, namely, $\{x_1, x_2, x_3, x_4\}$ and $\{x_3, x_4, x_5\}$. The joint probability distribution is given by

$$\mathrm{p}(x_1, x_2, x_3, x_4, x_5) = \underbrace{\mathrm{p}(x_1)\,\mathrm{p}(x_2|x_1)\,\mathrm{p}(x_3|x_1, x_2)\,\mathrm{p}(x_4|x_1, x_2, x_3)}_{\psi(x_1,x_2,x_3,x_4)}\underbrace{\mathrm{p}(x_5|x_3, x_4)}_{\psi_2(x_3,x_4,x_5)}.$$

Pair-wise MRF

An MRF is called a *pair-wise* MRF if all the cliques in the MRF are of size 2. In this case, the clique potentials are all functions of two variables. The clique potentials can then be denoted as $\psi_{i,j}(x_i, x_j)$, where x_i, x_j are variables connected by an edge in the MRF.

Consider a pair-wise MRF in which the x_is denote the underlying *hidden* variables on which the observed variables y_is are dependent. Let the dependence between the hidden variable x_i and the explicit variable y_i be represented by a *joint* compatibility function $\phi_i(x_i, y_i)$. This scenario is shown in Fig. 7.3. In such a scenario, the joint distribution of the hidden and explicit variables is given by

$$\mathrm{p}(\mathbf{x}, \mathbf{y}) \propto \prod_{i,j} \psi_{i,j}(x_i, x_j) \prod_i \phi_i(x_i, y_i). \tag{7.6}$$

7.1.3 Factor graphs

Factor graphs are graphical models that are popular in the context of parity check codes [2],[6]. These are bipartite graphs, which means there are two kinds of nodes in a factor graph, namely, variable nodes and factor nodes (or function nodes). An edge is allowed to be drawn only between a variable node and a function node. Variable nodes are usually denoted as circles and function nodes are usually denoted as squares. A factor graph explicitly depicts the factorization

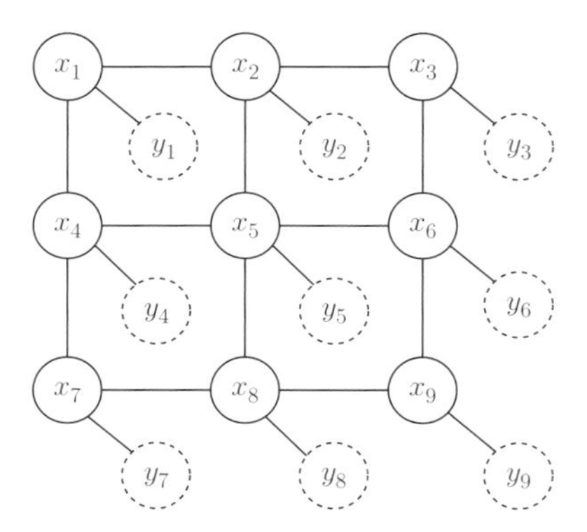

Figure 7.3 An example of a pair-wise MRF with observed (explicit) variables and hidden (implicit) variables.

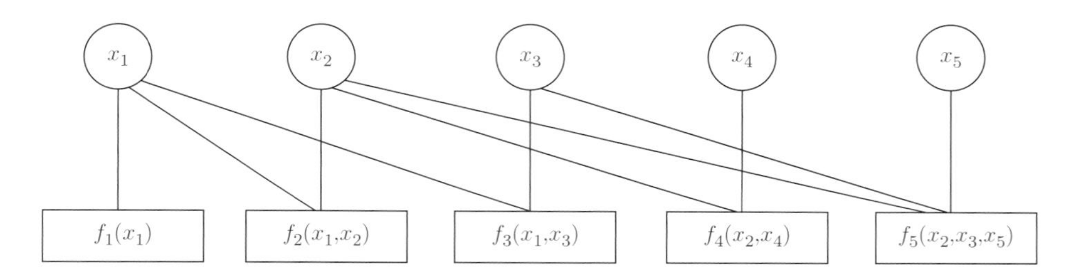

Figure 7.4 An example of a factor graph.

of a function – typically a probability distribution – into several *local functions*. Each local function depends on a subset of the set of all variables. For example, consider a function $f(\mathbf{x})$ that factorizes into N_F local functions as follows:

$$f(\mathbf{x}) \propto \prod_{j=1}^{N_F} f_j(\mathbf{x}_j) , \tag{7.7}$$

where f_js are the local functions and $\mathbf{x}_j$s are the sets of variables on which the local functions f_js depend. Every variable node in the factor graph of such a function denotes the variables in $\mathbf{x}$, whereas every function node represents a local function.

As an example, again consider the joint probability distribution as given by (7.1). Let us define the local functions as

$$f_1(x_1) = \mathrm{p}(x_1), \quad f_2(x_1, x_2) = \mathrm{p}(x_2|x_1), \quad f_3(x_1, x_3) = \mathrm{p}(x_3|x_1),$$
$$f_4(x_2, x_4) = \mathrm{p}(x_4|x_2), \quad f_5(x_2, x_3, x_5) = \mathrm{p}(x_5|x_2, x_3).$$

Then, the factor graph that demonstrates the factorization of the joint distribution in (7.1) is obtained as shown in Fig. 7.4.

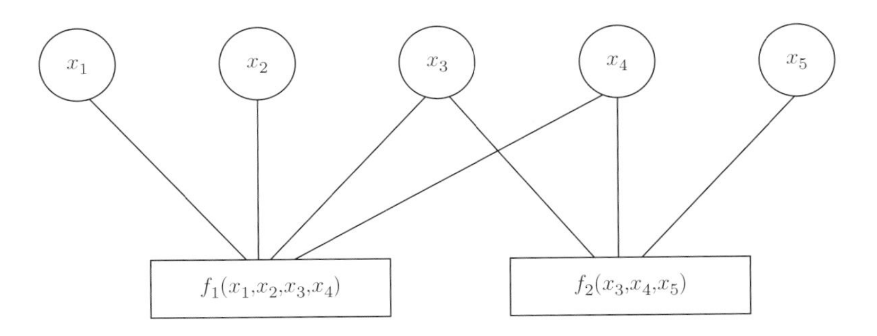

Figure 7.5 A factor graph obtained from the MRF of Fig. 7.2.

Inter-conversion of graphical models

The graphical models presented in the sections above can be inter-converted and are, in that sense, equivalent to each other. As can be seen from the examples in Fig. 7.1 and Fig. 7.4, a factor graph can be obtained from a Bayesian belief network by replacing every directed edge of the Bayesian network with a function node. Now, we see, through an example, how to convert an MRF into a factor graph.

Consider, for example, the MRF of Fig. 7.2. To convert such an MRF into a factor graph, a function node is created for every clique in the MRF, and the compatibility function for that clique represents the function pertaining to the function node. Therefore, the factor graph that represents the MRF of Fig. 7.2 is as shown in Fig. 7.5. In Fig. 7.5, the local functions are defined as

$$f_1(x_1, x_2, x_3, x_4) = \mathrm{p}(x_1)\,\mathrm{p}(x_2|x_1)\,\mathrm{p}(x_3|x_1, x_2)\,\mathrm{p}(x_4|x_1, x_2, x_3)\ ,$$
$$f_2(x_3, x_4, x_5) = \mathrm{p}(x_5|x_3, x_4)\,.$$

7.2 BP

BP is a technique that solves probabilistic inference problems using graphical models. The idea is to detect a hidden input to a system from its observed output. The system can be represented as a graphical model and the detection of the system input is equivalent to carrying out inference on the corresponding graph. More precisely, BP is an algorithm used to compute the marginalization of functions by passing messages on a graphical model. Pearl's BP algorithm [7], which has long been well known in the artificial intelligence community, was initially formalized for trees, and, in the case of trees, is known to solve the inference problem exactly. However, it is also empirically found to work well on many loopy graphs. It is generally believed that the BP algorithm gives good results on sparse graphs. Also, BP can be carried out at low computational cost if the underlying graph is sparse.

The generalized distributive law (GDL) is an algorithm that solves the marginalization of a product function problem by message passing on junction trees [8]. The sum–product algorithm is another general message passing algorithm that achieves marginalization of a global function on a factor graph [6]. The BP algorithm, the Viterbi algorithm, the Bahl–Cocke–Jelenik–Raviv (BCJR) algorithm and the turbo decoding algorithm are known to be special cases of GDL and the sum–product algorithm [6],[9],[10].

7.2.1 BP in communication problems

BP is a simple, yet highly effective, technique that has been successfully employed in a variety of applications including computational biology, statistical signal/image processing, data mining, etc. The BP algorithm is now widely recognized to be an efficient tool that can be used to solve several problems in communications as well [2]. The first mention of BP in a communications paper is in MacKay and Neal [11], where BP was used on Bayesian belief networks for decoding codes having a very sparse structure. BP-like algorithms are found in earlier works by Gallager [12] and Tanner [13]. McEliece *et al.*, in [9], showed that the famous turbo decoding algorithm of Berrou *et al.* in [14] is an instance of Pearl's BP algorithm. Though Pearl's BP algorithm was proven to compute exact marginals only for the case of graphs without loops, applying it on the belief network of a turbo code, which is loopy, results in very good performance in terms of nearness to capacity, particularly when the code block length is large (block lengths in the thousands) [14]. This makes BP an attractive general method for devising low complexity iterative decoding algorithms for several coding schemes including LDPC codes [15].

Another popular communication problem that has benefited from BP is MUD in CDMA [16]–[20]. The graph for the MUD problem is dense and loopy (fully connected); being dense makes the use of actual BP time consuming, and being loopy makes the resulting marginals approximate. Employing a Gaussian approximation of the other user interference has been shown to result in significant reduction in complexity; the achieved complexity of BP with Gaussian approximation is $O(K^2)$, where K is the number of users in the system [16]. Interestingly, this low complexity BP algorithm achieves near-single-user performance when the spreading factor N and number of users K are large (K and N in hundreds to thousands). In [17], PIC is derived as an approximate BP for the MUD problem. In [20], BP is employed for low complexity MUD in sparsely spread CDMA systems. BP has also been employed for equalization/detection in SISO and MIMO channels [21]–[29].

The applicability and success of BP in decoding turbo/LDPC codes with large block lengths and in CDMA MUD with large numbers of users/large spreading factors point to the possibility that it may work in large MIMO detection as well.

Taking this cue, BP techniques with suitable approximations for large MIMO detection have been devised and reported in the literature [30],[31].

7.2.2 BP algorithm on factor graphs

The BP algorithm on factor graphs can be viewed as a special case of the sum–product algorithm [6]. On a factor graph that represents the joint probability distribution $\mathrm{p}(\mathbf{x})$ as a global function that is factorized into a number of local functions, the BP algorithm becomes a special case of the sum–product algorithm. Message passing with the messages defined in [6] then implements the BP algorithm too. In particular, the message from variable node x to function node f is defined as

$$m_{x \to f}(x) = \prod_{h \in \mathcal{N}(x) \setminus \{f\}} m_{h \to x}(x) . \tag{7.8}$$

The message from the function node f to the variable node x is defined as

$$m_{f \to x}(x) = \sum_{\mathbf{x} \setminus \{x\}} \left[f(\mathbf{x}_f) \prod_{y \in \mathcal{N}(f) \setminus \{x\}} m_{y \to f}(y) \right] , \tag{7.9}$$

in which the outer summation is a marginalization over the variable x. The final belief at variable node x is calculated as a product of all incoming messages at the variable node, i.e.,

$$b(x) = \prod_{h \in \mathcal{N}(x)} m_{h \to x}(x) . \tag{7.10}$$

7.2.3 BP algorithm on pair-wise MRFs

Consider a situation similar to that of Fig. 7.3, where x_is are the hidden variables and y_is are the observed variables. If we consider the y_is to be fixed and write $\phi_i(x_i, y_i)$ simply as $\phi_i(x_i)$, then, from (7.6), the joint distribution for the hidden variables can be written as [3]

$$\mathrm{p}(\mathbf{x}) \propto \prod_{i,j} \psi_{i,j}(x_i, x_j) \prod_i \phi_i(x_i) . \tag{7.11}$$

A *message* from node j to node i, denoted as $m_{j,i}(x_i)$, is a vector of length equal to the number of values that the discrete variable x_i can possibly take. It is defined in such a way that on a cycle-free MRF, the belief at node i about the state of x_i is given by

$$\mathrm{b}_i(x_i) \propto \phi_i(x_i) \prod_{j \in \mathcal{N}(i)} m_{j,i}(x_i) . \tag{7.12}$$

In particular, the messages are defined as [3]

$$m_{i,j}(x_j) = \sum_{x_i} \phi_i(x_i) \psi_{i,j}(x_i, x_j) \prod_{k \in \mathcal{N}(i) \setminus j} m_{k,i}(x_i) . \tag{7.13}$$

Equation (7.13) actually constitutes an iteration, as the message is defined in terms of the other messages. Therefore, BP essentially involves computing the outgoing messages from a node to each of its neighbors using the local joint-compatibility function and the incoming messages and transmitting them.

7.2.4 Loopy BP

As mentioned before, BP computes the exact beliefs about the variables on cycle-free graphs. Moreover, in a cycle-free graph, the message passing needs to be carried out only once, to compute the belief about a variable. However, in a graph with cycles, one can start with an initial set of beliefs, probably a set of a priori beliefs about the variables, and iterate the message update rule of BP. Though there is no guarantee that this will yield the correct beliefs in a cyclic graph, it has been observed that in many practical cases, loopy BP does give satisfactory results [9],[32]. It is noted that graphical models of MIMO systems are fully/densely connected (loopy graphs). Still, as we will see later in the performance results section, running BP on the loopy graphs of MIMO systems gives very good performance.

7.2.5 Damped BP

In systems characterized by fully/densely connected graphical models, the BP algorithm may fail to converge, and if it does converge, the estimated marginals may be far from exact [33],[34]. However, in the literature there are several methods that include *damping*, [32],[35],[36], which can be used to improve things if BP does not converge (or converges too slowly). Double loop methods [37],[38] have also been shown to improve the convergence of BP.

In the damping method, messages to be passed are computed as a weighted average of the previous message (i.e., message in the previous iteration) and the current message (i.e., message in the current iteration) [32]. As an example, consider BP on a pair-wise MRF. Let $\widetilde{m}_{i,j}^{(t)}(x_j)$ denote the current message in iteration t, i.e.,

$$\widetilde{m}_{i,j}^{(t)}(x_j) \propto \sum_{x_i} \phi_i(x_i)\, \psi_{i,j}(x_i, x_j) \prod_{k \in \mathcal{N}(i) \setminus j} m_{k,i}^{(t-1)}(x_i). \tag{7.14}$$

The damped message to be passed from node i to node j in iteration t, denoted by $m_{i,j}^{(t)}(x_j)$, is computed as a convex combination of the previous message and the current message as

$$m_{i,j}^{(t)}(x_j) = \alpha\, m_{i,j}^{(t-1)}(x_j) + (1-\alpha)\, \widetilde{m}_{i,j}^{(t)}(x_j), \tag{7.15}$$

where $\alpha \in [0, 1)$ is referred to as the *damping factor*. This simple damping of messages has been shown to be very effective in improving BP dynamics and performance [32]. It will be seen later, in the performance results section, that damping can improve BP convergence and performance significantly in MIMO detection.

7.3 Application of BP in MIMO – an example

An illustration of the BP algorithm applied on factor graphs for equalization in MIMO-ISI channels is presented in this section.

7.3.1 MIMO-ISI system model

Consider a V-BLAST MIMO system with n_t transmit antennas and n_r receive antennas. Let $\mathbf{x}_n = \left(x_n^1, x_n^2, \ldots, x_n^{n_t}\right)^\mathsf{T}$ be the $n_t \times 1$ column vector of symbols transmitted from the n_t transmit antennas at time n. Assume $x_n^m \in \{+1, -1\}$, i.e., a BPSK modulated symbol is transmitted from the mth transmit antenna at time n. Assume a frequency-selective multipath fading channel model, where there are L independent paths between any two transmit and receive antennas with path delays of $0, 1, \ldots, L-1$. The ensemble of all paths having delay l between all transmit and receive antennas is represented by the $n_r \times n_t$ channel coefficient matrix $\mathbf{H}^{(l)}$ whose entries are independent zero mean complex Gaussian random variables with unit variance. The channel model between the ith transmit and jth receive antenna is illustrated in Fig. 7.6. In Fig. 7.6, $H_{j,i}^{(l)}$ is the coefficient at the jth row, ith column in the matrix $\mathbf{H}^{(l)}$. The received signal at time n is represented by the $n_r \times 1$ column vector $\mathbf{y}_n = \left(y_n^1, y_n^2, \ldots, y_n^{n_r}\right)^\mathsf{T}$, where y_n^m is the signal received at the mth receive antenna at time n. The received signal can be written as

$$\mathbf{y}_n = \sum_{l=0}^{L-1} \mathbf{H}^{(l)} \mathbf{x}_{n-l} + \mathbf{n}_n, \tag{7.16}$$

where $\mathbf{n}_n \sim \mathcal{CN}\left(\mathbf{0}, N_0 \, \mathbf{I}_{n_r}\right)$ is the additive complex Gaussian noise vector at time n, $\mathbf{I}_{n_r}$ is the $n_r \times n_r$ identity matrix, and $N_0/2$ is the noise power spectral density per real dimension. The channel matrix $\mathbf{H}^{(l)}$ is considered to be constant for a fixed quasi-static interval. As a special case, we can consider the quasi-static interval to be equal to one symbol duration.

7.3.2 Detection using BP

In order to use BP for MIMO detection over an ISI channel, the MIMO system needs to be represented as a graphical model. A convenient graphical model in this case is a factor graph. The messages that need to be passed between the function and the variable nodes of the factor graph are formulated. The graphical model and message passing schemes are as proposed in [26].

Graphical model

The multipath fading channel considered is essentially an ISI channel. Such a channel can conveniently be represented as a factor graph. The variable nodes

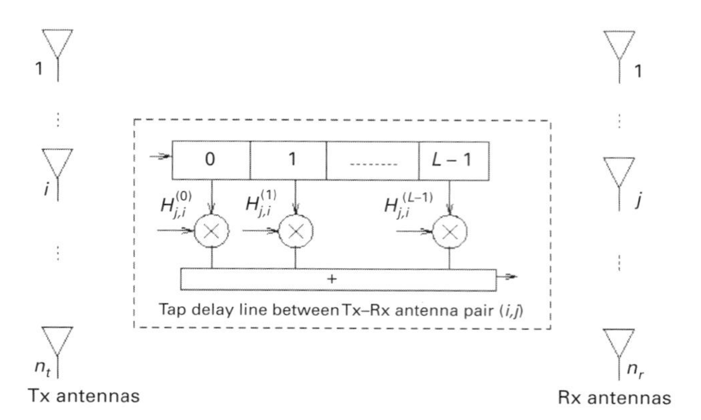

Figure 7.6 MIMO-ISI channel model.

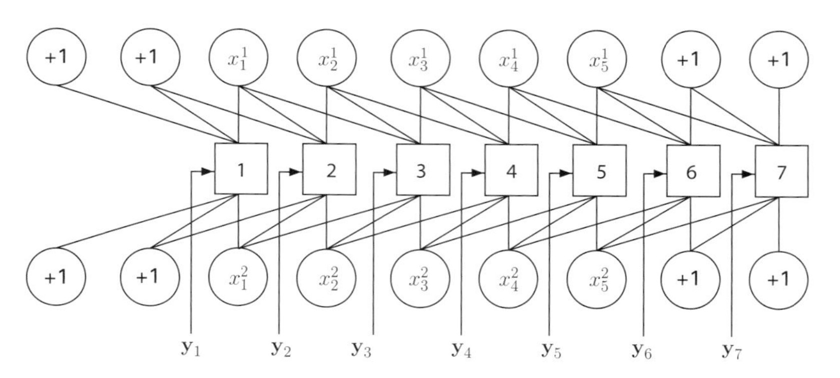

Figure 7.7 Factor graph model of a $(n_t, n_r, L) = (2, 4, 3)$ MIMO system for a block length of $N = 5$.

correspond to the transmitted symbols, which essentially are the variables that we intend to infer. Each function node in the graph corresponds to a time k. Since the received signal at any time k depends on the past L symbols transmitted from every transmit antenna, and since there are n_t transmit antennas, every function node is connected to Ln_t variable nodes. For detection, consider a block size of N bits. It is assumed that the $L - 1$ bits transmitted before and after a block are known, and, without loss of generality, these bits are assumed to be $+1$. An example of such a factor graph for $(n_t, n_r, L) = (2, 4, 3)$, and block length of 5 is shown in Fig. 7.7.

Let $\mathcal{X}_p \triangleq \{x_{p-t}^m : m = 1, 2, \ldots, n_t, t = 0, 1, \ldots, L - 1\}$ be the set of variables on which the vector received at time p depends. Then, in Fig. 7.7, the local function at the function node p is given by

$$f(\mathcal{X}_p) = \mathrm{p}(\mathbf{y}_p | \mathcal{X}_p), \tag{7.17}$$

which is the probability distribution of receiving $\mathbf{y}_p$ given that a variable pattern $\mathcal{X}_p$ was transmitted.

BP algorithm

BP is carried out on the factor graph by passing messages between the variable and function nodes. Let us denote the message passed by the function node p to the variable node pertaining to x_n^m as $R_{p \to n}^m(q)$. Also, let $Q_{n \to p}^m(q)$ be the message passed by the variable node pertaining to x_n^m to the function node p. The messages at function and variable nodes are computed as follows.

Computation at function nodes

$R_{p \to n}^m(q)$ is a quantity proportional to the conditional probability that the symbol transmitted from the mth transmit antenna at time n is q, given the receive vector $\mathbf{y}_p$, i.e., $\mathrm{P}(x_n^m = q|\mathbf{y}_p)$ for $q \in \{+1, -1\}$. This probability can be obtained by summing the joint probability $\mathrm{P}\left[x_n^m = q, \boldsymbol{\rho}_p(j)|\mathbf{y}_p\right]$ over all the $2^{(n_t L-1)}$ possible transmit patterns in which x_n^m is q. Here, $\boldsymbol{\rho}_p(j)$ is the noise-free received vector that will be received if the jth transmit pattern is transmitted. In other words,

$$
\begin{aligned}
\mathrm{P}(x_n^m = q|\mathbf{y}_p) &= \sum_{j=1}^{2^{(n_t L-1)}} \mathrm{P}\left[x_n^m = q, \boldsymbol{\rho}_p(j)|\mathbf{y}_p\right] \\
&= \sum_{j=1}^{2^{(n_t L-1)}} \frac{\mathrm{p}\left[x_n^m = q, \boldsymbol{\rho}_p(j), \mathbf{y}_p\right]}{\mathrm{p}(\mathbf{y}_p)} \\
&= \sum_{j=1}^{2^{(n_t L-1)}} \frac{\mathrm{p}\left[x_n^m = q, \boldsymbol{\xi}_p(j), \mathbf{y}_p\right]}{\mathrm{p}(\mathbf{y}_p)} \\
&\propto \sum_{j=1}^{2^{(n_t L-1)}} \underbrace{\mathrm{p}\left[\mathbf{y}_p|x_n^m = q, \boldsymbol{\xi}_p(j)\right]}_{\text{posterior probability}} \underbrace{\mathrm{P}\left[x_n^m = q, \boldsymbol{\xi}_p(j)\right]}_{\text{prior probability}} \\
&\propto \sum_{j=1}^{2^{(n_t L-1)}} \left[A_{posterior}\right]\left[A_{prior}\right] \\
&\triangleq R_{p \to n}^m(q),
\end{aligned}
$$

where $\boldsymbol{\xi}_p(j)$ is the jth transmit pattern pertaining to function node p.

$R_{p \to n}^m(q)$ is computed at function nodes. In order to do so, the posterior probability part can be computed as the density of a complex Gaussian random vector with mean $\boldsymbol{\rho}_p(j)$ and covariance matrix $N_0 \mathbf{I}_{n_r}$, assuming the value $\mathbf{y}_p$, i.e.,

$$
\left[A_{posterior}\right] \propto \exp\left[\frac{1}{N_0}\left\{\mathbf{y}_p - \boldsymbol{\rho}_p(j)\right\}^{\mathrm{H}}\left\{\mathbf{y}_p - \boldsymbol{\rho}_p(j)\right\}\right]. \tag{7.18}
$$

In the first iteration, the prior probability part can be computed by assuming equally likely transmit patterns. In subsequent iterations, the prior probability part is updated to reflect the latest belief about the variable node as follows:

$$[A_{\text{prior}}] = \prod_{m=1}^{n_t} \prod_{l \in \mathcal{N}(p) \backslash \{x_n^m\}} Q_{l \to p}^m \left(\triangle_{p,j}(m,l)\right), \tag{7.19}$$

where $\mathcal{N}(p)$ is the set of variable nodes connected to the function node p and $\triangle_{p,j}(m,l)$ is the bit in the transmit pattern corresponding to the pth function node, jth pattern index, mth transmit antenna, and lth variable node. It can be noted that the function to variable node messages are as expected as per (7.8).

Computation at variable nodes

The variable nodes compute and send the probabilities

$$Q_{n \to p}^m (q) = \frac{\prod_{l \in \mathcal{N}(n) \backslash p} R_{l \to n}^m (q)}{\prod_{l \in \mathcal{N}(n) \backslash p} R_{l \to n}^m (+1) + \prod_{l \in \mathcal{N}(n) \backslash p} R_{l \to n}^m (-1)}; \tag{7.20}$$

equation (7.20) is just a normalized extrinsic product of messages received from function nodes, as expected from (7.9).

Message passing schedule

The message passing schedule used is the *flooding* schedule. In the first iteration, function nodes begin the algorithm by computing and passing the function to variable node messages assuming equally likely transmit patterns. After all the variable nodes have received messages from connected function nodes, they compute the variable to function node messages and pass them to the connected function nodes. This complete process constitutes one iteration. From the second iteration onwards, the function nodes update their prior probabilities using the messages received during the previous iteration to compute the messages to be passed in the current iteration.

Final decision

After iterating on the graph for the required number of times, the variable node pertaining to the symbol x_n^m computes the *belief* about the event $x_n^m = q$ as

$$\mathrm{b}\left(x_n^m = q\right) = \frac{\prod_{l \in \mathcal{N}(n)} R_{l \to n}^m (q)}{\prod_{l \in \mathcal{N}(n)} R_{l \to n}^m (+1) + \prod_{l \in \mathcal{N}(n)} R_{l \to n}^m (-1)}, \tag{7.21}$$

which is the soft output of the detector and can be hard-limited to get the binary decision output. In the case of a coded system, the soft output can be directly fed to the decoder.

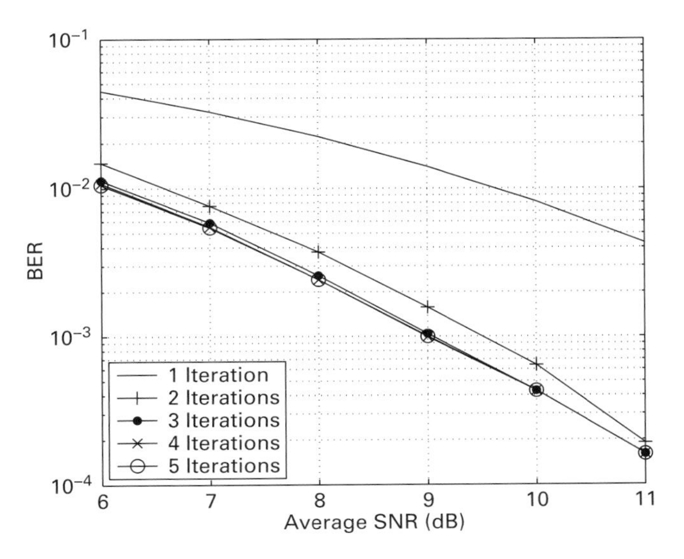

Figure 7.8 Performance of the BP detector in a MIMO-ISI channel as a function of the number of BP iterations. $(n_t, n_r, L) = (2, 2, 3)$, $N = 1000$.

7.3.3 Performance and complexity

The BER performance of the BP based detector described above, evaluated through simulations, is presented in Figs. 7.8–7.10. A MIMO system with $(n_t, n_r, L) = (2, 2, 3)$ and $N = 1000$ is considered. The effect of varying the number of BP iterations in this system is shown in Fig. 7.8, where the BER performance is seen to improve from one iteration to the next, and the improvement becomes saturated at around the fifth iteration. Figure 7.9 shows the performance for MIMO systems with channels with parameters (n_t, n_r, L) of $(2, 2, 2)$, $(2, 3, 2)$, $(2, 2, 3)$, $(2, 4, 2)$, and $(2, 2, 4)$. The block length is $N = 1000$ bits, and the number of BP iterations is 5. From the results shown in Fig. 7.9, the diversity gain achieved in the various systems is measured. The measured diversity gain is found to be 2.948 for the $(2, 2, 2)$ case, against the theoretically expected $Ln_r - n_t + 1 = 3$. For the $(2, 3, 2)$ case, the measured diversity gain is 4.2255 against the theoretically expected diversity of 5. For the $(2, 4, 2)$ case, the measured diversity gain is 6.4127 against the theoretically expected value of 7. Also, as expected, the performance of the $(2, 2, 3)$ and $(2, 2, 4)$ systems is very close to that of the $(2, 3, 2)$ and $(2, 4, 2)$ systems, respectively, in terms of achieved diversity. In Fig. 7.10, the performance of the BP detector is compared with the performance of the Viterbi algorithm in the $(2, 2, 2)$, $(2, 3, 2)$, and $(2, 4, 2)$ systems. It can be seen that the performance of the BP detector is very close to that of the Viterbi algorithm, and the BP detection can therefore be said to achieve near-ML performance.

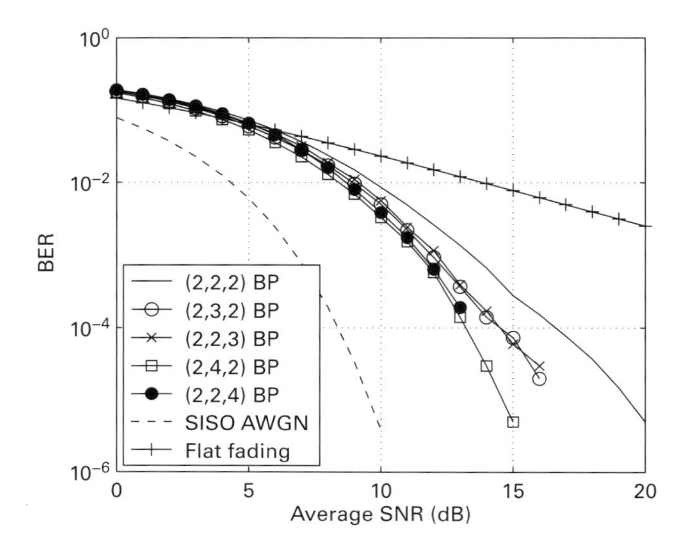

Figure 7.9 Performance of the BP detector in MIMO-ISI channels with parameters $(n_t, n_r, L) = (2, 2, 2), (2, 3, 2), (2, 2, 3), (2, 4, 2), (2, 2, 4)$.

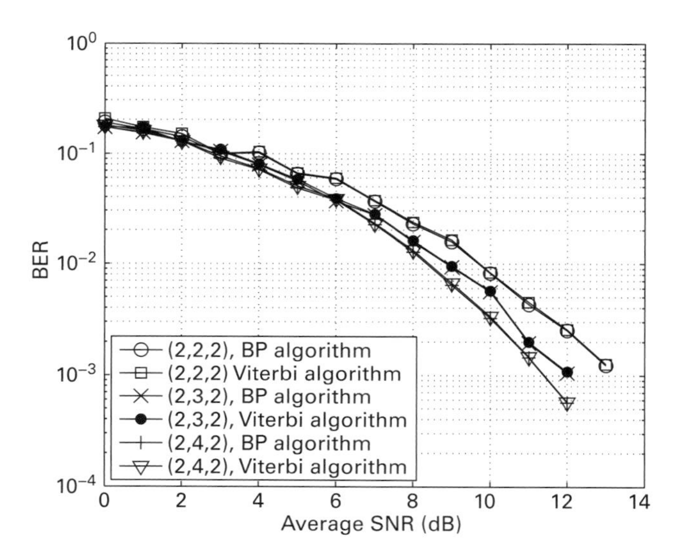

Figure 7.10 Performance comparison between the BP detector and the Viterbi algorithm in MIMO-ISI channels with parameters $(n_t, n_r, L) = (2, 2, 2), (2, 3, 2), (2, 4, 2)$.

Graph density and convergence

In Fig. 7.10, the BP algorithm is seen to achieve close to ML performance. The factor graph considered in this case has length-4 cycles. Even with this factor graph with cycles, BP has close to ML performance for the following reasons. (1) The weights of the edges of the factor graph are determined by the random channel coefficients. It is unlikely that all four edges that form a cycle have large weights, and, therefore, a performance loss due to such a length-4 cycle occurs with only a low probability [26]. (2) The second reason could be the sparsity in the variable-node graph for large block lengths. Consider the connectivity matrix, whose row and column headers are variable nodes of the corresponding factor graph. An entry of the matrix can be either a 1 or a 0. An entry of 1 at a location in the matrix indicates that the variables indicated by the row and column headers are connected by an edge in the variable node graph, which is essentially an MRF. For example, the connectivity matrix for the example of Fig. 7.7 is given below.

	x_1^1	x_1^2	x_2^1	x_2^2	x_3^1	x_3^2	x_4^1	x_4^2	x_5^1	x_5^2
x_1^1	1	1	1	1	1	1	0	0	0	0
x_1^2	1	1	1	1	1	1	0	0	0	0
x_2^1	1	1	1	1	1	1	1	1	0	0
x_2^2	1	1	1	1	1	1	1	1	0	0
x_3^1	1	1	1	1	1	1	1	1	1	1
x_3^2	1	1	1	1	1	1	1	1	1	1
x_4^1	0	0	1	1	1	1	1	1	1	1
x_4^2	0	0	1	1	1	1	1	1	1	1
x_5^1	0	0	0	0	1	1	1	1	1	1
x_5^2	0	0	0	0	1	1	1	1	1	1

As can be seen from this example, the connectivity matrix is a banded matrix, the band being $n_t(2L-1)$ thick. For larger and larger block lengths, this matrix becomes more and more sparse. At the block length of 1000 considered in the simulations, the matrix is quite sparse, and this could be another reason why the performance of the BP based detector is very close to that of the ML detector.

Complexity

The complexity of the BP algorithm is dominated by the computation of function to variable node messages. Therefore, we can ignore the complexity of computing the variable to function node messages. Every function node is connected to n_tL variable nodes. Therefore, every function node has to compute n_tL messages at every iteration, which contributes a factor of n_tL to the complexity. To compute each of these messages, it is necessary to carry out a summation over $2^{(n_tL-1)}$ terms. The prior probability part $[A_{prior}]$ of (7.19) needs a product of n_tL-1 terms to be computed. Therefore, the overall computational complexity is of the order of $n_t^2L^22^{(n_tL-1)}$. At the end of the required number of iterations, the BP algorithm detects n_t symbols. Therefore, the complexity is $O\left(n_tL^22^{(n_tL-1)}\right)$ per symbol per iteration. In view of the high complexity (exponential complexity in n_t and L) of this detector, it is not attractive for large MIMO systems, even though its performance is close to ML performance. In the following, two BP based approaches are presented, one based on a pair-wise MRF and another based on factor graph with scalar Gaussian approximation of interference, that scale very well (quadratic and linear per-symbol complexity in n_t) for large MIMO systems while achieving near-optimal performance.

7.4 Large MIMO detection using MRF

This section presents a BP based detector that employs message passing on an MRF and achieves a per-symbol complexity of $O(n_t n_r)$. Consider a V-BLAST MIMO communication system with n_t transmit antennas and n_r receive antennas. Let $\mathbf{x} = (x_1, x_2, \ldots, x_{n_t})^{\mathsf{T}}$ be the $n_t \times 1$ vector corresponding to the symbols transmitted over the n_t transmit antennas, $\mathbf{x} \in \{+1, -1\}^{n_t}$. Assume the channel to be a frequency-flat fading channel, characterized by an $n_r \times n_t$ channel matrix $\mathbf{H}$. The elements of $\mathbf{H}$ are all independent complex Gaussian random variables with zero mean and unit variance. The noise sample at any receive antenna is assumed to be a complex Gaussian random variable with zero mean and variance $\sigma^2 = n_t E_s/\gamma$, where E_s is the average energy of the transmitted symbols and γ is the average received SNR per receive antenna.

The signal received at the receiver is represented by the $n_r \times 1$ vector $\mathbf{y} = (y_1, y_2, \ldots, y_{n_r})^{\mathsf{T}}$. The received signal is given by

$$\mathbf{y} = \mathbf{H}\mathbf{x} + \mathbf{n}, \tag{7.22}$$

where $\mathbf{n} \sim \mathcal{CN}\left(\mathbf{0}, \sigma^2 \mathbf{I}_{n_r}\right)$ is an $n_r \times 1$ vector that represents the additive complex Gaussian noise.

Graphical model

The considered V-BLAST MIMO system can be conveniently represented as an undirected graph with a node for every symbol. Since every transmit antenna is used to transmit a separate symbol, there are n_t nodes in such a graph. The edges of the graph represent the conditional dependence between the symbols. Since every transmitted symbol interferes with every other transmitted symbol in V-BLAST, the graph is fully connected. A graphical model for a V-BLAST MIMO system with $n_t = 8$ is shown in Fig. 7.11.

7.4.1 MRF BP based detection algorithm

The MAP detector takes the joint posterior distribution

$$p(\mathbf{x} \mid \mathbf{y}, \mathbf{H}) \propto p(\mathbf{y} \mid \mathbf{x}, \mathbf{H})\, p(\mathbf{x}), \tag{7.23}$$

and marginalizes out each variable as $p(x_i|\mathbf{y}, \mathbf{H}) = \sum_{x_{-i}} p(\mathbf{x}|\mathbf{y}, \mathbf{H})$, where x_{-i} stands for all entries of $\mathbf{x}$ except x_i. The MAP estimate of the bit x_i, $i = 1, \ldots, n_t$,

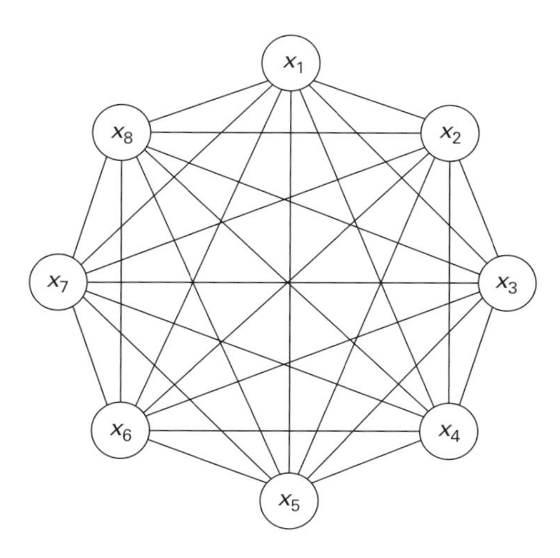

Figure 7.11 Graphical model of a V-BLAST MIMO system with eight transmit antennas. The variable node x_i represents the symbol transmitted from the ith antenna.

is then given by

$$\widehat{x}_i = \underset{a \in \{\pm 1\}}{\text{argmax}} \; p\big(x_i = a \mid \mathbf{y}, \mathbf{H}\big), \tag{7.24}$$

whose complexity is exponential in n_t. A low complexity BP based detection algorithm is described below.

7.4.2 MRF potentials

Given $\mathbf{x}$ and $\mathbf{H}$, $\mathbf{y}$ is a complex Gaussian random vector with mean $\mathbf{Hx}$ and covariance $\sigma^2 \mathbf{I}_{n_r}$. So,

$$p\left(\mathbf{y} | \mathbf{x}, \mathbf{H}\right) \propto \exp\Big(\frac{-\|\mathbf{y} - \mathbf{Hx}\|^2}{2\sigma^2}\Big). \tag{7.25}$$

Assuming that the symbols in $\mathbf{x}$ are all independent,

$$p\left(\mathbf{x}\right) = \prod_i p\left(x_i\right). \tag{7.26}$$

From (7.25), (7.26), and (7.23)

$$\begin{aligned} p(\mathbf{x} \mid \mathbf{y}, \mathbf{H}) &\propto \exp\Big(\frac{-\|\mathbf{y} - \mathbf{Hx}\|^2}{2\sigma^2}\Big) \exp\big(\ln p(\mathbf{x})\big) \\ &= \exp\Big(\frac{-1}{2\sigma^2}(\mathbf{y} - \mathbf{Hx})^H(\mathbf{y} - \mathbf{Hx})\Big) \prod_i \exp\big(\ln p(x_i)\big) \\ &\propto \exp\Big(\frac{-1}{2\sigma^2}\big(\mathbf{x}^H\mathbf{H}^H\mathbf{Hx} - 2\Re\{\mathbf{x}^H\mathbf{H}^H\mathbf{y}\}\big)\Big) \prod_i \exp\big(\ln p(x_i)\big). \end{aligned} \tag{7.27}$$

Defining $\mathbf{R} \triangleq (1/\sigma^2)\mathbf{H}^H\mathbf{H}$ and $\mathbf{z} \triangleq (1/\sigma^2)\mathbf{H}^H\mathbf{y}$, (7.27) can be written as

$$p(\mathbf{x} \mid \mathbf{y}, \mathbf{H}) \propto \exp\Big(-\sum_{i<j} \Re\{x_i^* R_{ij} x_j\}\Big) \exp\Big(\sum_i \Re\{x_i^* z_i\}\Big) \prod_i \exp\big(\ln p(x_i)\big)$$

$$= \Big(\prod_{i<j} \exp\Big(-x_i \Re\{R_{ij}\} x_j\Big)\Big)\Big(\prod_i \exp\Big(x_i \Re\{z_i\} + \ln p(x_i)\Big)\Big), \quad (7.28)$$

where z_i and R_{ij} are the elements of $\mathbf{z}$ and $\mathbf{R}$, respectively, and $\Re\{.\}$ denotes the real part of a complex number. Comparing (7.28) and (7.11), it is seen that the MRF of the MIMO system has pair-wise interactions with the following potentials:

$$\psi_{i,j}(x_i, x_j) = \exp\Big(-x_i \Re\{R_{ij}\} x_j\Big), \quad (7.29)$$

$$\phi_i(x_i) = \exp\Big(x_i \Re\{z_i\} + \ln p(x_i)\Big). \quad (7.30)$$

7.4.3 Message passing

The values of ψ and ϕ given by (7.29) and (7.30) define, respectively, the edge and self potentials of an undirected graphical model to which message passing algorithms, such as BP, can be applied to compute the marginal probabilities of the variables. BP attempts to estimate the marginal probabilities of all the variables by way of passing messages between the local nodes.

A *message* from node j to node i is denoted $m_{j,i}(x_i)$, and belief at node i is denoted $\mathrm{b}_i(x_i)$, $x_i \in \{\pm 1\}$. The belief $\mathrm{b}_i(x_i)$ depends on how likely it is x_i was transmitted. On the other hand, $m_{ji}(x_i)$ depends on how likely it is that node j thinks x_i was transmitted. The message from node i to a neighboring node j is then given by [3]

$$m_{i,j}(x_j) = \sum_{x_i} \phi_i(x_i)\, \psi_{i,j}(x_i, x_j) \prod_{k \in \mathcal{N}(i) \setminus j} m_{k,i}(x_i), \quad (7.31)$$

where $\mathcal{N}(i)$ denotes the set of all nodes neighboring the node i. Equation (7.31) actually constitutes an iteration, as the message is defined in terms of the other messages. So, BP essentially involves computing the outgoing messages from a node to each of its neighbors using the local joint compatibility function and the incoming messages and transmitting them. The algorithm terminates after a fixed number of iterations. Damping of messages as described in Section 7.2.5 can be carried out in each iteration. The final belief about the variable x_i is computed as

$$\mathrm{b}_i(x_i) \propto \phi_i(x_i) \prod_{j \in \mathcal{N}(i)} m_{j,i}(x_j), \quad (7.32)$$

which is the soft output of the detector that can be hard-limited to get the binary decision output. A complete listing of the MRF BP algorithm described above is listed in Table 7.1. In the case of a coded system, the soft output of the algorithm can be directly fed to the decoder.

Table 7.1. *MRF BP algorithm for large MIMO detection*

Initialization

1. $m_{i,j}^{(0)}(x_j) = 0.5, \quad p(x_i = 1) = p(x_i = -1) = 0.5, \quad \forall i, j = 1, \ldots, n_t$
2. $\widetilde{m}_{i,j}^{(0)}(x_j) = 0.5, \quad \forall i, j = 1, \ldots, n_t$
3. $\mathbf{z} = (1/\sigma^2)\,\mathbf{H}^H\mathbf{y}; \quad \mathbf{R} = (1/\sigma^2)\,\mathbf{H}^H\mathbf{H}$
4. for $i = 1$ to n_t
5. $\quad \phi_i(x_i) = \exp\left(x_i \Re\{z_i\} + \ln(p(x_i))\right)$
6. end for
7. for $i = 1$ to n_t
8. $\quad$ for $j = 1$ to $n_t, \quad j \neq i$
9. $\quad\quad \psi_{i,j}(x_i, x_j) = \exp(-x_i \Re\{R_{i,j}\} x_j)$
10. $\quad$ end for
11. end for

Iterative update of messages

12. for $t = 1$ to num_iter

Damped message calculation

13. $\quad$ for $i = 1$ to n_t
14. $\quad\quad$ for $j = 1$ to $n_t, \quad j \neq i$
15. $\quad\quad\quad \widetilde{m}_{i,j}^{(t)}(x_j) \propto \sum_{x_i} \phi_i(x_i)\psi_{i,j}(x_i, x_j) \prod_{k \in \mathcal{N}(i)\setminus j} m_{k,i}^{(t-1)}(x_i)$
16. $\quad\quad\quad m_{i,j}^{(t)}(x_j) = \alpha\, m_{i,j}^{(t-1)}(x_j) + (1-\alpha)\, \widetilde{m}_{i,j}^{(t)}(x_j)$
17. $\quad\quad$ end for
18. $\quad$ end for
19. end for; End of for loop starting at line 12

Belief calculation

20. for $i = 1$ to n_t
21. $\quad \mathrm{b}_i(x_i) \propto \phi_i(x_i) \prod_{j \in \mathcal{N}(i)} m_{j,i}^{(num\_iter)}(x_i)$
22. end for

Detection of data bits

23. $\widehat{x}_i = \underset{x_i \in \{\pm 1\}}{\operatorname{argmax}}\ \mathrm{b}_i(x_i), \quad \forall i = 1, \ldots, n_t$
24. Terminate

7.4.4 Performance

Figure 7.12 shows the BER performance of the above MRF BP detector with no message damping ($\alpha = 0$) as a function of average received SNR for various $n_t = n_r$, BPSK modulation, and five BP iterations. For comparison purposes, the BER performance over a SISO AWGN channel as well as on a SISO frequency-flat fading channel are plotted. From Fig. 7.12, it is seen that, as $n_t = n_r$ is increased, the BER performance improves and gets closer to SISO AWGN performance for $n_t = n_r$. When $n_t = n_r = 300$, the BER performance of the detector at high SNRs is close to that of SISO AWGN, illustrating the detector's near-optimality in large MIMO systems. The effect of message damping on the performance is illustrated in Fig. 7.13. $\alpha = 0$ corresponds to the case of no damping. In Fig. 7.13, damping is seen to significantly improve MRF

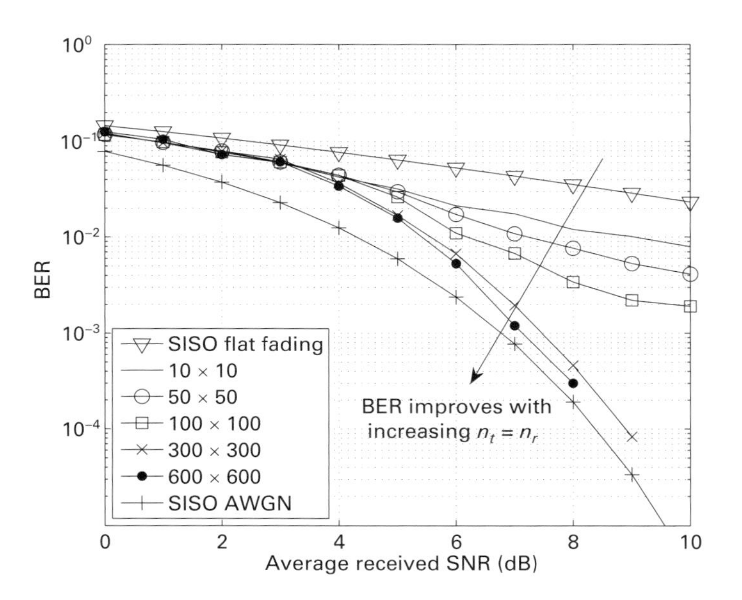

Figure 7.12 BER performance of MRF BP detector in V-BLAST MIMO systems for different $n_t = n_r$, $\alpha = 0$, and five BP iterations.

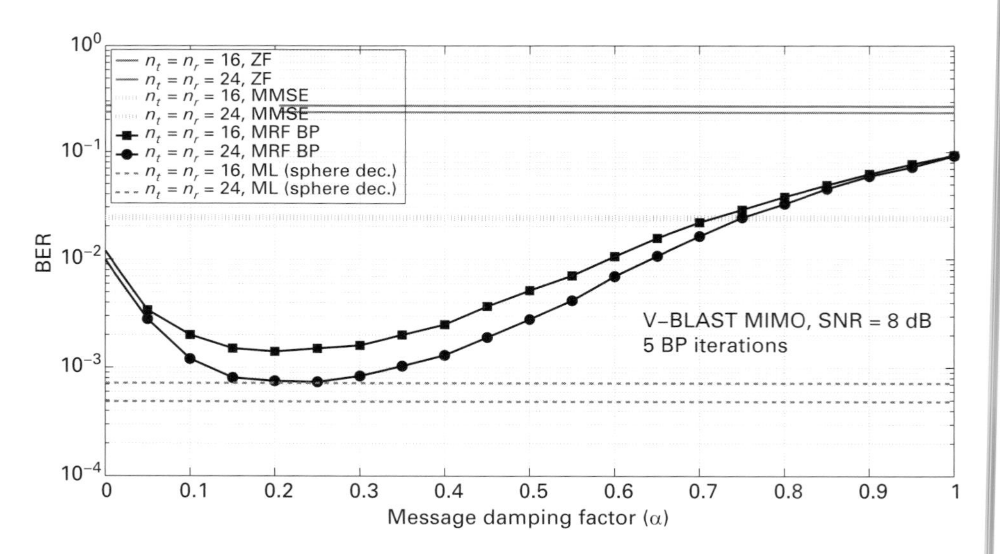

Figure 7.13 Effect of message damping on the performance of MRF BP detector. $n_t = n_r = 16, 24$, SNR $= 8$ dB, five BP iterations.

BP performance to the extent that close to ML performance (SD performance) is achieved at a damping factor of $\alpha = 0.2$ even with tens of antennas ($n_t = n_r = 16, 24$). Figure 7.14 shows the BER performance of the MRF BP detector as a function of SNR for different $n_t = n_r$ at a damping factor of $\alpha = 0.2$ and seven BP iterations.

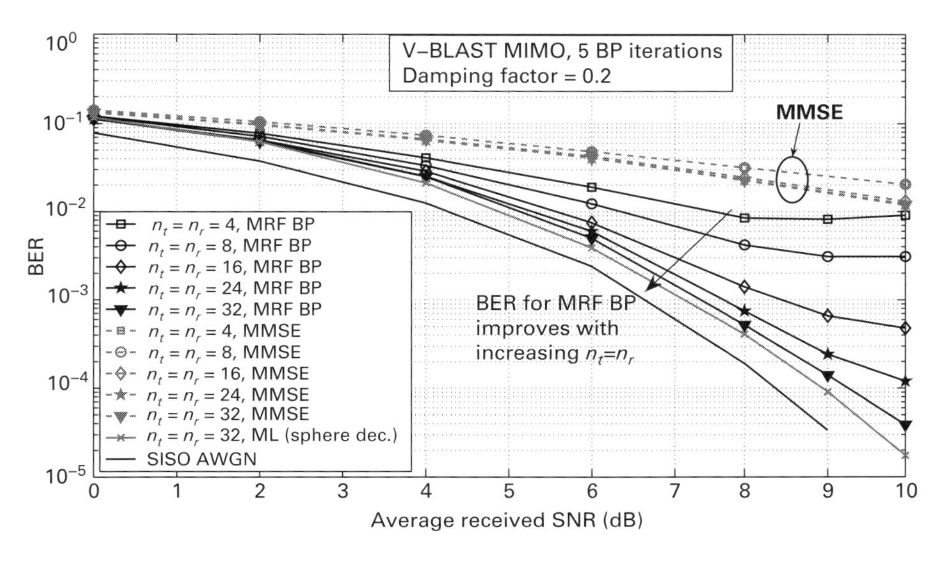

Figure 7.14 BER performance of the MRF BP detector as a function of SNR in V-BLAST MIMO for different $n_t = n_r$, $\alpha_m = 0.2$ and five BP iterations.

7.4.5 Complexity

In every iteration, each of the n_t nodes computes the message to be sent to the other nodes. As seen from (7.31), computing one such message involves multiplying incoming messages from the other $n_t - 1$ nodes. This is repeated for both $x_i = +1$ and $x_i = -1$ and after multiplying with $\phi_i(x_i)\,\psi_{i,j}(x_i, x_j)$, the results are added. This involves a complexity of $O(n_t^2)$ per iteration. Since n_t symbols are detected at a time, the complexity is $O(n_t)$ per iteration per symbol. However, the computations of $\mathbf{z}$ and $\mathbf{R}$ involve complexities of $O(n_t n_r)$ and $O(n_t^2 n_r)$, respectively. These computations need be carried out only once before the first iteration. Again, since n_t symbols are detected at a time, the complexities per symbol amount to $O(n_r)$ and $O(n_t n_r)$, respectively. Assuming the number of iterations is much less than n_r, the complexity order is given by $O(n_t, n_r)$ per symbol. Damping of messages does not increase the order of complexity.

7.5 Large MIMO detection using a factor graph

Consider the MIMO system model in (7.22). Treat each entry of the observation vector $\mathbf{y}$ as a function node (observation node) in a factor graph, and each transmitted symbol as a variable node. The received signal y_i can be written as

$$y_i = \sum_{j=1}^{n_t} h_{ij}x_j + n_i \quad = \quad h_{ik}x_k + \underbrace{\sum_{j=1, j\neq k}^{n_t} h_{ij}x_j + n_i}_{\text{Interference}}. \tag{7.33}$$

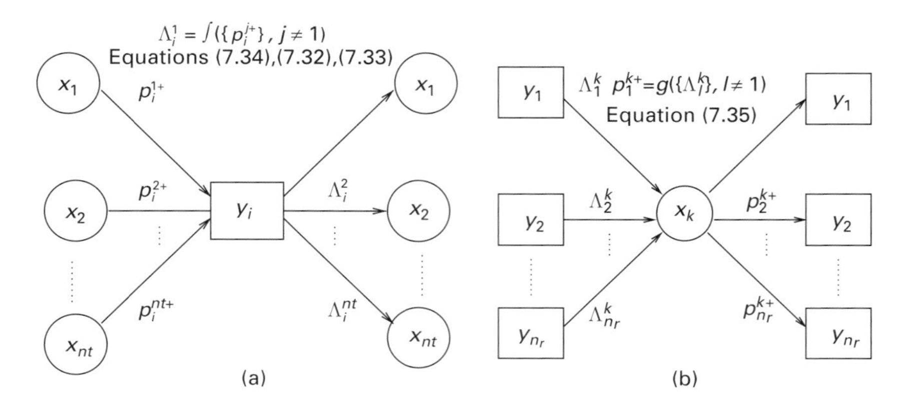

Figure 7.15 Message passing in the FG BP algorithm with SGA: (a) messages received at and sent by an observation node; (b) messages received at and sent by a variable node.

With the motivation of reducing complexity, when computing the message from the ith observation node to the kth variable node, make the following scalar Gaussian approximation (GA) of the interference (GAI):

$$y_i = h_{ik}x_k + \underbrace{\sum_{j=1, j\neq k}^{n_t} h_{ij}x_j + n_i}_{\triangleq\, z_{ik}}, \tag{7.34}$$

where the interference-plus-noise term, z_{ik}, is modeled as $\mathbb{C}\mathcal{N}(\mu_{z_{ik}}, \sigma^2_{z_{ik}})$ with

$$\mu_{z_{ik}} = \sum_{j=1, j\neq k}^{n_t} h_{ij}\mathbb{E}(x_j), \tag{7.35}$$

$$\sigma^2_{z_{ik}} = \sum_{j=1, j\neq k}^{n_t} |h_{ij}|^2 \, \mathrm{Var}(x_j) + \sigma^2. \tag{7.36}$$

For BPSK signaling, the log-likelihood ratio (LLR) of the symbol $x_k \in \{+1, -1\}$ at observation node i, denoted by Λ_i^k, can be written as

$$\begin{aligned}\Lambda_i^k &= \log \frac{p(y_i|\mathbf{H}, x_k = 1)}{p(y_i|\mathbf{H}, x_k = -1)} \\ &= \frac{4}{\sigma^2_{z_{ik}}} \Re\left(h^*_{ik}(y_i - \mu_{z_{ik}})\right).\end{aligned} \tag{7.37}$$

The LLR values computed at the observation nodes are passed to the variable nodes (Fig. 7.15(a)). Using these LLRs, the variable nodes compute the probabilities

$$\begin{aligned}p_i^{k+} &\triangleq p_i(x_k = +1|\mathbf{y}) \\ &= \frac{\exp(\sum_{l=1, l\neq i}^{n_r} \Lambda_l^k)}{1 + \exp(\sum_{l=1, l\neq i}^{n_r} \Lambda_l^k)},\end{aligned} \tag{7.38}$$

Table 7.2. *FG BP algorithm with SGA for large MIMO detection*

Initialization
1. $\Lambda_i^k = 0,\ p_i^{k+} = 0.5,\ s_{\Lambda^k} = 0,\quad \mu_{z_{ik}} = \sigma_{z_{ik}}^2 = s_{\mu_{z_i}} = 0,$
   $s_{\sigma_{z_i}^2} = 0,\quad \forall i = 1,\ldots,n_r,\ \forall k = 1,\ldots,n_t$
2. for $t = 1$ to num_iter
   Computation of LLRs at observation nodes
3. for $i = 1$ to n_r
4. $s_{\mu_{z_i}} = \sum_{j=1}^{n_t} h_{ij}\,(2p_i^{j+} - 1)$
5. $s_{\sigma_{z_i}^2} = 4\sum_{j=1}^{n_t} |h_{ij}|^2 p_i^{j+}(1 - p_i^{j+})$
6. for $k = 1$ to n_t
7. $\mu_{z_{ik}} = s_{\mu_{z_i}} - h_{ik}\,(2p_i^{k+} - 1)$
8. $\sigma_{z_{ik}}^2 = s_{\sigma_{z_i}^2} - 4|h_{ik}|^2 p_i^{k+}\,(1 - p_i^{k+}) + \sigma^2$
9. $\Lambda_i^k = (4/\sigma_{z_{ik}}^2)\Re\,(h_{ik}^*(r_i - \mu_{z_{ik}}))$
10. end for
11. end for
    Computation of probabilities at variable nodes
12. for $k = 1$ to n_t
13. $s_{\Lambda^k} = \sum_{l=1}^{n_r} \Lambda_l^k$
14. for $i = 1$ to n_r
15. $p_i^{k+} = \dfrac{exp(s_{\Lambda^k} - \Lambda_i^k)}{1 + exp(s_{\Lambda^k} - \Lambda_i^k)}$
16. end for
17. end for
18. end for; End of for loop starting at line 2
    Detection of data bits
19. for $k = 1$ to n_t
20. $\widehat{x}_k = \mathrm{sgn}\big(\sum_{i=1}^{n_r} \Lambda_i^k\big)$
21. end for
22. Terminate

and pass them back to the observation nodes (Fig. 7.15(b)). This message passing is carried out for a certain number of iterations. Messages can be damped as described in Section 7.2.5 and then passed. Finally, x_k is detected as

$$\widehat{x}_k = \mathrm{sgn}\Big(\sum_{i=1}^{n_r} \Lambda_i^k\Big). \tag{7.39}$$

The algorithm listing is given in Table 7.2. As will be seen from the complexity discussion next, approximating the interference as scalar Gaussian in (7.34) greatly simplifies the computation of messages. Because of this scalar Gaussian approximation (SGA), the algorithm is referred to as *FG BP-SGA algorithm.*

7.5.1 Computation complexity

The computation complexity of the FG BP-SGA algorithm in the above involves (i) LLR calculations at the observation nodes as per (7.37), which has $O(n_t n_r)$ complexity, and (ii) calculation of probabilities at variable nodes as per (7.38), which also requires $O(n_t n_r)$ complexity. A naive implementation of (7.37) would require a summation over $n_t - 1$ variable nodes for each message, amounting to a complexity of order $O(n_t^2 n_r)$. However, the summation over $n_t - 1$ variables in (7.35) can be written in the form $\sum_{j=1}^{n_t} h_{ij}\mathbb{E}(x_j) - h_{ik}\mathbb{E}(x_k)$, where the computation of the full summation from $j = 1$ to $j = n_t$ (which is independent of the variable index k) requires $n_t - 1$ additions. In addition, one subtraction operation for each k is required. The makes the complexity order for computing (7.35) only $O(n_t n_r)$. A similar argument holds for computation of the variance in (7.36), and hence the complexity of computing the LLR in (7.37) becomes $O(n_t n_r)$. Likewise, a similar rewriting of the summation in (7.38) leads to a complexity of $O(n_t n_r)$. Hence, the overall complexity of the algorithm is $O(2n_t n_r)$ for detecting n_t transmitted symbols. So, the per-symbol complexity is just $O(n_t)$ for $n_t = n_r$. Note that this complexity is one order less than that of the MRF BP approach in the previous section. Because of its linear complexity in n_t, the FG BP-SGA algorithm is very attractive for large MIMO systems. In addition, the BER performance achieved by the algorithm in large dimensions is very good (as will be seen in the performance results next).

7.5.2 Performance

Figure 7.16 shows the BER performance of the FG BP-SGA algorithm in $n_t \times n_r$ V-BLAST MIMO with $n_t = n_r = 8, 16, 32$ and BPSK. The number of BP iterations and the message damping factor used are 10 and 0.4, respectively. It is seen that, like the MRF BP algorithm, the performance of the FG BP-SGA algorithm also approaches SISO AWGN performance for increasing n_t. Importantly, the FG BP-SGA algorithm achieves this very good performance with $O(n_t)$ per-symbol complexity compared to the $O(n_t^2)$ per-symbol complexity of MRF BP. This illustrates that the SGA is very attractive in both complexity and performance for signal detection in large dimensions.

7.5.3 Vector GA (VGA) in PDA versus SGA in FG BP

At this point, it is interesting to compare the PDA algorithm presented in Chapter 6 and the FG BP algorithm presented in this section. Both are iterative algorithms of a similar kind, which start with some initial prior probabilities of the symbols, compute updated conditional probabilities of these symbols, pass them as prior probabilities for the next iteration, and carry out this procedure for several iterations. Importantly, both the algorithms make GAs. PDA makes a VGA of interference, whereas FG BP makes an SGA of interference. For a

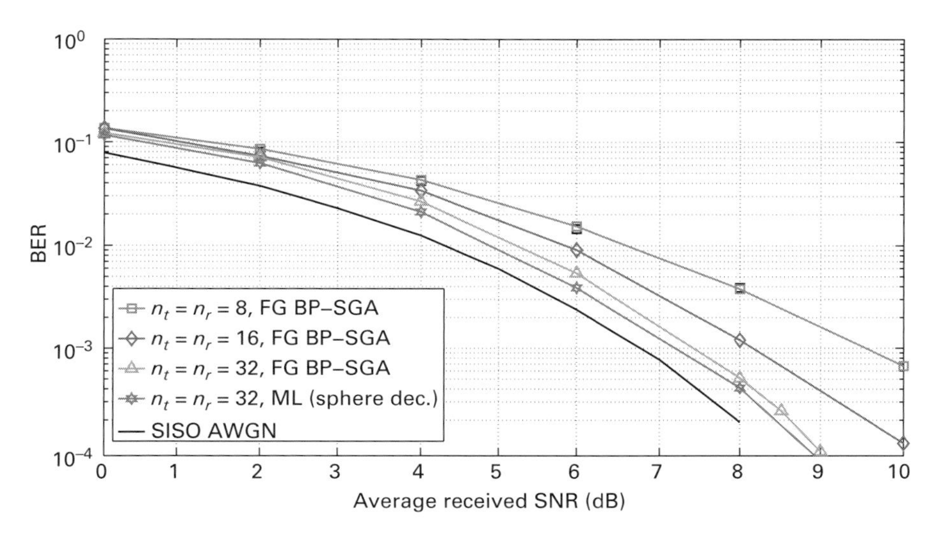

Figure 7.16 BER performance of FG BP-SGA algorithm in V-BLAST MIMO with $n_t = n_r = 8, 16, 32$, $\alpha_m = 0.4$, and ten BP iterations.

V-BLAST MIMO system with BPSK, the VGA adopted in PDA is

$$\mathbf{y} = \mathbf{h}_i x_i + \underbrace{\sum_{\substack{j=0 \\ j \neq i}}^{n_t-1} \mathbf{h}_j x_j + \mathbf{n}}_{\triangleq \tilde{\mathbf{n}}}, \tag{7.40}$$

where $\tilde{\mathbf{n}}$ is the interference-plus-noise vector, which is approximated to be a Gaussian vector. The covariance matrix $\mathbf{C}_i$ of $\mathbf{y}$ given x_i is

$$\mathbf{C}_i = \sigma^2 \mathbf{I}_{n_r} + \sum_{\substack{j=0 \\ j \neq i}}^{n_t-1} \mathbf{h}_j \mathbf{h}_j^T \, 4p_j^+ (1 - p_j^+), \tag{7.41}$$

where $p_j^+ = p(x_j = +1)$. The second term on the right-hand side of (7.41) contains the off-diagonal elements of the covariance matrix, which is proportional to $\mathbf{HH}^H$ matrix. It is noted that, as the size of $\mathbf{H}$ increases, the diagonal entries of $\mathbf{HH}^H$ become increasingly larger in magnitude than the off-diagonal entries, showing the channel hardening behavior discussed in Section 2.2. This can be observed in Fig. 7.17, where the intensity plots of $\mathbf{HH}^H$ matrices for $n_t = n_r = 8, 16, 32, 64$ are shown. It can be seen that the diagonal entries of $\mathbf{HH}^H$ are relatively more prominent than their off-diagonal counterparts in a 64×64 MIMO channel than in an 8×8 MIMO channel. Therefore, one can ignore the off-diagonal terms for large values of $n_t = n_r$ without much penalty.

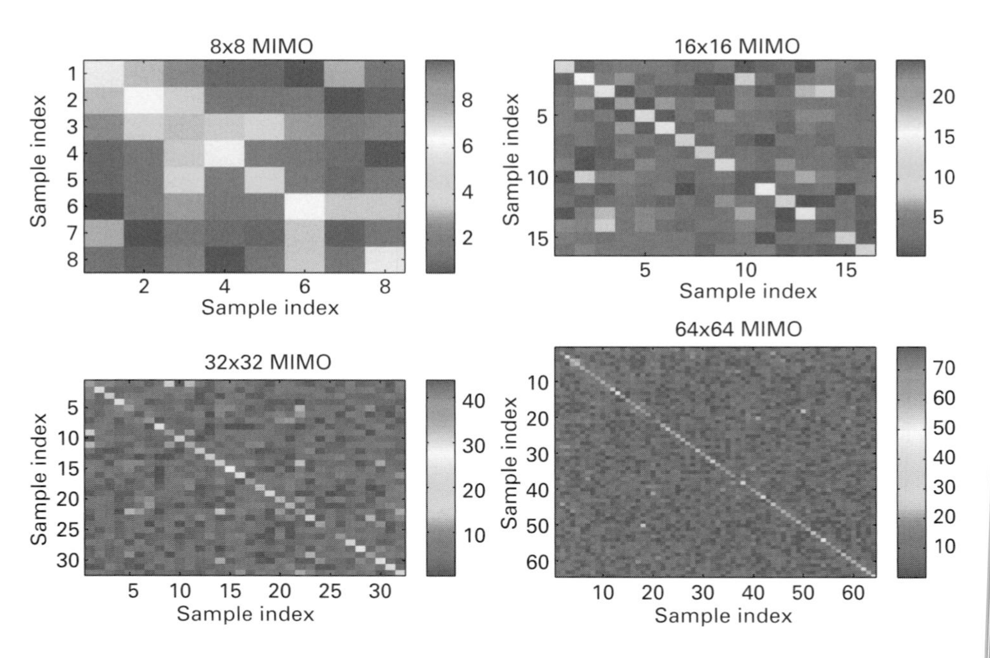

Figure 7.17 Intensity plots of $\mathbf{HH}^H$ matrices for 8×8, 16×16, 32×32, and 64×64 MIMO channels.

What the SGA in the FG BP algorithm does in (7.34) can be viewed as essentially amounting to ignoring the off-diagonal terms in the covariance matrix as a further approximation, in which case only variance computation is needed. This in essence gives one order complexity advantage for the SGA in FG BP compared to the VGA in PDA. What is more interesting is that this complexity advantage comes without much price in terms of performance loss when the number of antennas is large. This is illustrated in Fig. 7.18, where there is a noticeable performance gap between VGA and SGA in 8×8 MIMO (due to strong off-diagonal terms in 8×8), whereas in 64×64 MIMO there is almost no performance gap between SGA and VGA (due to weak off-diagonal terms in 64×64 MIMO). This shows that the FG BP algorithm with SGA is very attractive for large MIMO detection.

7.6 BP with the Gaussian tree approximation (GTA)

The MRF and FG BP algorithms presented above perform well when the modulation alphabet is BPSK. However, these algorithms do not perform quite so well in higher-order QAM. An attempt to design an MRF based BP detection algorithm for higher order QAM is reported in [39]. This algorithm uses a *Gaussian tree approximation* (GTA) to convert the fully-connected graph representing the

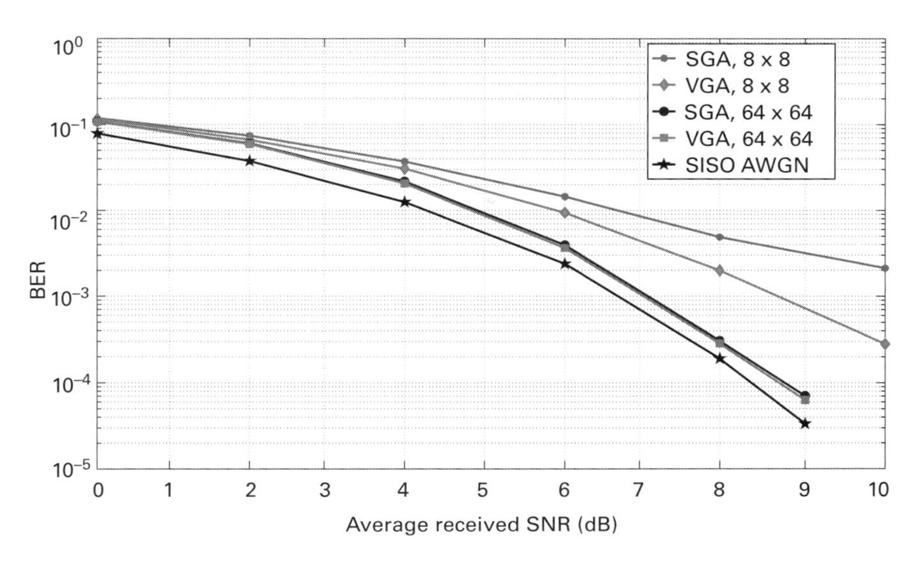

Figure 7.18 BER performance of FG BP with SGA versus PDA with VGA in 8×8 and 64×64 V-BLAST MIMO systems.

MIMO system into a tree, and to carry out BP on the resultant approximated tree. The GTA BP algorithm uses the real-valued system model and works as follows.

The GTA BP algorithm is based on an approximation of the exact probability function

$$p(\mathbf{x}|\mathbf{y}, \mathbf{H}) \propto \exp\left(-\frac{1}{2\sigma^2}\|\mathbf{H}\mathbf{x} - \mathbf{y}\|^2\right), \quad \mathbf{x} = (x_1, x_2, \ldots, x_{2n_t})^\mathsf{T} \in \mathbb{A}^{2n_t}, \quad (7.42)$$

where $\mathbb{A}$ is the underlying PAM alphabet corresponding to a QAM alphabet. The graph corresponding to (7.42) is fully connected. Applying BP on a fully-connected graph is exponentially complex and may fail to produce convergence. Finding an *optimal tree approximation* of a fully-connected graph would be useful, since BP has been proved to be optimal for trees. In [40], Chow and Liu proposed a method to find an optimal tree approximation of a given distribution that has the minimal Kullback–Leibler (KL) divergence to the true distribution. They showed that the optimal tree can be learned efficiently via a *maximum spanning tree* whose edge weights represent the mutual information between the two variables corresponding to the edge's end points. The GTA BP algorithm uses the approximation technique in [40] to construct an approximate tree from the fully-connected graph corresponding to (7.42), and applies BP on the resulting approximated tree. The listing of the GTA BP algorithm is given in Table 7.3. The per-bit complexity of the GTA BP algorithm is $O(n_t^2)$. The BER performance of the algorithm in 12×12 V-BLAST MIMO is shown in Fig. 7.19 for 16-QAM. It can be seen that the performance of the GTA BP algorithm is far from that of the SD, and that further improvements are possible.

Table 7.3. *Listing of the GTA BP algorithm*

Input: A constrained linear least square problem: $\mathbf{y} = \mathbf{H}\mathbf{x} + \mathbf{n}$, a noise level σ^2 and a finite symbol set $\mathbb{A}$ whose mean symbol energy is denoted by e.

Algorithm:

1. Compute: $\mathbf{z} = (\mathbf{H}^T\mathbf{H} + (\sigma^2/e)\mathbf{I})^{-1}\mathbf{H}^T\mathbf{y}$ and $\mathbf{C} = \sigma^2(\mathbf{H}^T\mathbf{H} + (\sigma^2/e)\mathbf{I})^{-1}$

2. Denote:

$$f(x_i; \mathbf{z}, \mathbf{C}) = \exp\left(-\frac{1}{2}\frac{(x_i - z_i)^2}{2C_{ii}}\right)$$

$$f(x_i \mid x_j; \mathbf{z}, \mathbf{C}) = \exp\left(-\frac{1}{2}\frac{((x_i - z_i) - C_{ij}/C_{jj}(x_j - z_j))^2}{C_{ii} - C_{ij}^2/C_{jj}}\right)$$

3. Compute the maximum spanning tree of the n-node graph where the weight of the edge between the nodes i and j is the square of the correlation coefficient, $\rho^2 = C_{ij}^2/(C_{ii}C_{jj})$. Assume the tree is rooted at node x_1 and denote the parent of node i by π_i.

4. Apply BP on the tree approximation using the loop-free Gaussian distribution:
$$p(x_1, \ldots, x_n \mid \mathbf{y}) \propto f(x_1; \mathbf{z}, \mathbf{C}) \prod_{i=2}^{n} f(x_i \mid x_{\pi_i}; \mathbf{z}, \mathbf{C})$$
as follows to find the most likely configuration.

5. Downward BP message:
Message from variable x_i to its parent variable x_{π_i} is computed based on all the messages x_i received from its children as
$$m_{i\to\pi_i}(x_{\pi_i}) = \sum_{x_i \in \mathbb{A}} f(x_i \mid x_{\pi_i}; \mathbf{z}, \mathbf{C}) \prod_{j \mid \pi_j = i} m_{j\to i}(x_i)$$
If x_i is a leaf node in the tree then the message is simply
$$m_{i\to\pi_i}(x_{\pi_i}) = \sum_{x_i \in \mathbb{A}} f(x_i \mid x_{\pi_i}; \mathbf{z}, \mathbf{C})$$

6. Upward BP message:
Message from a parent variable x_{π_i} to its child variable x_i is computed based on the messages from its parent $x_{\pi_{\pi_i}}$ and from downward messages received from all the siblings of x_i
$$m_{\pi_i\to i}(x_i) = \sum_{x_{\pi_i} \in \mathbb{A}} f(x_i \mid x_{\pi_i}; \mathbf{z}, \mathbf{C})\, m_{\pi_{\pi_i}\to\pi_i}(x_{\pi_i}) \prod_{j \mid j \neq i, \pi_j = \pi_i} m_{j\to\pi_i}(x_{\pi_i})$$
If x_{π_i} is the root of the tree then the message is simply:
$$m_{\pi_i\to i}(x_i) = \sum_{x_{\pi_i} \in \mathbb{A}} f(x_i \mid x_{\pi_i}; \mathbf{z}, \mathbf{C}) \prod_{j \mid j \neq i, \pi_j = \pi_i} m_{j\to\pi_i}(x_{\pi_i})$$

7. Belief calculation:
For root node belief is: $\mathrm{b}(x_i) = f(x_i; \mathbf{z}, \mathbf{C}) \prod_{j \mid \pi_j = i} x_i$
For other nodes belief is: $\mathrm{b}(x_i) = m_{\pi_i\to i}(x_i) \prod_{j \mid \pi_j = i} x_i$

8. Symbol decoding:
$$\widehat{x}_i = \underset{a \in \mathbb{A}}{\mathrm{argmax}}\ \mathrm{b}(a)$$

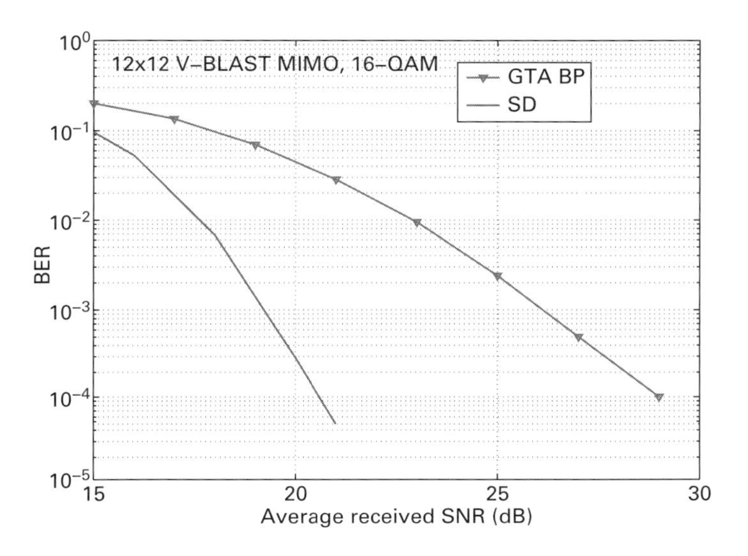

Figure 7.19 BER performance of the GTA BP algorithm in 12 × 12 V-BLAST MIMO with 16-QAM.

7.7 BP based joint detection and LDPC decoding

Often, detection and decoding in communications receivers are carried out as two independent functions. However, processing detection and decoding functions jointly can lead to improved performance [41]. Also, turbo equalization that performs detection and decoding in an iterative manner is known to give coded performance at low complexities [42]. In [42], a receiver that performs detection and decoding is (*i*) independently referred to as a Type-B receiver, (*ii*) iteratively (between detection and decoding) referred to as a Type-C receiver, and (*iii*) jointly (optimal) referred to as a Type-A receiver, as shown in Fig. 7.20.

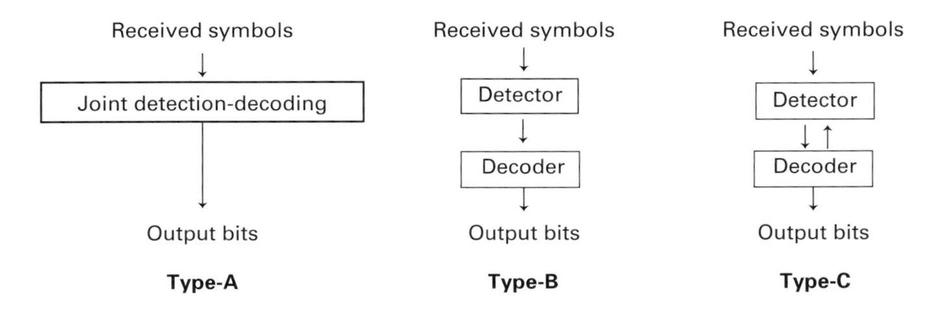

Figure 7.20 Different types of receivers for coded systems.

Receivers of the Type-C kind that iterate extrinsic information between a detector and a decoder are the predominantly studied ones. However, given that large MIMO detection can be done using message passing on factor graphs at low complexities and that decoding of LDPC codes is also done using message passing on factor graphs, joint detection–decoding schemes based on message passing on an integrated factor graph that combines MIMO constraints as well as LDPC code constraints can be devised to achieve improved performance. The joint factor graph formulation would involve two sets of factor nodes, one representing the received vector over the MIMO channel and the other representing the LDPC check equations. The marginalization of the joint probabilities is done through message passing from the different sets of nodes and appropriately combining them at the variable nodes. A message passing scheme on a combined factor graph for joint detection–decoding in large MIMO systems and its coded BER performance are presented in the following subsections.

7.7.1 System model

Consider a MIMO system with n_t transmit and n_r receive antennas. A sequence $\mathbf{u}$ of k information bits is encoded by an LDPC code into a codeword $\mathbf{b}$ of n coded bits. The coded bits are then BPSK modulated, i.e., the modulation alphabet is[1] $\{+1, -1\}$. In one channel use n_t modulated symbols are sent on n_t transmit antennas using spatial multiplexing. So, the number of channel uses needed to send all n coded bits is $M = \lceil n/n_t \rceil$. Let $\mathbf{x}^{(m)}$, $m = 1, \ldots, M$, denote the $n_t \times 1$ sized modulated symbol vector sent in the mth channel use. Let $\mathbf{H}^{(m)} \in \mathbb{C}^{n_r \times n_t}$ denote the channel gain matrix in the mth channel use, whose entries are assumed to be iid Gaussian with zero mean and unit variance. The received vector in the mth channel use, $\mathbf{y}^{(m)}$, is given by

$$\mathbf{y}^{(m)} = \mathbf{H}^{(m)}\mathbf{x}^{(m)} + \mathbf{w}^{(m)}, \tag{7.43}$$

where $\mathbf{w}^{(m)}$ is the noise vector whose entries are modeled as iid $\mathcal{CN}(0, \sigma^2)$. Perfect knowledge of the channel gains at the receiver is assumed. Detection is done on a per-channel use basis, and decoding is done over all n coded bits from M channel uses. For MIMO detection, we need to compute the MAP, given by

$$p(\mathbf{x}^{(m)} \mid \mathbf{y}^{(m)}, \mathbf{H}^{(m)}) \propto p(\mathbf{y}^{(m)} \mid \mathbf{x}^{(m)}, \mathbf{H}^{(m)}) p(\mathbf{x}^{(m)}), \tag{7.44}$$

whose exact computation requires exponential complexity in n_t.

7.7.2 Individual detection and decoding

In Type-B receiver architecture, detection and decoding are done individually. In the considered large MIMO system, MIMO detection can be independently carried out using the BP algorithm on a factor graph based on the scalar GAI (FG-SGA algorithm) presented in Section 7.5, whose per-symbol complexity is $O(n_t)$. Similarly, the decoding of the LDPC code can be carried out independently using another instance of the BP algorithm. The LDPC decoding algorithm is briefly summarized here before describing the joint detection–decoding algorithm.

[1] 4-QAM modulation can be considered likewise by vectorizing the $\{+1, -1\}$ values in the real and imaginary components and working with the real-valued system model.

LDPC decoding

The LDPC decoding algorithm is a message passing algorithm, which gives the APPs of the coded bits. The LDPC decoder graph is described by the parity check matrix, $\mathbf{F}$, of dimension $n \times (n-k)$. If $\mathbf{b}$ is a valid codeword, then $\mathbf{bF} = \mathbf{0}$. The graph over which the messages are passed is a bipartite graph, consisting of n variable nodes corresponding to the coded bits in a block and $n-k$ check nodes corresponding to the check equations. The message passing algorithm for LDPC decoding can be briefly described as follows.

- *Initialization* Initialize the variable node to check node messages V_{lj}s, $l = 1, \ldots, n$, $j = 1, \ldots, n-k$ with initial probabilities $p(b_l = a)$, $a \in \{0, 1\}$, where b_l is the lth element in the coded bit vector $\mathbf{b}$.
- *Step (1)* Compute check node to variable node messages $U_{jl} = p(S_j \mid b_l = a)$, where S_j is the event of the jth check equation being satisfied. U_{jl} is a function of all V_{rj}, $r \in \mathcal{N}(j) \setminus l$, where $\mathcal{N}(j)$ denotes the set of all variable nodes connected to check node j.
- *Step (2)* Compute $V_{lj} \propto p(b_l = a \mid S_{\mathcal{N}(l) \setminus j}, \mathbf{y}) p(b_l = a)$, where $S_{\mathcal{N}(l) \setminus j}$ is the event that all check equations which involve bit b_l except the jth check equation are satisfied. V_{lj} is a function of all U_{rl}, $r \in \mathcal{N}(l) \setminus j$.
- *Step (3)* Compute $p(b_l = a \mid S, \mathbf{y})$, where S is the event of all check equations being satisfied.

Steps (1) and (2) are repeated until $\mathbf{bF} = \mathbf{0}$ or a certain number of iterations are completed. At the end, decisions on the bits are made based on the probabilities in Step (3).

7.7.3 Joint detection and decoding

The message passing algorithm can be used to jointly detect and decode the received symbols in large MIMO systems [43]. When jointly performing detection and decoding operations, there is a transfer of extrinsic information from the decoder to the detector and vice versa, which results in efficient usage of information at the receiver to marginalize the joint probability of the received symbols. In a joint detection and decoding scenario, the objective is to compute

$$\begin{aligned} p(\mathbf{x} \mid S, \mathbf{y}) &\propto p(\mathbf{x}, S, \mathbf{y}) \\ &= p(S \mid \mathbf{x}) \mathbf{p}(\mathbf{y} \mid \mathbf{x}) \mathbf{p}(\mathbf{x}), \end{aligned} \tag{7.45}$$

where

$$p(S \mid \mathbf{x}) = \prod_{j=1}^{n-k} p(S_j \mid \mathbf{x}). \tag{7.46}$$

A graph whose joint probability factorizes according to the above equation can be constructed, and marginalization on this graph gives the probability of the

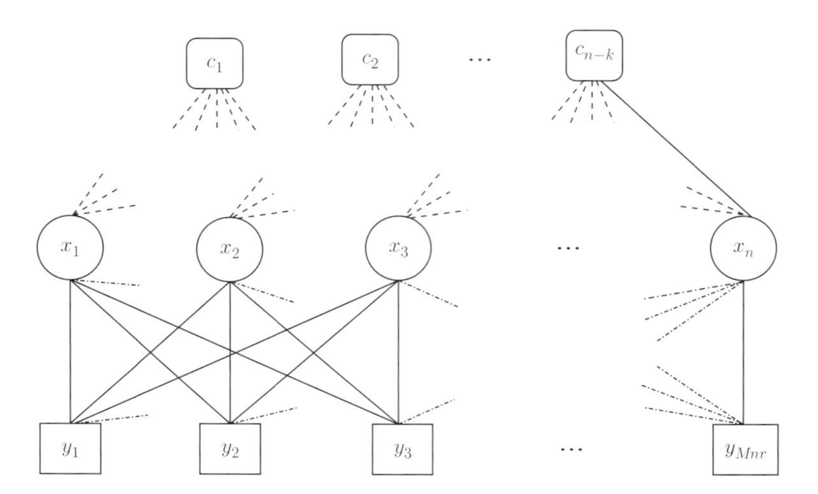

Figure 7.21 Illustration of the joint graph with observation nodes y_i, variable nodes x_l, and check nodes c_j.

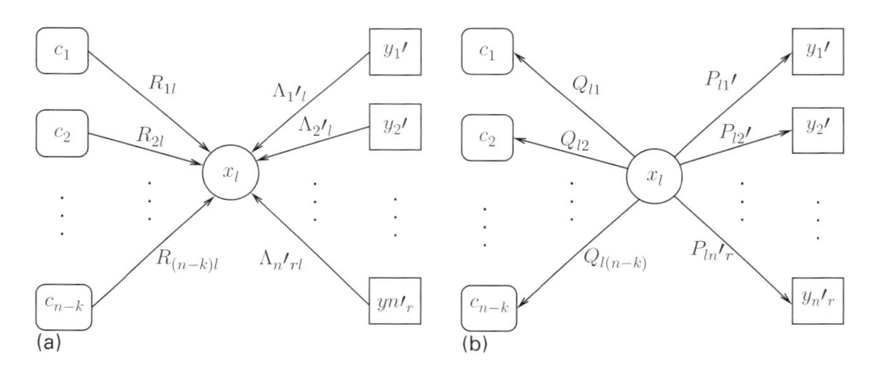

Figure 7.22 Message passing in joint detection–decoding: (a) messages passed to variable node x_l; (b) messages passed from variable node x_l.

received symbols. The constructed graph consists of three sets of nodes, namely, the variable nodes set, the observation nodes set, and the check nodes set. There are Mn_r observation nodes corresponding to the received vectors, $Mn_t = n$ variable nodes corresponding to the transmitted coded symbol vectors over M channel uses, and $n-k$ check nodes corresponding to the check equations of the LDPC code. Figures 7.21 and 7.22 illustrate the joint graph and the messages passed over it.

The messages passed over the graph are: (i) P_{li}, the message from variable node x_l to observation node y_i; (ii) Q_{lj}, the message from variable node x_l to check node c_j; (iii) R_{jl}, the message from check node c_j to variable node x_l; and (iv) Λ_{il}, the message from observation node y_i to variable node x_l, where $l = 1, \ldots, n$, $i = 1, \ldots, Mn_r$, $j = 1, \ldots, n-k$, $m = 1, \ldots, M$, and n is chosen such that $n = Mn_t$. It can be observed that $m = \lceil i/n_r \rceil = \lceil l/n_t \rceil$. The message

Λ_{il} is computed as in (7.37) with only the corresponding $\mathbf{H}^{(m)}$ and $\mathbf{x}^{(m)}$. Thus, Λ_{il} is a function of all P_{ri}, where $r \in \{l' \mid \lceil l'/n_t \rceil = m\} \setminus l$.

$$R_{jl} = \ln\left(\frac{p(S_j \mid x_l = +1)}{p(S_j \mid x_l = -1)}\right) = \ln\left(\frac{1 + \prod_{r \in \mathcal{N}_v(j)\setminus l}(1 - 2p(b_r = 1))}{1 - \prod_{r \in \mathcal{N}_v(j)\setminus l}(1 - 2p(b_r = 1))}\right), \tag{7.47}$$

$$P_{li} = p(x_l = +1 \mid \mathbf{y}_{\mathcal{N}_o(l)\setminus i}, S_{\mathcal{N}_c(l)}) = \frac{\exp(\sum_{r \in \mathcal{N}_o(l)\setminus i} \Lambda_{rl} + \sum_{r \in \mathcal{N}_c(l)} R_{rl})}{1 + \exp(\sum_{r \in \mathcal{N}_o(l)\setminus i} \Lambda_{rl} + \sum_{r \in \mathcal{N}_c(l)} R_{rl})}, \tag{7.48}$$

$$Q_{lj} = \ln\left(\frac{p(x_l = +1 \mid \mathbf{y}_{\mathcal{N}_o(l)}, S_{\mathcal{N}_c(l)\setminus j})}{p(x_l = -1 \mid \mathbf{y}_{\mathcal{N}_o(l)}, S_{\mathcal{N}_c(l)\setminus j})}\right) = \sum_{r \in \mathcal{N}_o(l)} \Lambda_{rl} + \sum_{r \in \mathcal{N}_c(l)\setminus j} R_{rl}, \tag{7.49}$$

where $\mathcal{N}_c(l)$ is the set of check nodes connected to x_l, $\mathcal{N}_v(j)$ is the set of variable nodes connected to c_j, and $\mathcal{N}_o(l)$ is the set of observation nodes connected to x_l, $\mathcal{N}_o(l) = \{i' \mid \lceil i'/n_r \rceil = m\}$. In typical LDPC decoding, the computation of R_{jl} is simplified by the use of the tanh(.) function. The messages P_{li} and Q_{lj} are computed as given by (7.45) and only the extrinsic information from one set of nodes is passed to the other. The LLRs or the beliefs of the symbols at the end of an iteration are given by

$$L_l = \sum_{i \in \mathcal{N}_o(l)} \Lambda_{il} + \sum_{j \in \mathcal{N}_c(l)} R_{jl}. \tag{7.50}$$

The iterations are continued till $\mathbf{bF} = \mathbf{0}$ or till a certain number of iterations is completed, upon which bit decisions are made using the final LLRs. The convergence of the algorithm can be improved by damping of beliefs, where, in each iteration, the beliefs are taken to be a weighted average of the beliefs in the previous and current iterations [35].

7.7.4 Performance and complexity

The coded BER performance of the above joint message passing algorithm for detection and decoding is shown in Fig. 7.23. The average received SNR is $\gamma = n_t E_s/\sigma^2$, where E_s is the average symbol energy. Figure 7.23 shows the coded BER performance in a 64×64 V-BLAST MIMO system using a rate-1/2 LDPC code with block length $n = 2640$. The performance of this LDPC coded system with individual detection and decoding (i.e., a Type-B receiver), and iterative detection and decoding (i.e., a Type-C receiver, also called the turbo equalizer) are plotted for comparison. For the Type-B receiver, the numbers of local iterations in detection and decoding, respectively, are 5 and 10. For the

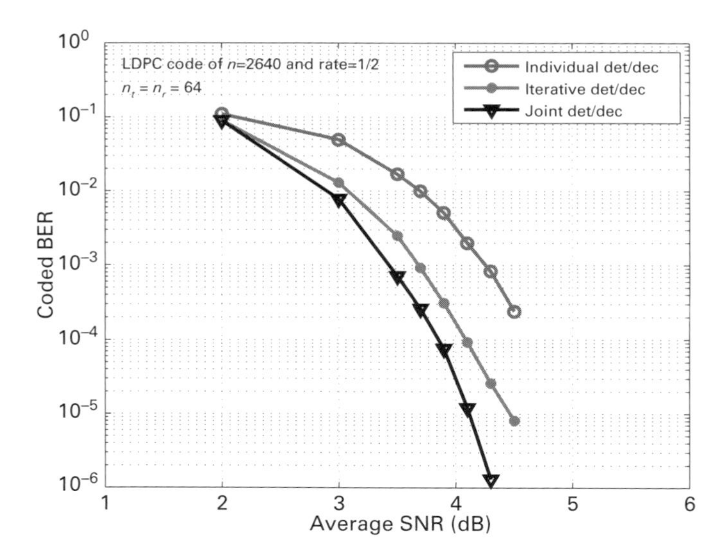

Figure 7.23 Coded BER performance of joint detection–decoding in a 64×64 V-BLAST MIMO system with rate-1/2 LDPC code and block length $n = 2640$ bits.

Type-C receiver, the numbers of local iterations in detection and decoding, respectively, are 1 and 10, with two global (turbo) iterations between detection and decoding. For detection and decoding, the number of iterations is 10. Belief damping with a damping factor of 0.2 is used. It can be observed that the turbo equalizer and joint detection–decoding perform significantly better than the individual detection and decoding scheme, and that the joint detection–decoding performance is better than the turbo equalizer.

The total complexity of the factor graph based detection scheme is $O(n_t n_r)$ for both the computation of the messages at observation nodes and at variable nodes [31]. The complexity of the LDPC decoding algorithm requires $O(cn)$ additions for variable node messages and $O(r(n-k))$ multiplications for check node messages, where c and r are the column and row weights of the parity check matrix, respectively. The joint message passing algorithm requires the same complexity for the computation of messages at observation nodes and check nodes. The complexity of variable node messages computation is $O(n_t M(n_r M + c))$. The overall complexity therefore scales well for large numbers of antennas.

7.8 Irregular LDPC codes design for large MIMO

The feasibility of the low complexity joint detection–decoding approach in the previous section allows one to design good LDPC codes for large MIMO systems. In particular, LDPC codes suited for large MIMO systems can be designed using the EXIT behavior of the joint detection–decoding receiver [44]. In large MIMO systems, these specially designed LDPC codes can outperform off-the-shelf LDPC codes designed for non-fading AWGN channels.

An LDPC code is a linear block code for which the $m \times n$ parity check matrix of interest has a low density of ones. A regular LDPC code is one for which the parity check matrix of interest has a fixed number (w_c) of ones in every column and a fixed number (w_r) of ones in every row. Each code bit is involved with w_c parity constraints and each parity constraint involves w_r bits. Low density means $w_c \ll m$ and $w_r \ll n$. The number of ones in the parity matrix is $w_c n = w_r m$. So, $m \gg n - k$ means the code rate $r = k/n \geq 1 - (w_c/w_r)$, and thus $w_c < w_r$. An irregular LDPC code is one in which the row weights and/or column weights of the parity check matrix are not constant. Irregular LDPC codes are known to generally perform better than regular LDPC codes [45].

A way to construct irregular LDPC codes is to optimize the degree distribution using either density evolution [46] or EXIT charts [47]. One requirement in LDPC code design using the EXIT chart approach is knowledge of the EXIT characteristics of the detector of interest on a given channel. For simple detectors and channels, the EXIT behavior can be analytically characterized in closed form [48]. When such analytical characterization is not tractable, one can resort to Monte Carlo simulations to obtain the EXIT curves [47].

One can design irregular LDPC codes for large MIMO systems by first characterizing the EXIT characteristics of the FG BP-SGA detector in Section 7.5, then obtaining the EXIT curves for the combination of the detector and decoder using knowledge of the EXIT characteristics of the detector, and finally constructing LDPC codes by matching the EXIT chart of the combined detector–decoder with the EXIT chart of the LDPC check node set. That is, the ability to generate EXIT curves for the joint message passing detector–decoder for a large number of antennas can be exploited for the purpose of designing efficient LDPC codes for large MIMO systems. Such an EXIT chart based LDPC code design approach applied in a large MIMO system model in Section 7.7.1 with QPSK modulation is described in the following subsections.

7.8.1 EXIT chart analysis

EXIT chart analysis was first introduced in [48] for analyzing iterative decoders. The EXIT function of a decoder describes the relation between the input a priori information and the output extrinsic information. Let I_A be the average mutual information between the coded bits and the a priori input (a priori information) and I_E be the average mutual information between the coded bits and the extrinsic output. The function $f(I_A) = I_E$ is the EXIT function of the decoder. This function $f(.)$ completely characterizes the information transfer in the decoder. The EXIT curves for the FG BP-SGA based detector are first obtained, and these are then utilized to obtain the EXIT curves of the joint detector–decoder.

EXIT characteristics of the FG BP-SGA detector

The FG BP-SGA detector is introduced and explained in detail in Section 7.5. In brief, the detector has two sets of nodes, namely, the observation nodes and the variable nodes. The observation nodes are initialized with $\mathbf{y}_r^{(l)}$, and after a few iterations of message passing, the soft values or LLRs of the transmitted $\mathbf{x}_r^{(l)}$ are obtained from the variable nodes. In computing the messages, the interference from other antennas is approximated to be Gaussian, and its mean and variance are obtained in closed form and used in the message computation. Denoting the extrinsic information at the output of the FG BP-SGA detector by $I_{E,det}$ and the a priori information at the input of the detector by $I_{A,det}$, we have

$$I_{E,det} = f\left(\gamma, n_t, n_r, I_{A,det}\right), \tag{7.51}$$

where γ is the average SNR at the receiver. To obtain the EXIT curve, the input a priori LLR value A is modeled as $A \sim \mathcal{N}(x\mu_A, \sigma_A^2)$ such that $\mu_A = \sigma_A^2/2$ and $x \in \{+1, -1\}$. Then $I_{A,det}$ can be written as [48]

$$\begin{aligned} I_{A,det}(\sigma_A) &= I(X, A) \\ &= \frac{1}{2}\sum_{x=-1,1}\int_{-\infty}^{+\infty} P_A(z|X=x)\log_2\frac{2P_A(z|X=x)}{P_A(z|X=1)+P_A(z|X=-1)}dz \\ &\triangleq J(\sigma_A). \end{aligned} \tag{7.52}$$

Therefore, $\sigma_A = J^{-1}(I_{A,det})$. Similarly, the output extrinsic information $I_{E,det}$ is given by

$$\begin{aligned} I_{E,det}(\sigma_A) = \frac{1}{2}\sum_{x=-1,1}\int_{-\infty}^{+\infty} & P_E(z|X=x) \\ & \times \log_2\frac{2P_E(z|X=x)}{P_E(z|X=1)+P_E(z|X=-1)}dz. \end{aligned} \tag{7.53}$$

The above extrinsic information depends on the detector used. For the considered system and the FG BP-SGA detector, the analytical evaluation of $I_{E,det}$ in (7.53) is a difficult task. For simplicity, one can resort to the computation of $P_E(z|X = x)$ through Monte Carlo simulations. The EXIT curves can be computed by simulating (7.53) in conjunction with the simulation of the FG BP-SGA detection algorithm for various values of received SNRs. Figure 7.24(a) shows the EXIT curves obtained for the FG BP-SGA detector for a 16×16 MIMO system for different SNR values, and Fig. 7.24(b) shows the EXIT curves of the FG BP-SGA detector for $n_t = n_r = 16, 32, 64, 128, 256$ MIMO configurations at 3 dB SNR.

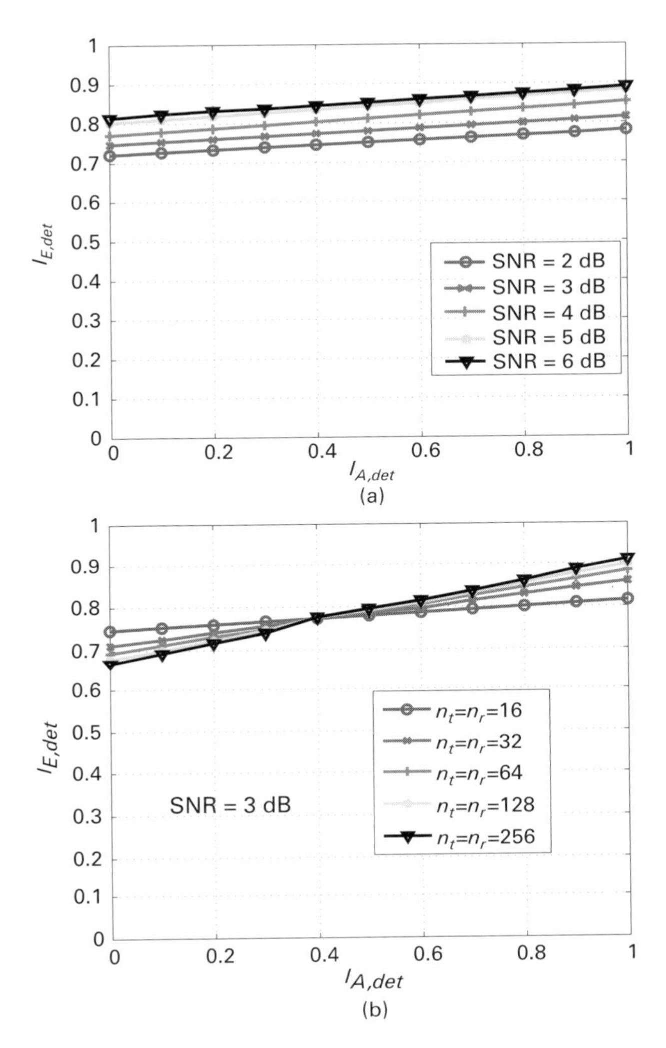

Figure 7.24 $I_{E,det}$ versus $I_{A,det}$ EXIT curves of FG BP-SGA detector in large MIMO systems: (a) different SNRs; (b) different $n_t = n_r$.

Analysis of the joint detection–decoding receiver

Let d_v and d_c denote the variable node and check node (CN) degrees of the LDPC decoder. $I_{E,joint}(\gamma, d_v, I_{A,joint})$ is the extrinsic information at the variable nodes of the LDPC decoder, where $I_{A,joint}$ is the a priori information at the variable nodes. Likewise, $I_{A,CN}(d_c, I_{E,CN})$ and $I_{E,CN}$ are the a priori and extrinsic information at the CNs of the LDPC decoder. For the joint detector–decoder, a single graph that combines the variable nodes of the FG BP-SGA MIMO detector and the variable nodes of the LDPC decoder is used. Each node in the variable nodes set of the detector and decoder represents a coded bit. The edges from this combined variable nodes set are appropriately interleaved and connected to the CN set of the decoder. The EXIT curve of the combined detector–decoder is evaluated as [47]

$$I_{E,joint}(\gamma, d_v, I_{A,joint}, I_{E,det}) = J\left(\sqrt{(d_v - 1)(J^{-1}(I_{A,joint}))^2 + (J^{-1}(I_{E,det}))^2}\right), \quad (7.54)$$

where $I_{E,Joint}$ is the extrinsic information and $I_{A,joint}$ is the a priori information at the block formed by the combination of the variable nodes of the FG BP-SGA detector and LDPC decoder. Figure 7.25 illustrates the transfer of mutual information between the combined observation node/variable node (ON/VN) block and the LDPC check node block of the joint detector/decoder.

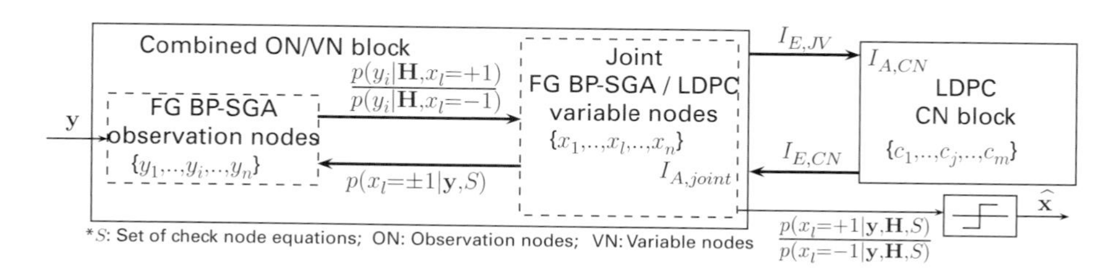

Figure 7.25 Transfer of mutual information in the joint detector–decoder.

The EXIT curves for the combined detector–decoder for $n_t = n_r$ =16, 32, 64, 128, 256 MIMO systems at 3 dB SNR evaluated using the EXIT curves of the FG BP-SGA detector and the LDPC variable nodes are shown in Fig. 7.26(a). The decoding trajectory for $d_v = 3$, $d_c = 6$, $n_t = n_r = 16$ and 3 dB SNR is plotted in Fig. 7.26(b). It is necessary that the EXIT curve of the CN set lies below the EXIT curve of the combined variable nodes set for the decoder to converge, which must be satisfied by the code design.

7.8.2 LDPC code design

To approach the capacity of the channel, the EXIT curves of the CN set and the combined variable nodes set should be matched [49]. To match these curves, the degree distribution of both the variable nodes (d_v) and the CNs (d_c) can be varied. Alternatively, only d_v can be varied for a fixed d_c [47]. The LDPC codes so obtained will have a non-uniform degree distribution (i.e., the codes are irregular codes). The design methodology that varies d_v for a fixed d_c and the methodology that varies d_v and d_c are explained below.

Let $\{d_v^i\}$ and $\{p_v^i\}$, respectively, denote the variable node degrees and the probabilities of occurrence of each d_v^i, $i = 1, \ldots, D_v$, where D_v is the number of variable node degrees considered for the design. An example is as follows: $D_v = 4$ and $\{d_v^1, d_v^2, d_v^3, d_v^4\} = \{2, 4, 5, 7\}$. The average variable node degree is

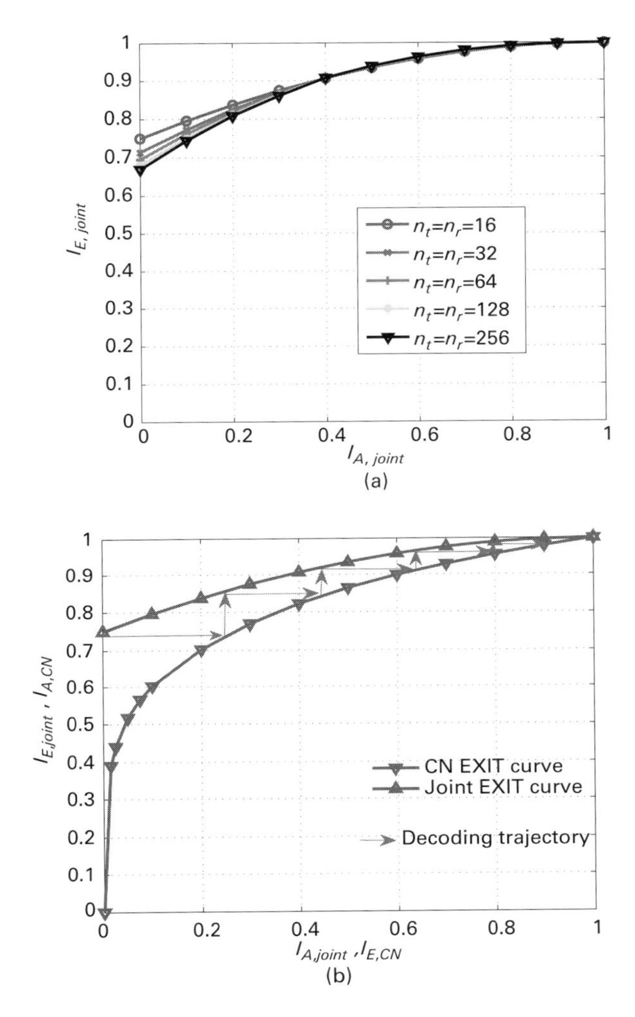

Figure 7.26 (a) EXIT curves $d_v = 3$, different $n_t = n_r$ and (b) mutual information trajectory of joint detector–decoder for large MIMO systems at SNR = 3 dB, $d_v = 3, d_c = 6, 16 \times 16$ MIMO.

then given by

$$\bar{d_v} = \mathbb{E}(d_v) = \sum_{i=1}^{D_v} p_v^i d_v^i = \mathbf{p}_v^T \mathbf{d}_v, \tag{7.55}$$

where $\mathbf{d}_v = [d_v^1 \; d_v^2 \; \cdots \; d_v^{D_v}]$, $\mathbf{p}_v = [p_v^1 \; p_v^2 \; \cdots \; p_v^{D_v}]$, and $\|\mathbf{p}_v\|_1 = 1$. Since the number of edges incident at the variable nodes is the same as the number of edges incident at the CNs, $n\bar{d_v} = (n-k)d_c$ for a fixed d_c. Thus, $\bar{d_v} = (1-R)d_c$. The probability that an edge is connected to a variable node of degree d_v^i is $p_{ve}^i = np_v^i d_v^i / n\bar{d_v}$. For a fixed d_c, let $\mathbf{D} = (1/(1-R)d_c)\,\mathrm{diag}(d_v^1, \ldots, d_v^{D_v})$. Hence,

$\mathbf{p}_{ve} \triangleq [p_{ve}^1 \, p_{ve}^2 \cdots p_{ve}^{D_v}] = \mathbf{D}\mathbf{p}_v$, and $\|\mathbf{p}_{ve}\|_1 = 1$. Since the EXIT curve of a mixture of codes is the same as the average of the individual codes, the effective EXIT curve of the mixture of codes with varying variable node degree is

$$I_{E,joint}^{eff}(\gamma, I_{A,joint}) = \sum_{i=1}^{D_v} p_{ve}^i \; I_{E,joint}(\gamma, d_v^i, I_{A,joint},). \tag{7.56}$$

Since a closed-form analytical expression for $I_{E,det}$, and hence for $I_{E,joint}$, is unavailable, $I_{E,joint}$ is evaluated at N different points, using which (7.56) is written in the form

$$\mathbf{q}_{E,joint} = \mathbf{Q}_{E,joint} \; \mathbf{p}_{ve}, \tag{7.57}$$

where $\mathbf{q}_{E,joint}$ is an $N \times 1$ vector and $\mathbf{Q}_{E,joint}$ is an $N \times D_v$ matrix whose element in the gth row and ith column is $I_{E,JV}(\gamma, d_v^i, I_{A,joint}^g)$, where $I_{A,joint}^g$ is the gth $I_{A,joint}$ value, $g = 1, \ldots, N$. The aim is to find a vector $\mathbf{p}_v$ such that $(\mathbf{Q}_{E,joint}\mathbf{D}\mathbf{p}_v - \mathbf{q}_{A,CN})$ is minimized. This ensures that the EXIT curve of the LDPC check nodes set matches the EXIT curve of the joint detector–decoder, by varying only the variable node degrees in the LDPC code. That is, we need to find a vector $\mathbf{p}_v$ such that

$$\widehat{\mathbf{p}}_v = \underset{\mathbf{p}_v}{\operatorname{argmin}} \, \{\mathbf{Q}_{E,joint}\mathbf{D}\mathbf{p}_v - \mathbf{q}_{A,CN}\}, \tag{7.58}$$

subject to $\|\mathbf{p}_v\|_1 = 1$, $\|\mathbf{D}\mathbf{p}_v\|_1 = 1$, and $p_v^i \geq 0$, where $\mathbf{q}_{A,CN}$ is the $N \times 1$ vector consisting of the N evaluated values of $I_{A,CN}$ for a fixed d_c. For a fixed d_c, (7.58) can be solved using well-known quadratic programming optimization methods like interior point methods. The EXIT curves are matched by varying the CN degrees also. For this, as defined for the variable nodes, define $\{d_c^j\}$ and $\{p_c^j\}$, $j = 1, \ldots, D_c$ for the CN degrees so that the average CN degree is $\bar{d}_c = \mathbf{p}_c^T \mathbf{d}_c$, where $\mathbf{d}_c = [d_c^1 \; d_c^2 \; \cdots \; d_c^{D_c}]$, $\mathbf{p}_c = [p_c^1 \; p_c^2 \; \cdots \; p_c^{D_c}]$, and $\|\mathbf{p}_c\|_1 = 1$. The probability that an edge is connected to a CN of degree d_c^j is

$$p_{ce}^j = \frac{(n-k)p_c^j d_c^j}{(n-k)\bar{d}_c}, \quad \mathbf{p}_{ce} \triangleq [p_{ce}^1 \; p_{ce}^2 \; \cdots \; p_{ce}^{D_c}]. \tag{7.59}$$

Let $R' = 1 - R$. So $\bar{d}_v$ becomes $\bar{d}_v = R'\bar{d}_c$. To start the optimization, fix either $\bar{d}_v$ or $\bar{d}_c$. The effective EXIT curve for a varying CN degree is

$$I_{A,CN}^{eff}(I_{E,CN}) = \sum_{j=1}^{D_c} p_{ce}^j I_{A,CN}(d_c^j, I_{E,CN}). \tag{7.60}$$

Let $\mathbf{Q}_{A,CN}$ be an $N \times D_c$ matrix whose element in the gth row and jth column is $I_{A,CN}(d_c^j, I_{E,CN}^g)$, where $I_{E,CN}^g$ is the gth $I_{E,CN}$ value, $g = 1, \ldots, N$. Write (7.60) in the form

$$\mathbf{q}_{A,CN} = \mathbf{Q}_{A,CN}\mathbf{p}_{ce}, \tag{7.61}$$

Table 7.4. *Degree profile distribution of LDPC code for 16 × 16 MIMO*

d_v^i	2	8	10	16	18	24
p_v^i	0.3928	0.1990	0.2160	0.0505	0.1244	0.0173

d_c^i	8	12	24	64
p_c^i	0.6216	0.2072	0.0604	0.1108

where $\mathbf{q}_{A,CN}$ is an $N \times 1$ vector. Let $\mathbf{V} = \text{diag}(d_v^1, \ldots, d_v^{D_v})$, and $\mathbf{C} = \text{diag}(d_c^1, \ldots, d_c^{D_c})$. Then, $\mathbf{p}_{ve} = K_v \mathbf{V} \mathbf{p}_v$, $\mathbf{p}_{ce} = K_c \mathbf{C} \mathbf{p}_c$, where the scalars K_v and K_c are given by $K_v = 1/\bar{d}_v$ (for fixed $\bar{d}_v$) and $K_v = 1/R'\bar{d}_c$ (for fixed $\bar{d}_c$), and $K_c = R'/\bar{d}_v$ (for fixed $\bar{d}_v$) and $K_c = \frac{1}{\bar{d}_c}$ (for fixed $\bar{d}_c$). $\mathbf{q}_{A,CN}$ and $\mathbf{q}_{E,joint}$ need to be matched. Thus, the optimization problem is to find $\mathbf{p}_v$ and $\mathbf{p}_c$ such that $\{K_v \mathbf{Q}_{E,joint} \mathbf{V} \mathbf{p}_v - K_c \mathbf{Q}_{A,CN} \mathbf{C} \mathbf{p}_c\}$ is minimized, subject to $\|\mathbf{p}_v\|_1 = 1$, $\|\mathbf{p}_c\|_1 = 1$, $\|\mathbf{V}\mathbf{p}_v\|_1 = 1/K_v$, $\|\mathbf{C}\mathbf{p}_c\|_1 = 1/K_c$, $p_c^i \geq 0$, and $p_v^i \geq 0$. Let

$$\mathbf{p} \triangleq \begin{bmatrix} \mathbf{p}_v \\ \mathbf{p}_c \end{bmatrix}.$$

The optimization problem can then be reduced to

$$\widehat{\mathbf{p}} = \underset{\mathbf{p}}{\text{argmin}} \; \{[K_v \mathbf{Q}_{E,JV} \mathbf{V} - K_c \mathbf{Q}_{A,CN} \mathbf{C}] \, \mathbf{p}\}, \tag{7.62}$$

subject to the linear constraint

$$\begin{bmatrix} 1 \; 1 \; \cdots \; 1 & \mathbf{0} \\ \mathbf{0} & 1 \; 1 \; \cdots \; 1 \\ d_v^1 \; d_v^2 \; \cdots \; d_v^{D_v} & \mathbf{0} \\ \mathbf{0} & d_c^1 \; d_c^2 \; \cdots \; d_c^{D_c} \end{bmatrix} \mathbf{p} = \begin{bmatrix} 1 \\ 1 \\ 1/K_v \\ 1/K_c \end{bmatrix},$$

and each element of $\mathbf{p}$ is ≥ 0. The solution $\widehat{\mathbf{p}}$ is the required degree profile distribution for the irregular LDPC code. An example irregular LDPC code constructed to match the EXIT curves at an average receiver SNR of 5 dB for 16 × 16 MIMO systems has the degree profile distribution shown in Table 7.4.

7.8.3 Coded BER performance

The coded BER performance of the irregular LDPC codes designed based on the EXIT chart approach presented in the above is shown in Fig. 7.27. The performance of the designed irregular LDPC codes is compared with those of (*i*) the (3,6)-regular LDPC code in [50], (*ii*) the irregular LDPC code in [45], and (*iii*) the irregular LDPC code in WiMax standard [51]. 4-QAM is used.

Figure 7.27(a) shows the coded BER performances of (*i*) the designed irregular code, (*ii*) the regular code in [50], and (*iii*) the irregular code in [45] for 16 × 16 and 64 × 64 MIMO configurations with rate-1/2 codes and block length n = 4000. It is seen that in 16 × 16 MIMO, the designed irregular LDPC code has

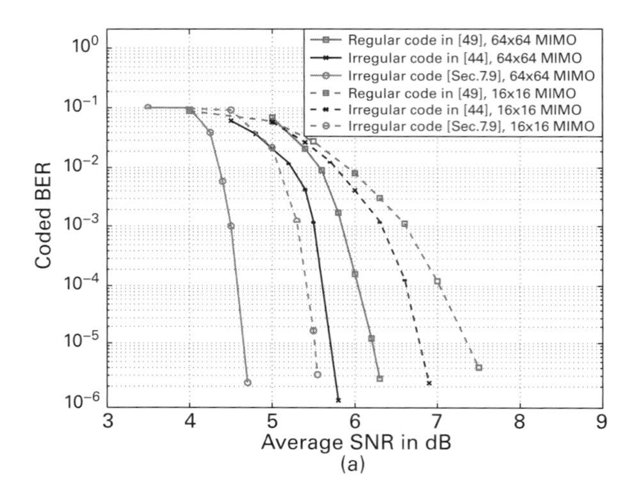

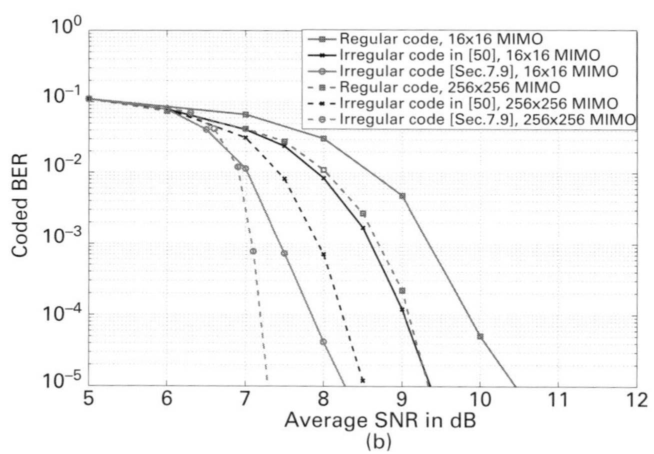

Figure 7.27 Coded BER of (a) rate-1/2 LDPC codes with $n = 4000$ for 16×16 and 64×64 MIMO systems with the joint detector–decoder, (b) rate-3/4 LDPC codes with $n = 2048$ for 16×16 and 256×256 MIMO systems with the joint detector–decoder.

a performance improvement of about 1.8 dB over the regular code in [50] and 1.2 dB over the irregular code in [45], at a coded BER of 10^{-5}. Likewise, the performance improvement in 64×64 MIMO is 1.6 dB over the regular code in [50] and 1 dB over the irregular code in [45]. Figure 7.27(b) shows a performance comparison between the designed LDPC codes and the irregular LDPC code in WiMax standard [51] for rate-3/4 and $n = 2048$ in 16×16 and 256×256 MIMO. Here again, it is observed that the designed irregular LDPC codes perform better by about 2 dB than their regular counterparts and by 1 dB than the irregular code specified in WiMax, at 10^{-5} coded BER. These performance improvements are due to the matching of the EXIT charts of the combined detector–decoder and the LDPC check node set in the design.

References

[1] D. Koller and N. Friedman, *Probabilistic Graphical Models: Principles and Techniques.* Cambridge, MA: The MIT Press, 2009.

[2] B. J. Frey, *Graphical Models for Machine Learning and Digital Communication.* Cambridge, MA: MIT Press, 1998.

[3] J. S. Yedidia, W. T. Freeman, and Y. Weiss, "Understanding belief propagation and its generalizations," in *Exploring Artificial Intelligence in the New Millennium*, G. Lakemeyer and B. Nebel, Eds. San Mateo, CA: Morgan Kaufmann, 2002, ch. 8.

[4] D. Heckerman and M. P. Wellman, "Bayesian networks," *Commun. ACM*, vol. 38, pp. 27–30, 1990.

[5] D. Griffeath, Introduction to Markov Random Fields, in Denumerable Markov Chains, J. G. Kerney, J. L. Snell and A.W. Knupp, Eds., second edition. New York: Springer-Verlag, 1976, pp. 425–485.

[6] F. R. Kschischang, B. J. Frey, and H.-A. Loeliger, "Factor graphs and the sum-product algorithm," *IEEE Trans. Inform. Theory*, vol. 47, no. 2, pp. 498–519, Feb. 2001.

[7] J. Pearl, *Probabilistic Reasoning in Intelligent Systems: Networks of Plausible Inference.* San Mateo, CA: Morgan Kaufmann, 1988.

[8] S. M. Aji and R. J. McEliece, "The generalized distributive law," *IEEE Trans. Inform. Theory*, vol. 46, no. 2, pp. 325–343, Mar. 2000.

[9] R. J. McEliece, D. J. C. MacKay, and J.-F. Cheng, "Turbo decoding as an instance of Pearl's 'belief propagation' algorithm," *IEEE J. Sel. Areas Commun.*, vol. 16, no. 2, pp. 140–152, Feb. 1998.

[10] N. Wiberg, "Codes and decoding on general graphs," in Ph.D. dissertation, Linkoping University, 1996.

[11] D. J. C. MacKay and R. Neal, "Good codes based on very sparse matrices," in *5th IMA Conf. Cryptography and Coding*, vol. 1025. Berlin, Germany: Springer Lecture Notes in Computer Science. Berlin: Springer, 1995, pp. 100–111.

[12] R. G. Gallager, "Low density parity check codes," *IRE Trans. Inform. Theory*, vol. IT-8, no. 2, pp. 21–28, Jan. 1962.

[13] R. M. Tanner, "A recursive approach to low complexity codes," *IEEE Trans. Inform. Theory*, vol. IT-27, no. 5, pp. 533–547, Sep. 1981.

[14] G. Berrou, A. Glavieux, and P. Thitimajshima, "Near shannon limit error-correcting coding: Turbo codes," in *IEEE ICC'1993*, Geneva, May 1993, pp. 1064–1070.

[15] D. J. C. MacKay, "Good error-correcting codes based on very sparse matrices," *IEEE Trans. Inform. Theory*, vol. 45, no. 2, pp. 399–431, Feb. 1999.

[16] Y. Kabashima, "A CDMA multiuser detection algorithm on the basis of belief propagation," *J. Phys. A: Math. General*, Vol. 36, pp. 11 111–11 121, Oct. 2003.

[17] T. Tanaka and M. Okada, "Approximate belief propagation, density evolution, and statistical neurodynamics for CDMA multiuser detection," *IEEE Trans. Inform. Theory*, vol. 51, no. 2, pp. 700–706, Feb. 2005.

[18] A. Montanari, B. Prabhakar, and D. Tse, "Belief propagation based multiuser detection," arXiv:cs/0510044v2 [cs.IT], 22 May 2006.

[19] D. Bickson, D. Dolev, O. Shental, P. H. Siegel, and J. K. Wolf, "Gaussian belief propagation based multiuser detection," in *IEEE ISIT'2008*, Toronto, Jul. 2008, pp. 1878–1882.

[20] D. Guo and C.-C. Wang, "Multiuser detection of sparsely spread CDMA," *IEEE J. Sel. Areas. Commun.*, vol. 26, no. 3, pp. 421–431, Apr. 2008.

[21] H. Wymeersch, *Iterative Receiver Design.* Cambridge, UK: Cambridge University Press, 2007.

[22] M. Tutchler, R. Koetter, and A. C. Singer, "Graphical models for coded data transmission over inter-symbol interference channels," *Eur. Trans. Telecomm.*, vol. 5, no. 4, pp. 307–321, Jul./Aug. 2005.

[23] O. Shental, A. J. Weiss, N. Shental, and Y. Weiss, "Generalized belief propagation receiver for near-optimal detection of two-dimensional channels with memory," in *IEEE ITW'2004*, San Antonio, TX, Oct. 2004, pp. 225–229.

[24] G. Colavolpe and G. Germi, "On the application of factor graphs and the sum-product algorithm to ISI channels," *IEEE Trans. Commun.*, vol. 53, no. 5, pp. 818–825, May 2005.

[25] R. J. Drost and A. C. Singer, "Factor graph algorithms for equalization," *IEEE Trans. Signal Process.*, vol. 55, no. 5, pp. 2052–2065, May 2007.

[26] M. N. Kaynak, T. M. Duman, and E. M. Kurtas, "Belief propagation over MIMO frequency selective fading channels," in *Joint Intl. Conf. on Autonomic and Autonomous Systems and Int. Conf. on Networking and Services*, Papeeti, Oct. 2005.

[27] J. Soler-Garrido, R. J. Piechocki, K. Maharatna, and D. McNamara, "Analog MIMO detection on the basis of belief propagation," in *IEEE Mid-West Symposium on Circuits and Systems*, San Juan, Aug. 2006, pp. 50–54.

[28] X. Yang, Y. Xiong, and F. Wang, "An adaptive MIMO system based on unified belief propagation detection," in *IEEE ICC'2007*, Jun. 2007, pp. 4156–4161.

[29] T. Wo and P. A. Hoeher, "A simple iterative Gaussian detector for severely delay-spread MIMO channels," in *IEEE ICC'2007*, Glasgow, Jun. 2007, pp. 4598–4603.

[30] M. Suneel, P. Som, A. Chockalingam, and B. S. Rajan, "Belief propagation based decoding of large non-orthogonal STBCs," in *IEEE ISIT'2009*, Jun.-Jul. 2008, Sead, pp. 2003–2007.

[31] P. Som, T. Datta, N. Srinidhi, A. Chockalingam, and B. S. Rajan, "Low-complexity detection in large-dimension MIMO-ISI channels using graphical models," *IEEE J. Sel. Topics Signal Process.*, vol. 5, no. 8, pp. 1497–1511, Dec. 2011.

[32] K. Murphy, Y. Weiss, and M. Jordan, "Loopy belief propagation for approximate inference: An empirical study," in *15th Annual Conference on Uncertainty*

in Artificial Intelligence (UAI-99), K. Laskey and H. Prade, Eds., San Francisco, CA: Morgan Kaufmann, 1999, pp. 467–475.

[33] J. M. Mooij, Understanding and improving belief propagation. Ph.D Thesis, Radboud University Nijmegen, May 2008.

[34] J. M. Mooij and H. J. Kappen, "Sufficient conditions for convergence of the sum-product algorithm," *IEEE Trans. Inform. Theory*, vol. 53, no. 12, pp. 4422–4437, Dec. 2007.

[35] M. Pretti, "A message passing algorithm with damping," *J. Stat. Mech.: Theory and Practice*, Nov. 2005.

[36] T. Heskes, "On the uniqueness of loopy belief propagation fixed points," *Neural Computation*, vol. 16, no. 11, pp. 2379–2413, Nov. 2004.

[37] T. Heskes, K. Albers, and B. Kappen, "Approximate inference and constrained optimization," in *19th Uncertainty in AI*, Acapulco, Aug. 2003, pp. 313–320, San Francisco, CA: Morgan Kaufmann.

[38] A. L. Yuille, "A double-loop algorithm to minimize Bethe and Kikuchi free energies," in *EMMCVPR'2001*, Sophic Antipolis Sep. 2001. Berlin: Springer, 2001 pp. 3–18.

[39] J. Goldberger and A. Leshem, "MIMO detection for high-order QAM based on a Gaussian tree approximation," *IEEE Trans. Inform. Theory*, vol. 57, no. 8, pp. 4973–4982, Aug. 2011.

[40] C. K. Chow and C. N. Liu, "Approximating discrete probability distributions with dependence trees," *IEEE Trans. Inform. Theory*, vol. 14, no. 3, pp. 462–467, May 1968.

[41] B. M. Kurkoski, P. H. Siegel, and J. K. Wolf, "Joint message-passing decoding of LDPC codes and partial-response channels," *IEEE Trans. Inform. Theory*, vol. 48, no. 6, pp. 1410–1422, Jun. 2002.

[42] R. Koetter, A. C. Singer, and M. Tuchler, "Turbo equalization," *IEEE Signal Process. Mag.*, Vol. 21, no 1, pp. 67–80, Jan. 2004.

[43] T. L. Narasimhan, A. Chockalingam, and B. S. Rajan, "Factor graph based joint detection/decoding for LDPC coded large-MIMO systems," in *IEEE VTC'2012-Spring*, Yokohama, May 2012, pp. 1–5.

[44] T. L. Narasimhan and A. Chockalingam, "EXIT chart based design of irregular LDPC codes for large-MIMO systems," *IEEE Comm. Lett.*, vol. 17, no. 1, pp. 115–118, Jan. 2013.

[45] T. Richardson, A. Shokrollahi, and R. Urbanke, "Design of capacity-approaching irregular codes," *IEEE Trans. Inform. Theory*, vol. 47, no. 2, pp. 619–637, Feb. 2001.

[46] S.-Y. Chung, J. G. D. Forney, T. Richardson, and R. Urbanke, "On the design of low-density parity-check codes within 0.0045 dB of the Shannon limit," *IEEE Commun. Lett.*, vol. 5, no. 2, pp. 58–60, Feb. 2001.

[47] S. ten Brink, G. Kramer, and A. Ashikhmin, "Design of low-density parity-check codes for modulation and detection," *IEEE Trans. Commun.*, vol. 52, no. 4, pp. 670–678, Apr. 2004.

[48] S. ten Brink, "Convergence behavior of iteratively decoded parallel concatenated codes," *IEEE Trans. Commun.*, vol. 49, no. 10, pp. 1727–1737, Oct. 2001.

[49] A. Ashikhmin, G. Kramer, and S. ten Brink, "Extrinsic information transfer functions: A model and two properties," in *Conf. Inform. Sci. and Sys. (CISS'2002)*, Prinston, Mar. 2002, pp. 742–747.
[50] D. J. C. MacKay, "Encyclopedia of sparse graph codes," Online: http://www.inference.phy.cam.ac.uk/mackay/codes/data.html.
[51] "Air interface for fixed and mobile broadband systems," in IEEE P802.16e Draft, 2005.

8 Detection based on MCMC techniques

In large systems where the physical laws governing system behavior are inherently probabilistic and complicated, traditional methods of obtaining closed-form analytic solutions may not be adequate for the level of detailed study needed. In such situations, one could simulate the behavior of the system in order to estimate the desired solution. This approach of using randomized simulations on computers, which came to be called the Monte Carlo methods, is powerful, elegant, flexible, and easy to implement. From the early days of their application in simulating neutron diffusion evolution in fissionable material in the late 1940s, Monte Carlo methods have found application in almost every discipline of science and engineering. Central to the Monte Carlo approach is the generation of a series of random numbers, often a sequence of numbers between 0 and 1 sampled from a uniform distribution, or, in several other instances, a sequence of random numbers sampled from other standard distributions (e.g., the normal distribution) or from more general probability distributions that arise in physical models. More sophisticated techniques are needed for sampling from more general distributions, one such technique being acceptance–rejection sampling. These techniques, however, are not well suited for sampling large-dimensional probability distributions. This situation can be alleviated through the use of Markov chains, in which case the approach is referred to as the Markov chain Monte Carlo (MCMC) method [1]. Typically, MCMC methods refer to a collection of related algorithms, namely, the Metropolis–Hastings algorithm, simulated annealing, and Gibbs sampling. Before introducing these MCMC algorithms, a brief introduction to Monte Carlo integration and Markov chains is presented in the next two sections. Subsequently, large MIMO detection algorithms based on MCMC techniques are presented.

8.1 Monte Carlo integration

The Monte Carlo approach was originally developed to use random number generation to compute complex integrals. The basic idea is to view integration of a function as an expectation over a certain distribution. Suppose we are interested in computing the complex integral

$$\int_a^b h(x)dx. \tag{8.1}$$

If $h(x)$ can be decomposed as a product of a function $f(x)$ and a PDF $p(x)$ defined over the interval (a, b), then the integral can be expressed as an expectation of $f(x)$ over the density $p(x)$, i.e.,

$$\int_a^b h(x)dx = \int_a^b f(x)p(x)dx = \mathbb{E}_{p(x)}[f(x)]. \tag{8.2}$$

Often, $m = \mathbb{E}_{p(x)}[f(x)]$ is analytically intractable, i.e., the integration or summation required is too complicated. In such cases, a Monte Carlo estimate of m is obtained by

1. drawing N random variables from the density $p(x)$, say $X_1, X_2, \dots, X_N$, and
2. taking the average of $f(X_1), f(X_2), \dots, f(X_N)$, i.e.,

$$m \approx \frac{1}{N}\sum_{n=1}^{N} f(X_n). \tag{8.3}$$

As N (the number of samples) becomes large, the estimate converges to the true expectation m. This is referred to as Monte Carlo integration. Monte Carlo integration can also be used to approximate posterior distributions. For example, the integral

$$\phi(y) = \int f(y|x)p(x)dx \tag{8.4}$$

can be approximated by

$$\widehat{\phi}(y) = \frac{1}{N}\sum_{n=1}^{N} f(y|X_n), \tag{8.5}$$

where X_ns are draws from $p(x)$.

In situations where estimating properties of a particular distribution is of interest, but only samples from a different distribution rather than the distribution of interest can be generated, the importance sampling technique can be used. Suppose the density $p(x)$ roughly approximates the density of interest $q(x)$. Then

$$\int f(x)q(x)dx = \int f(x)\left(\frac{q(x)}{p(x)}\right)p(x)dx = \mathbb{E}_{p(x)}\left[\frac{f(x)q(x)}{p(x)}\right]. \tag{8.6}$$

This forms the basis of the importance sampling technique with

$$\int f(x)q(x)dx \approx \frac{1}{N}\sum_{n=1}^{N}\frac{f(X_n)q(X_n)}{p(X_n)}, \tag{8.7}$$

where X_ns are draws from the approximating density $p(x)$. Likewise, $\int f(y|x)$ $q(x)dx$ can be approximated by

$$\int f(y|x)q(x)dx \approx \frac{1}{N}\sum_{n=1}^{N}\frac{f(y|X_n)q(X_n)}{p(X_n)}. \tag{8.8}$$

Importance sampling can be used for variance reduction in Monte Carlo methods. Variance reduction procedures are aimed at increasing the accuracy of the estimates that can be achieved for a given number of iterations. The idea behind importance sampling is that certain values of the input random variables in a simulation have more impact than others on the parameter being estimated. The variance of the estimator can be reduced if these "important" values are emphasized in the simulations. This is done by choosing a distribution which encourages the important values. Choosing or designing a good biased distribution is key in importance sampling. The benefit of a good distribution can be huge simulation run-time savings. Direct use of such biased distributions results in a biased estimator. However, the simulation outputs can be weighted to correct for the use of the biased distribution to get an unbiased estimate.

8.2 Markov chains

Let X_t denote a random variable at time t, and the possible values that it can take form a finite or countable set called the state space $S = \{s_j : j = 1, 2, \ldots, K\}$, $K \leq \infty$. A Markov chain is a sequence of random variables $X_0, X_1, X_2, \ldots$ with the Markov property, namely,

$$p(X_{t+1} = s_j | X_t = s_i, X_{t-1} = s_{i_{t-1}}, \ldots) = p(X_{t+1} = s_j | X_t = s_i), \qquad (8.9)$$

for all $s_j, s_i, s_{i_{t-1}}, \ldots \in S$ and $t = 0, 1, 2, \ldots$, i.e., given the current state, the future and past states are independent. The probability

$$p_{ij} = p(X_{t+1} = s_j | X_t = s_i), \quad s_i, s_j \in S, \qquad (8.10)$$

is called the transition probability from state s_i to state s_j (the probability that the chain moves from state s_i to state s_j in a single step). For every i, $\sum_{j=1}^{K} p_{ij} = 1$. A state s_i is called an absorbing state if $p_{ii} = 1$, i.e., once the chain enters into this state, it will not come out of it.

Let $\pi_j(t) = p(X_t = s_j)$ denote the probability that the chain is in state s_j at time t, and $\boldsymbol{\pi}(t)$ denote the K-length vector of these state probabilities at time t. The chain is started by specifying a starting vector $\boldsymbol{\pi}(0)$ drawn from some initial distribution $\boldsymbol{\mu} = \{\mu_j : j = 1, 2, \ldots, K\}$, i.e., $p(X_0 = s_j) = \mu_j$ and $\sum_{j=1}^{K} \mu_j = 1$. The probability that the chain is in state s_j at time (or step) $t+1$ is given by

$$\begin{aligned}\pi_j(t+1) &= p(X_{t+1} = s_j) \\ &= \sum_i p(X_{t+1} = s_j | X_t = s_i)\, p(X_t = s_i) \\ &= \sum_i p_{ij}\, \pi_i(t). \qquad (8.11)\end{aligned}$$

Successive iterations of the above equation describe the evolution of the chain. Defining the transition probability matrix $\mathbf{P}$ as a $K \times K$ matrix whose (i, j)th element is p_{ij}, the evolution equation (8.11) can be compactly written in matrix form as

$$\boldsymbol{\pi}(t+1) = \boldsymbol{\pi}(t)\mathbf{P}. \qquad (8.12)$$

From this, it follows that

$$\boldsymbol{\pi}(t) = \boldsymbol{\pi}(0)\mathbf{P}^t. \tag{8.13}$$

Defining the n-step transition probability $p_{ij}^{(n)}$ as the probability that the chain is in state s_j given that n steps ago it was in stage s_i, i.e.,

$$p_{ij}^{(n)} = p(X_{t+n} = s_j | X_t = s_i), \tag{8.14}$$

it follows that $p_{ij}^{(n)}$ is just the (i, j)th element of $\mathbf{P}^n$.

A Markov chain is said to be *irreducible* if there exists a positive integer n_{ij} such that $p_{ij}^{(n_{ij})} > 0$, for all $i, j = 1, 2, \ldots, K$. That is, one can go from any state to any other state in S in some finite number of steps.

A state in a Markov chain is classified as *absorbing*, *transient*, or *recurrent* to characterize how often the state is visited or the time between visits. Let $f_{ij}^{(n)}$ denote the probability that the chain "first" visits state s_j at step n whereas it started in state s_i at step 0, i.e.,

$$f_{ij}^{(n)} = p(X_1 \neq s_j, X_2 \neq s_j, \ldots, X_{n-1} \neq s_j, X_n = s_j | X_0 = s_i), \tag{8.15}$$

with $f_{ii}^{(0)} = 1$ and $f_{ij}^{(0)} = 0$ for $j \neq i$. Further, define the sum of probabilities of first visiting times being $n = 1, 2, 3, \ldots$,

$$f_{ij} = \sum_{n=1}^{\infty} f_{ij}^{(n)}, \tag{8.16}$$

which is the probability that the chain visits state s_j in finite time if it starts in state s_i. In particular, f_{ii} is the probability of returning to the starting state s_i in finite time. A state s_j is said to be

- transient if $f_{jj} < 1$,
- recurrent if $f_{jj} = 1$, and
- absorbing if $p_{jj} = 1$.

If state s_j is recurrent, then it is said to be *positive recurrent* if the mean time between revisits is finite, i.e.,

$$\sum_{n=1}^{\infty} n f_{jj}^{(n)} < \infty. \tag{8.17}$$

Otherwise it is said to be *null recurrent*. A transient state is visited only finitely often. A recurrent state, after it is entered, is visited infinitely often. In addition, if the recurrent state is positive recurrent, then the mean time between visits is finite. If one state in an irreducible Markov chain is positive recurrent, then all the states are positive recurrent. The period of a state s_j is defined as

$$d_j = \gcd\left\{n \geq 0 \,|\, p_{j,j}^{(n)} > 0\right\}. \tag{8.18}$$

It can be shown that for an irreducible Markov chain, $d_j = d, \forall j$. If $d > 1$, the chain is said to be periodic with period d. If $d = 1$, then the chain is said to be *aperiodic*, which means that the chain is not forced into some cycle of fixed length between certain states. It can be seen that if $\mathbf{P}$ has no eigenvalues equal to -1 then the chain is aperiodic.

The limiting probability $\lim_{n\to\infty} p_{jj}^{(n)}$ may or may not converge. For a transient or null recurrent state s_j, $\lim_{n\to\infty} p_{jj}^{(n)} = 0$, i.e., the probability of the chain being in state s_j eventually goes to zero. If state s_j is positive recurrent and periodic, then $\lim_{n\to\infty} p_{jj}^{(n)}$ will not converge. If s_j is positive recurrent and aperiodic, then $\lim_{n\to\infty} p_{jj}^{(n)}$ will converge to steady state probability $\pi_j > 0$. A positive recurrent and aperiodic Markov chain reaches a stationary distribution $\boldsymbol{\pi}$, where the vector of probabilities of being in any particular given state is independent of the initial condition $\boldsymbol{\pi}(0)$. The stationary distribution satisfies

$$\boldsymbol{\pi} = \boldsymbol{\pi}\mathbf{P}. \tag{8.19}$$

A sufficient condition for a unique stationary distribution is that the detailed balance or time reversibility condition, namely, $\pi_i p_{ij} = \pi_j p_{ji}, \forall i, j$, is satisfied.

8.3 MCMC techniques

MCMC methods are a collection of related algorithms, including the Metropolis–Hastings algorithm, simulated annealing, and Gibbs sampling. These algorithms are introduced in the following subsections.

8.3.1 Metropolis–Hastings algorithm

Obtaining samples for Monte Carlo integration becomes an issue when sampling has to be done on complex probability distributions. MCMC methods essentially attempt to address this issue. These methods can be traced back to attempts by mathematical physicists to integrate very complex functions by random sampling [2],[3], and the resulting Metropolis–Hastings algorithm [4],[5]. Suppose our goal is to draw samples from some distribution $p(x)$, where $p(x) = f(x)/C$. The normalizing constant C may not be known and may be difficult to compute. The Metropolis algorithm [2],[3] generates a sequence of draws from this distribution as follows.

(1) Start with an initial value x_0 satisfying $f(x_0) > 0$.
(2) Using the current x value, sample a *candidate value* x^* from some *proposal distribution* $q(x_1, x_2)$, which is the probability of returning a value of x_2 given a previous value of x_1. In the Metropolis algorithm, the only restriction on the proposal distribution is that it is symmetric, i.e., $q(x_1, x_2) = q(x_2, x_1)$.
(3) Compute the ratio of densities evaluated at the current and previous values as

$$\alpha = \frac{p(x^*)}{p(x_{t-1})} = \frac{f(x^*)}{f(x_{t-1})}, \tag{8.20}$$

where the normalizing constant C cancels out in taking the ratio.

(4) If the jump increases the density (i.e., if $\alpha > 1$), then accept the candidate value (i.e., set $x_t = x^*$) with probability 1, and return to Step (2). If the jump decreases the density (i.e., if $\alpha < 1$), then accept the candidate value with probability α and reject with probability $1-\alpha$, and return to Step (2).

In summary, we see that Metropolis sampling is a procedure that computes

$$\alpha = \min\left(\frac{f(x^*)}{f(x_{t-1})}, 1\right) \tag{8.21}$$

in each step and accepts the candidate value x^* with probability α. This procedure generates a Markov chain $(x_0, x_1, x_2, \ldots,)$, as the transition probabilities from x_t to x_{t+1} depend only on x_t and not on $(x_0, x_1, \ldots, x_{t-1})$. After a sufficiently long burn-in period of, say, k steps, the chain approaches its stationary distribution and the samples from the vector $(x_{k+1}, \ldots, x_{k+N})$ are samples from $p(x)$. Hastings [4] generalized the Metropolis algorithm by using an arbitrary transition probability function $q(x_1, x_2) = \Pr(x_1 \to x_2)$ and taking the acceptance probability α as

$$\alpha = \min\left(\frac{f(x^*)q(x^*, x_{t-1})}{f(x_{t-1})q(x_{t-1}, x^*)}, 1\right). \tag{8.22}$$

This is the Metropolis–Hastings algorithm. The Metropolis algorithm results when the proposal distribution is symmetric, i.e., $q(x, y) = q(y, x)$. It can be shown that Metropolis–Hastings sampling generates a Markov chain whose stationary density is $p(x)$. The chain is said to be poorly mixing if the value of x remains flat for long periods in the evolution of the chain. For example, this could correspond to the situation where several consecutive x^* values are rejected in the accept–reject test. On the other hand, if the value of x varies significantly over iterations, then the chain is said to be mixing well.

Burn-in period, starting value, proposal distribution

A key issue in any MCMC sampler, including the Metropolis–Hastings sampler, is the burn-in period (number of steps/iterations until the chain approaches stationarity). Typically, the first several draws are discarded and then one of the various convergence tests (e.g., the Geweke test [6], the Raftery–Lewis test [7]) is used to assess whether stationarity has been reached. A poor choice of proposal distribution and/or starting value can greatly increase the burn-in period. One suggestion is to choose a starting value that is close to the center of the distribution, e.g., choose a value close to the distribution's mode.

Two approaches, namely, *random walk* and *independent chain sampling*, are the generally adopted approaches to choosing proposal distributions. In a proposal distribution based on a random walk approach, the new value is taken to be the sum of the current value and a random value. In the independent chain approach, on the other hand, the probability of moving to point y is independent of the current position x in the chain, i.e., $q(x, y) = g(y)$. That is, the candidate value is simply drawn from a distribution of interest independent of the current value. A number of standard distributions can be used for $g(y)$. Since $g(x)$ is not generally equal to $g(y)$, i.e., the proposal distribution in this case is not generally symmetric, Metropolis–Hastings sampling has to be used. The proposal distribution can be tuned to adjust the mixing (in particular, the accept probability). For example, this is generally done by (i) adjusting the variance/eigenvalues of the covariance matrix if a normal/multivariate normal distribution is used, (ii) changing the range if a uniform distribution is used, and (iii) changing the degrees of freedom if the chi-square distribution is used.

8.3.2 Simulated annealing

As mentioned earlier, a chain is poorly mixing if it stays in small regions of the parameter space for long periods of time. A poorly mixing chain can occur, for example, if the target distribution is multimodal and the choice of starting value leads to a trap near one of the modes. Two approaches to alleviate such situations are common. One approach is to run multiple chains with widely varying (e.g., randomly chosen) starting values. Another approach is to use simulated annealing on a single chain. Simulated annealing was developed for finding the maximum of complex functions which have multiple peaks, where standard hill-climbing techniques may become trapped in less than optimal peaks. The idea of simulated annealing is to allow a non-zero probability of a down-hill move (to encourage exploration in larger parts of the parameter space), and reducing this probability as time (iterations) progresses. The name simulated annealing is due to its analogy to the annealing of a crystal where initially there is a lot of movement which reduces further and further as the temperature cools. Simulated annealing is closely related to Metropolis sampling with the only difference being in the way the accept–reject probability α is defined. In simulated annealing, α is defined as

$$\alpha = \min\left(\left(\frac{f(x^*)}{f(x_{t-1})}\right)^{1/T(t)}, 1\right), \tag{8.23}$$

where T is called the temperature and $T(t)$ is called the cooling schedule. Metropolis sampling becomes a special case for $T(t) = 1, \forall t$. Typically, a cooling schedule with a geometric decrease in temperature, given by

$$T(t) = T_0\left(\frac{T_f}{T_0}\right)^{t/n}, \tag{8.24}$$

is used, where T_0 is the initial temperature and T_f is the final temperature at the nth iteration. If we want to cool to temperature T_f by iteration n and subsequently keep the temperature constant at T_f, then we can use

$$T(t) = \max\left(T_0\left(\frac{T_f}{T_0}\right)^{t/n}, T_f\right). \tag{8.25}$$

To cool down to Metropolis sampling, set $T_f = 1$ and the cooling schedule becomes

$$T(t) = \max\left(T_0{}^{(1-t/n)}, 1\right). \tag{8.26}$$

8.3.3 Gibbs sampling

Gibbs sampling is a special case of Metropolis–Hastings sampling with $\alpha = 1$ (i.e., the candidate values are always accepted). Gibbs sampling was introduced by Geman and Geman in 1984 in the context of image processing [8]. The Gibbs sampler generates samples from univariate conditional distributions (distributions in which all the random variables except one variable are assigned fixed values) rather than generating samples from the full joint distribution which can be difficult for large dimensions. That is, it simulates n random variables sequentially from the n univariate conditionals rather than generating a single n-dimensional vector using the full joint distribution. As an example, consider a bivariate random variable (x, y). Suppose we are interested in computing the marginals $p(x)$ and/or $p(y)$. Compared to obtaining the marginal by integrating the joint density $p(x, y)$, e.g., $p(x) = \int p(x, y)dy$, it is easier to compute marginals by sequentially sampling from conditional distributions $p(x|y)$ and $p(y|x)$ as follows.

- Start with an initial value x_0 for x, and obtain y_0 by generating a draw from the conditional distribution $p(y|x = x_0)$.
- For $t > 0$, the sampling proceeds as follows:

$$x_t \sim p(x|y = y_{t-1}), \tag{8.27}$$

$$y_t \sim p(y|x = x_t). \tag{8.28}$$

The above procedure generates a Gibbs sequence of length k. The points (x_t, y_t), $t = 1, 2, \ldots, k$ are taken as the simulated draws from the full joint distribution. Each of the operations in (8.27) and (8.28) is referred to as a "coordinate update." In this example, the two coordinate updates in (8.27) and (8.28) form one iteration. A similar sampling procedure for an n-dimensional distribution is straightforward. In an n-dimensional problem, each iteration consists of n coordinate updates, one update for each of the n dimensions. The Gibbs sequence converges to a stationary distribution that is independent of the starting values, and by construction this stationary distribution is the target distribution that we are trying to simulate. Sample points in the Gibbs sequence, after a sufficient burn-in period, say, B, can be used to compute any feature of the marginals. For example, using N Gibbs sampling draws from the post burn-in period, $x_{B+1}, \ldots, x_{B+N}$, the expectation of the function f of random variable x can be approximated as

$$\mathbb{E}[f(x)]_N \approx \frac{1}{N} \sum_{i=B+1}^{B+N} f(x_i). \tag{8.29}$$

As $N \to \infty$, $\mathbb{E}[f(x)]_N \to \mathbb{E}[f(x)]$. Likewise, Monte Carlo estimates of other moments can be computed using the Gibbs sequence. Also, approximate marginals can be obtained directly using the Gibbs sequence. Alternatively, the approximate marginals can be obtained from the average of the conditional densities $p(x|y = y_i)$. Since $p(x) = \int p(x|y)p(y)dy = \mathbb{E}_y[p(x|y)]$, the marginal can be approximated by

$$p(x) \approx \frac{1}{N} \sum_{i=B+1}^{B+N} p(x|y = y_i). \tag{8.30}$$

8.4 MCMC based large MIMO detection

In communication receivers, detecting the transmitted symbol vector $\mathbf{x}$ based on the observation vector $\mathbf{y}$ involves the computation of the APP distribution of each transmitted symbol

$$p(x_i = a|\mathbf{y}) = \sum_{\mathbf{x} \backslash x_i} p(\mathbf{x}|\mathbf{y}), \tag{8.31}$$

where $a \in \mathbb{A}$, the modulation alphabet. Note that the posterior density $p(\mathbf{x}|\mathbf{y}) \propto p(\mathbf{y}|\mathbf{x})p(\mathbf{x})$, where $p(\mathbf{x})$ is the prior distribution of $\mathbf{x}$. The computation of (8.31) is clearly prohibitive for large dimensions, in which case one can resort to Monte Carlo methods. Suppose we can generate samples $\mathbf{x}^{(1)}, \mathbf{x}^{(2)}, \ldots, \mathbf{x}^{(N)}$ from the distribution $p(\mathbf{x}|\mathbf{y})$. Then, we can approximate the marginal posterior $p(x_i = a|\mathbf{y})$ by the empirical distribution based on the corresponding component in the Monte Carlo sample, i.e., $x_i^{(1)}, x_i^{(2)}, \ldots, x_i^{(N)}$, and approximate the marginalization in (8.31) as

$$p(x_i = a|\mathbf{y}) \approx \frac{1}{N} \sum_{n=1}^{N} I(x_i^{(n)} = a), \tag{8.32}$$

where $I(.)$ is an indicator function. MCMC methods that use Gibbs sampling are an effective means to sample from the posterior distribution $p(\mathbf{x}|\mathbf{y})$. MCMC simulations are found useful in reducing the exponential complexity in (8.31) to polynomial complexity.

MCMC methods have been applied to design receivers in a number of digital communication applications including signal detection and decoding in AWGN channels, ISI channels, CDMA channels, and MIMO channels [9]–[14].

In the rest of this chapter, large MIMO detection algorithms based on MCMC techniques are presented. It will be seen that a careful choice of sampling distribution and stopping criteria is needed to simultaneously achieve both near-optimal performance as well as scalability to large dimensions.

8.4.1 System model

As mentioned in Chapter 1, the large MIMO detection algorithms in Chapters 5–7, though presented in the context of point-to-point MIMO systems, are applicable in multiuser MIMO systems on the uplink. MCMC based large MIMO detection algorithms are also applicable in both point-to-point as well as uplink multiuser MIMO systems. As an illustration to its applicability in multiuser MIMO settings, the MCMC based large MIMO detection algorithm in this section is presented by considering the following large-scale uplink multiuser MIMO system model.

Consider a large-scale multiuser MIMO system on the uplink consisting of a BS with N receive antennas and K uplink users with one transmit antenna each, $K \leq N$ (Fig. 8.1). N and K are in the range of tens to hundreds. All users transmit symbols from a modulation alphabet $\mathbb{B}$. Though single-antenna users are considered here, the detection schemes apply to a general setting where user k can have n_{t_k} transmit antennas and transmit n_{t_k} spatial streams of data subject to $\sum_k n_{t_k} = K$. It is assumed that synchronization and sampling procedures have been carried out, and that the sampled baseband signals are available at the BS receiver. Let $x_k \in \mathbb{B}$ denote the transmitted symbol from user k. Let $\mathbf{x}_c = [x_1, x_2, \ldots, x_K]^T$ denote the vector comprising the symbols transmitted simultaneously by all users in one channel use. Let $\mathbf{H}_c \in \mathbb{C}^{N\times K}$, given by $\mathbf{H}_c = [\mathbf{h}_1, \mathbf{h}_2, \ldots, \mathbf{h}_K]$, denote the channel gain matrix, where $\mathbf{h}_k = [h_{1k}, h_{2k}, \ldots, h_{Nk}]^T$ is the channel gain vector from user k to the BS, and the h_{jk} denotes the channel gain from the kth user to the jth receive antenna at the BS. Assuming rich scattering and adequate spatial separation between the BS antenna elements, $h_{jk}, \forall j$ are assumed to be independent Gaussian with zero mean and σ_k^2 variance such that $\sum_k \sigma_k^2 = K$. The imbalance in received powers from different users is modeled by σ_k^3, and $\sigma_k^2 = 1$ corresponds to the perfect power control scenario. The received signal vector at the BS in a channel use, denoted by $\mathbf{y}_c \in \mathbb{C}^N$, can be written as

$$\mathbf{y}_c = \mathbf{H}_c\mathbf{x}_c + \mathbf{n}_c, \tag{8.33}$$

where $\mathbf{n}_c$ is the noise vector whose entries are modeled as iid $\mathcal{CN}(0, \sigma^2)$. We will work with the real-valued system model corresponding to (8.33), given by

$$\mathbf{y}_r = \mathbf{H}_r\,\mathbf{x}_r + \mathbf{n}_r, \tag{8.34}$$

where $\mathbf{x}_r \in \mathbb{R}^{2K}$, $\mathbf{H}_r \in \mathbb{R}^{2N\times 2K}$, $\mathbf{y}_r \in \mathbb{R}^{2N}$, $\mathbf{n}_r \in \mathbb{R}^{2N}$ given by

$$\mathbf{H}_r = \begin{bmatrix} \Re(\mathbf{H}_c) & -\Im(\mathbf{H}_c) \\ \Im(\mathbf{H}_c) & \Re(\mathbf{H}_c) \end{bmatrix}, \quad \mathbf{y}_r = \begin{bmatrix} \Re(\mathbf{y}_c) \\ \Im(\mathbf{y}_c) \end{bmatrix},$$
$$\mathbf{x}_r = \begin{bmatrix} \Re(\mathbf{x}_c) \\ \Im(\mathbf{x}_c) \end{bmatrix}, \quad \mathbf{n}_r = \begin{bmatrix} \Re(\mathbf{n}_c) \\ \Im(\mathbf{n}_c) \end{bmatrix}. \tag{8.35}$$

Dropping the subscript r in (8.34) for notational simplicity, the real-valued

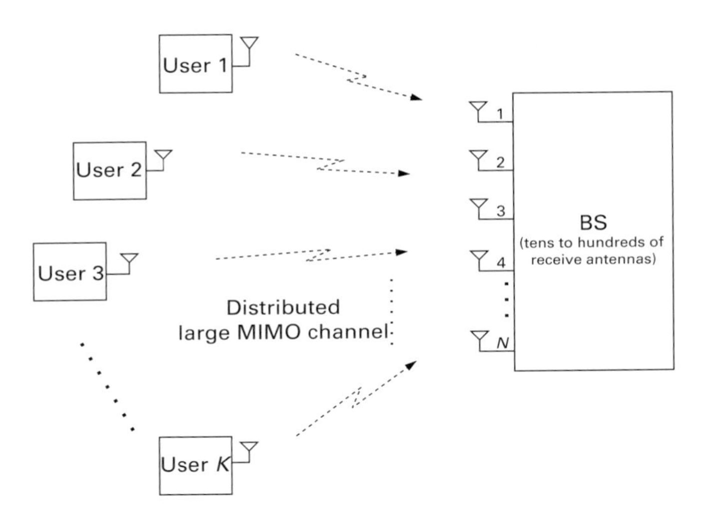

Figure 8.1 Large-scale multiuser MIMO system on the uplink.

system model is written as

$$\mathbf{y} = \mathbf{H}\mathbf{x} + \mathbf{n}. \tag{8.36}$$

For a QAM alphabet $\mathbb{B}$, the elements of $\mathbf{x}$ take values from the underlying PAM alphabet $\mathbb{A}$, i.e., $\mathbf{x} \in \mathbb{A}^{2K}$. The symbols from all the users are jointly detected at the BS. The ML decision rule is given by

$$\mathbf{x}_{ML} = \underset{\mathbf{x}\in\mathbb{A}^{2K}}{\operatorname{argmin}} \ \|\mathbf{y} - \mathbf{H}\mathbf{x}\|^2 = \underset{\mathbf{x}\in\mathbb{A}^{2K}}{\operatorname{argmin}} \ f(\mathbf{x}), \tag{8.37}$$

where $f(\mathbf{x}) \stackrel{\triangle}{=} \mathbf{x}^T\mathbf{H}^T\mathbf{H}\mathbf{x} - 2\mathbf{y}^T\mathbf{H}\mathbf{x}$ is the ML cost. While the ML detector in (8.37) is exponentially complex in K (which is prohibitive for large K), the MCMC based algorithms in the following subsections have a per-symbol complexity that is quadratic in K and they achieve near-ML performance as well.

The ML detection problem in (8.37) can be solved by using MCMC simulations [15]. First, consider the conventional Gibbs sampler, which is an MCMC method used for sampling from distributions of multiple dimensions. In the context of MIMO detection, the joint probability distribution of interest is

$$p(x_1, \ldots, x_{2K}|\mathbf{y}, \mathbf{H}) \propto \exp\Big(-\frac{\| \mathbf{y} - \mathbf{H}\mathbf{x} \|^2}{\sigma^2}\Big). \tag{8.38}$$

Assume perfect knowledge of channel gain matrix $\mathbf{H}$ at the BS receiver.

8.4.2 Conventional Gibbs sampling for detection

In conventional Gibbs sampling based detection, the algorithm starts with an initial symbol vector, denoted by $\mathbf{x}^{(t=0)}$. The initial vector can be a random vector or an output vector from known detectors like MF, the ZF, MMSE detectors. Let

t denote the iteration index and i denote the coordinate index, $i = 1, 2, \ldots, 2K$. Each iteration consists of $2K$ coordinate updates. In each iteration, $2K$ updates are carried out by sampling from distributions as follows:

$$\begin{aligned} x_1^{(t+1)} &\sim p(x_1 | x_2^{(t)}, x_3^{(t)}, \ldots, x_{2K}^{(t)}, \mathbf{y}, \mathbf{H}), \\ x_2^{(t+1)} &\sim p(x_2 | x_1^{(t+1)}, x_3^{(t)}, \ldots, x_{2K}^{(t)}, \mathbf{y}, \mathbf{H}), \\ x_3^{(t+1)} &\sim p(x_3 | x_1^{(t+1)}, x_2^{(t+1)}, x_4^{(t)}, \ldots, x_{2K}^{(t)}, \mathbf{y}, \mathbf{H}), \\ &\vdots \\ x_{2K}^{(t+1)} &\sim p(x_{2K} | x_1^{(t+1)}, x_2^{(t+1)}, \ldots, x_{2K-1}^{(t+1)}, \mathbf{y}, \mathbf{H}). \end{aligned} \tag{8.39}$$

The updated symbol vector at the end of each iteration is fed back to the next iteration for further coordinate updates. The algorithm is run for a certain number of iterations. The detected symbol vector is chosen to be that symbol vector which has the least ML cost in all the iterations.

A problem with the above conventional Gibbs sampling based detection is the stalling problem which results in BER floors at high SNRs [11]. This is illustrated in Fig. 8.2(a) for $K = N = 16$, 4-QAM, random initial vector, and 256 iterations, where the BER of the conventional Gibbs sampler is degraded for SNRs more than 8 dB. The reason for this flooring is that the algorithm becomes trapped in some poor local solutions for a long time (i.e, for many iterations). This can be observed in Fig. 8.2(b) which shows an evolution of the ML cost of the state vector in the nth iteration as a function of n for 12 dB SNR. Note that the ML cost of the state vector does not change much from iteration 4 to iteration 256, and that this trapped ML cost is quite poor compared to the ML cost of the SD solution. This leads to inferior performance compared to the SD. Although the chain is guaranteed to converge to the target distribution (8.38) asymptotically as $n \to \infty$, stalling occurs and degrades performance with finite number of iterations.

8.4.3 Motivation for mixed-Gibbs sampling (MGS)

One might think that the most natural target distribution for sampling is the posterior distribution itself, i.e., the distribution of $\mathbf{x}$, given $\mathbf{y}$ and $\mathbf{H}$ in (8.38). Gibbs sampling with this posterior distribution is indeed guaranteed to take us to this target distribution in the limit $n \to \infty$ [15]. However, this is not the appropriate distribution to sample from if one's goal is to minimize the expected number of iterations for finding the correct solution, as has been demonstrated in [16] (p. 5). This result was shown in the context of guessing passwords using MCMC. As per the result in [16] (p. 6), the correct target distribution with which one must sample to minimize the expected number of iterations for finding the correct solution is a tilted version of the posterior; specifically it must be proportional to the square root of the posterior, i.e., $\big(p(x_i | \mathbf{y}, \mathbf{H}, \mathbf{x}_{-i})\big)^{1/2}$ [16]. If there are only a

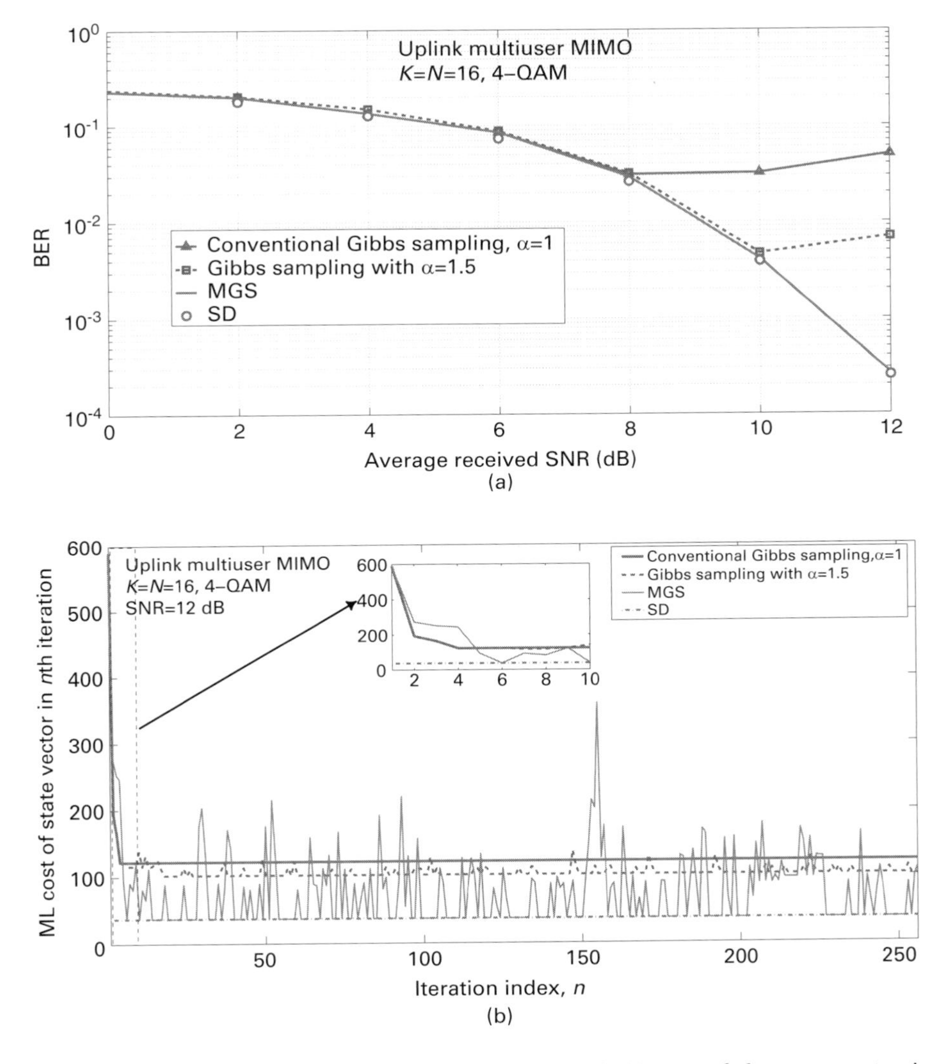

Figure 8.2 (a) BER performance and (b) evolution of ML cost of the state vector in a conventional Gibbs sampler, Gibbs sampler with $\alpha = 1.5$, and mixed-Gibbs sampler for $K = N = 16$ and 4-QAM.

finite number of iterations, and we need to maximize the probability of arriving at the correct solution within these iterations, a heuristic is to sample in such a way as to minimize the higher moments of the number of iterations for finding the correct solution (see [17],[18]). This can be achieved by choosing a temperature parameter $\alpha \geq 1$ and by sampling according to $\left(p(x_i|\mathbf{y}, \mathbf{H}, \mathbf{x}_{-i})\right)^{1/\alpha^2}$. The target distribution for sampling proposed in [19] for MIMO detection used such a parameter α, where the target distribution is taken as

$$p(x_1, \ldots, x_{2K}|\mathbf{y}, \mathbf{H}) \propto \exp\left(- \frac{\| \mathbf{y} - \mathbf{Hx} \|^2}{\alpha^2 \sigma^2} \right). \tag{8.40}$$

α represents a tunable positive parameter which controls the mixing time of the Markov chain; the larger the value of α, the shorter will be the mixing time [19]. Conventional Gibbs sampling results as a special case when $\alpha = 1$. A larger α speeds up the mixing and serves the purpose of reducing the higher moments of the number of iterations for finding the correct solution. However, the stalling problem persists even with large α. This is illustrated for Gibbs sampling with $\alpha = 1.5$ for 12 dB SNR in Fig. 8.2(b); the corresponding evolution of the ML cost of the state vector shows that the ML cost does not go below a certain value (which is well above the ML cost of the SD solution) from iteration 20 to iteration 256. Such poor local solutions, in turn, result in a degraded BER performance for SNRs of more than 10 dB as observed in Fig. 8.2(a) for Gibbs sampling with $\alpha = 1.5$. The above observations motivate the need to devise sampling strategies that can avoid local traps and alleviate the stalling problem significantly.

8.4.4 MGS

In order to break away from traps that lead to stalling, one needs to use a noisy version of the MCMC procedure. The noisiest is the one with infinite temperature (i.e., $\alpha = \infty$), which randomly and uniformly samples from all the possibilities. A simple, yet effective, approach is to use MGS which employs a mixture of (*i*) Gibbs sampling with the posterior in (8.38) (i.e., $\alpha = 1$) and (*ii*) random uniform sampling (i.e., $\alpha = \infty$) [20]. The idea behind the MGS approach is that, in each coordinate update, instead of updating $x_i^{(t)}$s as per the update rule in (8.39) with probability 1 as is done in conventional Gibbs sampling, they are updated as per (8.39) with probability $(1 - q)$ and a different update rule with probability q is used. The different update rule is as follows [21],[22]. Generate $|\mathbb{A}|$ probability values from uniform distribution as

$$p(x_i^{(t)} = j) \sim U[0, 1], \quad \forall j \in \mathbb{A},$$

such that $\sum_{j=1}^{|\mathbb{A}|} p(x_i^{(t)} = j) = 1$, and sample $x_i^{(t)}$ from this generated probability mass function (pmf). In other words, the mixture distribution for sampling is given by

$$p(x_1, \ldots, x_{2K}|\mathbf{y}, \mathbf{H}) \propto (1 - q)\psi(\alpha_1) + q\psi(\alpha_2), \tag{8.41}$$

where

$$\psi(\alpha) = \exp\left(- \frac{\| \mathbf{y} - \mathbf{Hx} \|^2}{\alpha^2 \sigma^2} \right),$$

and q is the mixing ratio. Different values for (α_1, α_2) can be chosen. Note that with $\alpha_1 = 1$ and $\alpha_2 = \infty$, the first and second distributions in (8.41) become the true distribution and the uniform distribution, respectively. That is, the

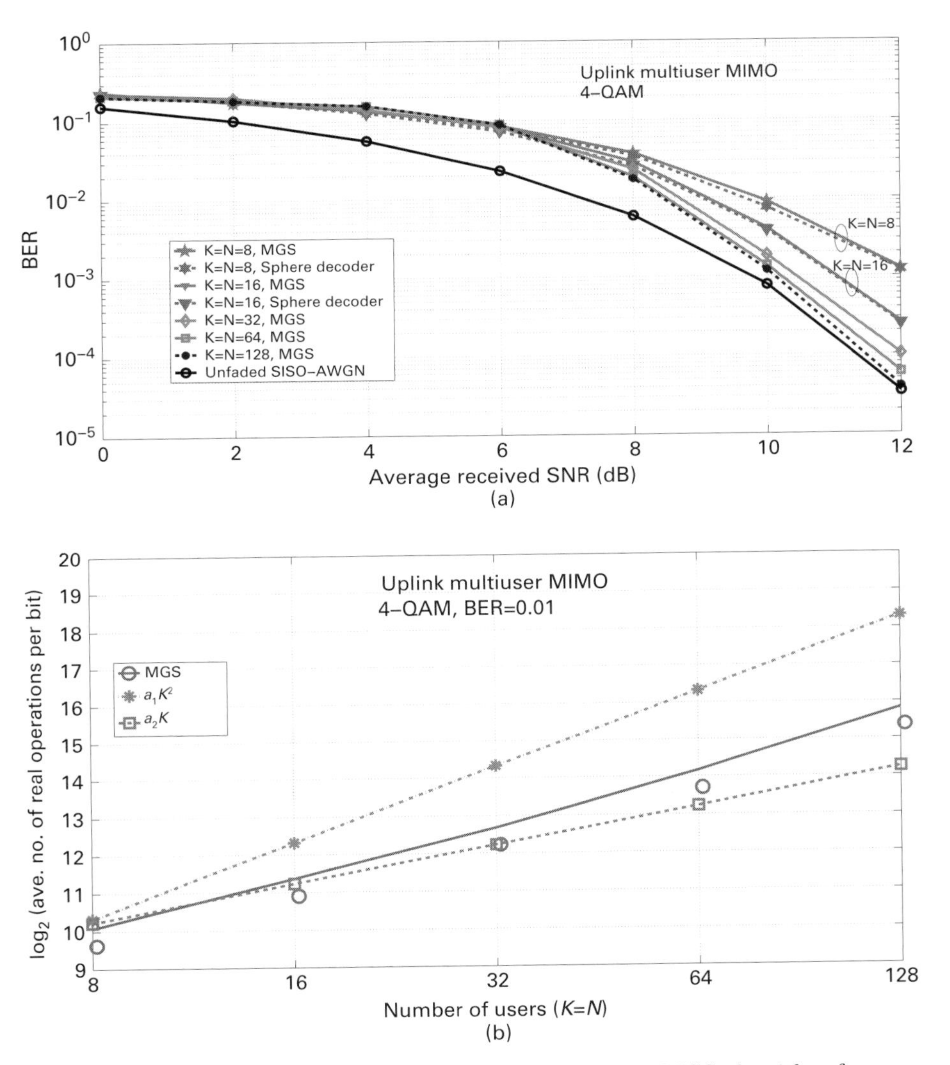

Figure 8.5 (a) BER performance and (b) complexity of the MGS algorithm for $K = N = 8, 16, 32, 64, 128$ and 4-QAM.

grows only quadratically in K (i.e., $O(K^2)$). Because of such low complexity, the MGS algorithm scales easily for $K = N = 32, 64, 128$, whose simulated BER performances are also shown in Fig. 8.5(a). Since SD simulation is prohibitive for such large dimensions, the unfaded SISO AWGN performance is plotted as a lower bound on ML performance for comparison. It can be seen that the MGS detector achieves a performance which is very close to the SISO AWGN performance for large $K = N$, e.g., close to within 0.5 dB at 10^{-3} BER for $K = N = 128$. This illustrates the ability of the MGS detector to achieve near-optimal performance in large-scale multiuser MIMO systems.

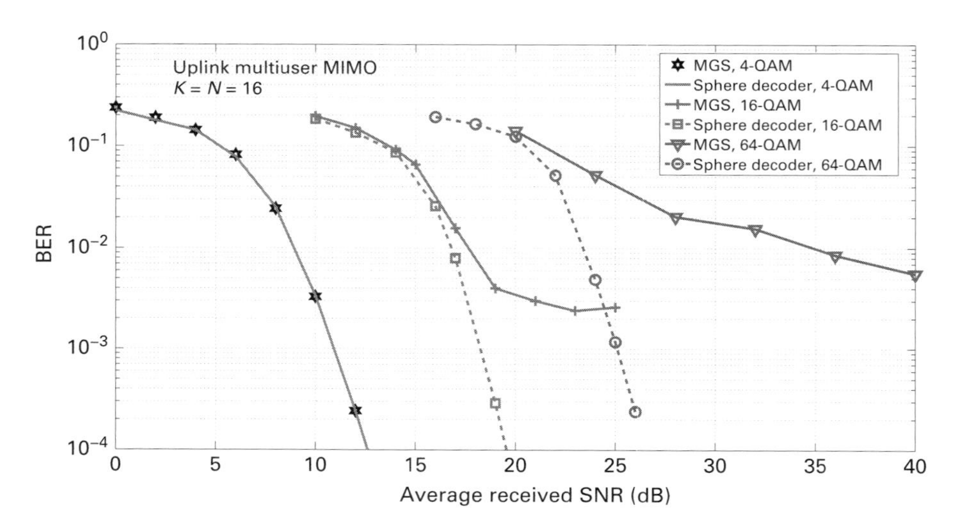

Figure 8.6 Comparison between MGS algorithm performance and SD performance in uplink multiuser MIMO with $K = N = 16$ and 4-/16-/64-QAM.

8.4.8 Multirestart MGS algorithm for higher-order QAM

Although the MGS algorithm is very attractive in terms of both performance and complexity for 4-QAM, its performance for higher-order QAM is far from optimal. This is illustrated in Fig. 8.6, where MGS is seen to achieve SD performance for 4-QAM, whereas for 16-QAM and 64-QAM it performs poorly compared to the SD. This observation motivates the need for ways to improve MGS performance in higher-order QAM. One approach to this is to do parallel explorations (i.e., multiple restarts). Multiple restarts, also referred to as running multiple parallel Gibbs samplers, have been tried with conventional and other variants of MCMC in [11],[23],[24]. It turns out that coupling multiple restarts with MGS is very effective in achieving near-ML performance in large systems with higher-order QAM.

8.4.9 Effect of multiple restarts

Figures 8.7(a) and (b) show the effect of multiple random restarts in MGS and conventional Gibbs sampling algorithms for 4-QAM and 16-QAM, respectively. For a given realization of $\mathbf{x}, \mathbf{H}$, and $\mathbf{n}$, both algorithms were run for three different random initial vectors, and the least ML cost up to the nth iteration as a function of n was plotted. Results are shown for multiuser MIMO with $K = N = 16$ at 11 dB SNR for 4-QAM and 18 dB SNR for 16-QAM (these SNRs give about 10^{-3} BER with sphere decoding for 4-QAM and 16-QAM, respectively). The true ML vector cost (obtained through SD simulation for the same realization) is also plotted. It is seen that MGS achieves much better least ML cost than conventional Gibbs sampling. This is because conventional Gibbs

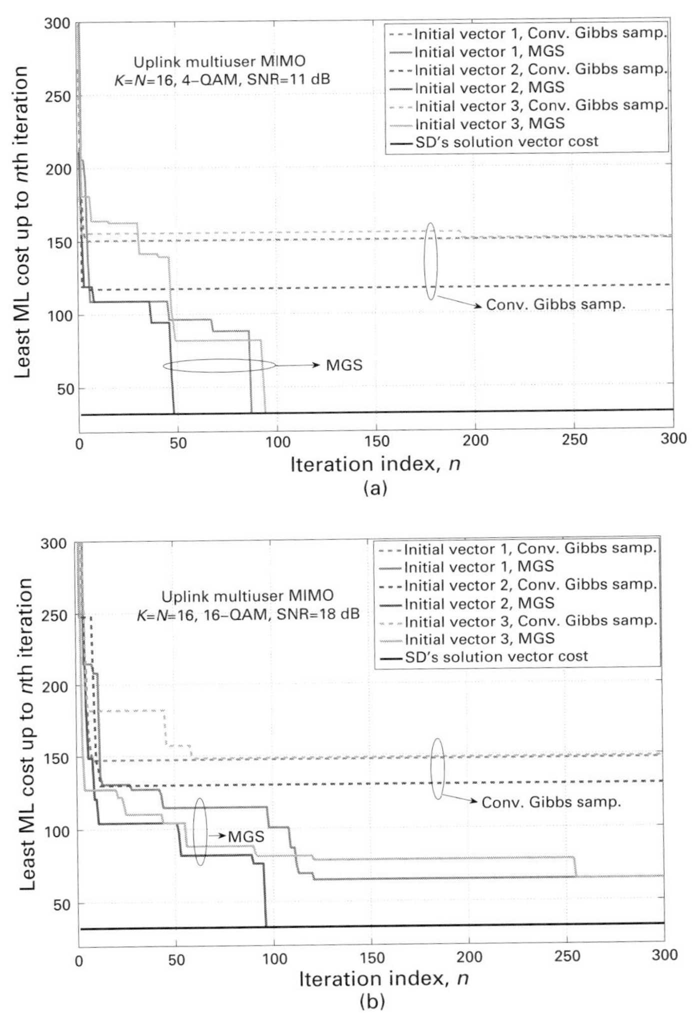

Figure 8.7 Least ML cost up to the nth iteration versus n in conventional Gibbs sampling and MGS for different initial vectors in multiuser MIMO with $K = N = 16$: (a) 4-QAM, SNR=11 dB, (b) 16-QAM, SNR=18 dB.

sampling becomes locked in some state (with very low state transition probability) for a long time without any change in ML cost in subsequent iterations, whereas the mixed sampling strategy is able to exit from such states quickly and give improved ML costs in subsequent iterations. This shows that MGS is preferred over conventional Gibbs sampling. More interestingly, comparing the least ML costs of 4-QAM and 16-QAM (in Figs. 8.7(a) and (b), respectively), it is seen that all the three random initializations could converge almost to the true ML vector cost for 4-QAM within 100 iterations, whereas only initial vector 3 converges to near true ML cost for 16-QAM, while initial vectors 1 and 2 do

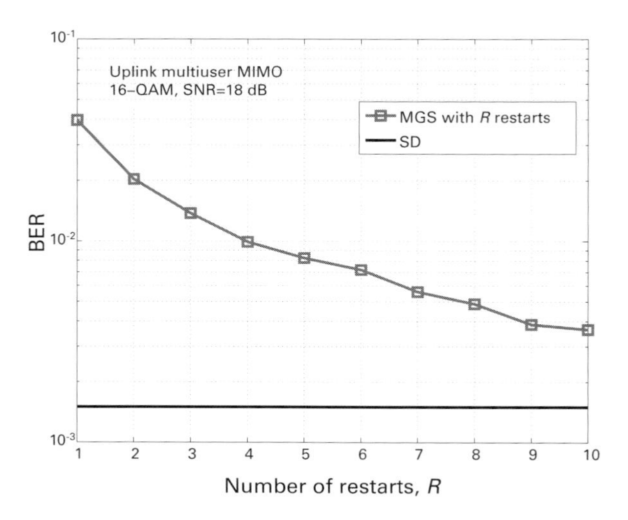

Figure 8.8 BER performance of MGS algorithm as a function of number of restarts in multiuser MIMO with $K = N = 16$ and 16-QAM at SNR = 18 dB.

not. Since any random initialization works well with 4-QAM, MGS is able to achieve near-ML performance without multiple restarts for 4-QAM. However, it can be seen that 16-QAM performance is more sensitive to the initialization, which explains the poor performance of MGS without restarts in higher-order QAM. An MMSE vector can be used as an initial vector, but it is not a good initialization for all channel realizations. This points to the possibility of achieving good initializations through multiple restarts to improve the performance of MGS in higher-order QAM.

8.4.10 MGS with multiple restarts

In MGS with multiple restarts, the basic MGS algorithm is run multiple times, each time with a different random initial vector, and the vector with the least ML cost at the end is chosen as the solution vector. Figure 8.8 shows the improvement in the BER performance of MGS as the number of restarts (R) is increased in multiuser MIMO with $K = N = 16$ and 16-QAM at SNR = 18 dB. Three hundred iterations were used in each restart. It can be observed that, though BER improves with increasing R, a gap still remains between SD performance and MGS performance even with $R = 10$. A larger R could bring the performance of MGS close to that of SD, but at the cost of increased complexity. While a small R results in poor performance, a large R results in high complexity. So, instead of arbitrarily fixing R, a good restart criterion that can significantly enhance the performance without incurring much increase in complexity is needed. One such criterion is described below.

8.4.11 Restart criterion

At the end of each restart, a decision has to be made as to whether to terminate the algorithm or to go for another restart. To do that, one can use

- the standardized ML costs (given by (8.42)) of solution vectors, and
- the number of repetitions of the solution vectors.

The closeness of the ML costs obtained so far to the error-free ML cost in terms of its statistics may allow the algorithm to approach the ML solution. Checking for repetitions allows the number of restarts, and hence the complexity, to be restricted. The minimum standardized ML cost obtained so far and its number of repetitions are used to decide the credibility of the solution. An integer threshold (P) is defined for the best ML cost obtained so far for the purpose of comparison with the number of repetitions. The number of repetitions needed for termination (P, the integer threshold) is chosen as per the following expression [20]:

$$P = \lfloor \max\left(0, c_2 \phi(\tilde{\mathbf{x}})\right) \rfloor + 1, \tag{8.45}$$

where $\tilde{\mathbf{x}}$ is the solution vector with minimum ML cost so far, and c_2 is a constant chosen depending on the QAM size; a larger value of c_2 is chosen for larger QAM size. Now, denoting R_{max} to be the maximum number for restarts, the MGS with multiple restarts algorithm (referred to as the MGS-MR algorithm) can be stated as follows.

- ***Step (1)*** Choose an initial vector.
- ***Step (2)*** Run the basic MGS algorithm in Section 8.4.3.
- ***Step (3)*** Check if R_{max} restarts are completed. If yes, go to Step (5); else go to Step (4).
- ***Step (4)*** For the solution vector with minimum ML cost obtained so far, find the required number of repetitions needed using (8.45). Check if the number of repetitions of this solution vector so far is less than the required number of repetitions computed in Step (4). If yes, go to Step (1), else go to Step (5).
- ***Step (5)*** Output the solution vector with the minimum ML cost so far as the final solution.

Note that the output solution vectors from the MGS and MGS-MR algorithms are hard-decision outputs. Soft-decision values for channel decoding can be generated from these hard-decision output vectors following the method proposed in Section 5.1.4.

8.4.12 Performance and complexity of the MGS-MR algorithm

The BER performance and complexity of the MGS-MR algorithm evaluated through simulations are presented here. The simulation parameters of MGS and MGS-MR are: $c_{min} = 10$, $c_1 = 10 \log_2 M$ (i.e., $c_1 = 20, 40, 60$ for 4-/16-/64-QAM, respectively), MAX-ITER $= 8K\sqrt{M}$, $R_{max} = 50$, $c_2 = 0.5 \log_2 M$,

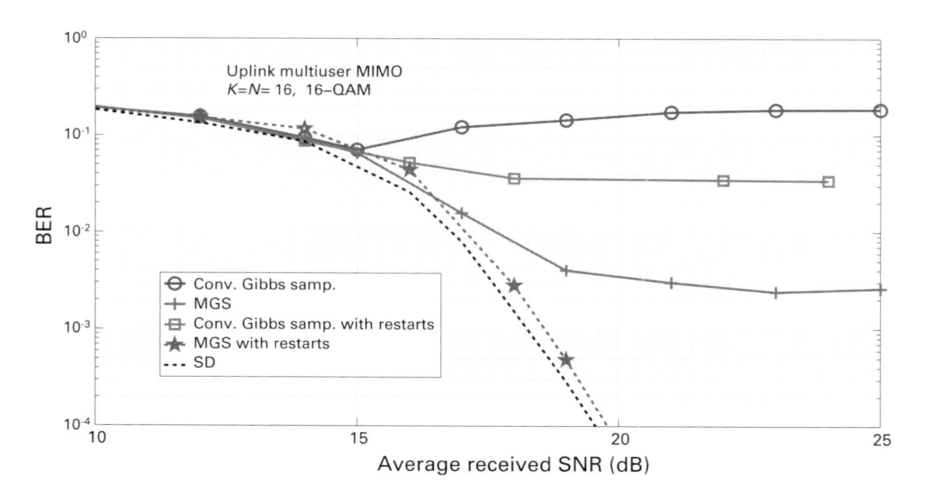

Figure 8.9 BER performance comparison between conventional Gibbs sampling (without and with restarts), MGS (without and with restarts), and SD in uplink multiuser MIMO with $K = N = 16$ and 16-QAM.

$q = 1/2K$. In Fig. 8.9, the BER performance of conventional Gibbs sampling, MGS, MGS-MR and SD in multiuser MIMO with $K = N = 16$ and 16-QAM are compared. In the first start, the MMSE solution vector is used as the initial vector. In the subsequent restarts, random initial vectors are used. For 64-QAM, the mixed sampling is applied only to the one-symbol-away neighbors of the previous iteration index; this helps to reduce complexity in 64-QAM. From Fig. 8.9, it can be seen that the performance of conventional Gibbs sampler, both without and with restarts, is quite poor. That is, using restarts in conventional Gibbs sampling is not of much help. This shows the persistence of the stalling problem. The performance of MGS (without restarts) is better than conventional Gibbs sampling with and without restarts, but its performance is still far from SD performance. This shows that MGS alone (without restarts) is inadequate to alleviate the stalling problem in higher-order QAM. However, the MGS when used with restarts (i.e., MGS-MR) gives strikingly improved performance. In fact, the proposed MGS-MR algorithm achieves almost SD performance (within 0.4 dB at 10^{-3} BER). This points to the important observations that application of any one of the two features, namely, mixture sampling and restarts, to the conventional algorithm is not adequate, and that simultaneous application of both these features is needed to alleviate the stalling problem and achieve near-ML performance in higher-order QAM.

Figure 8.10(a) shows that the MGS-MR algorithm is able to achieve almost SD performance for 4-/16-/64-QAM in multiuser MIMO with $K = N = 16$. Similar performance plots for 4-/16-/64-QAM for $K = N = 32$ are shown in Fig. 8.10(b), where the performance of MGS-MR algorithm is seen to be quite

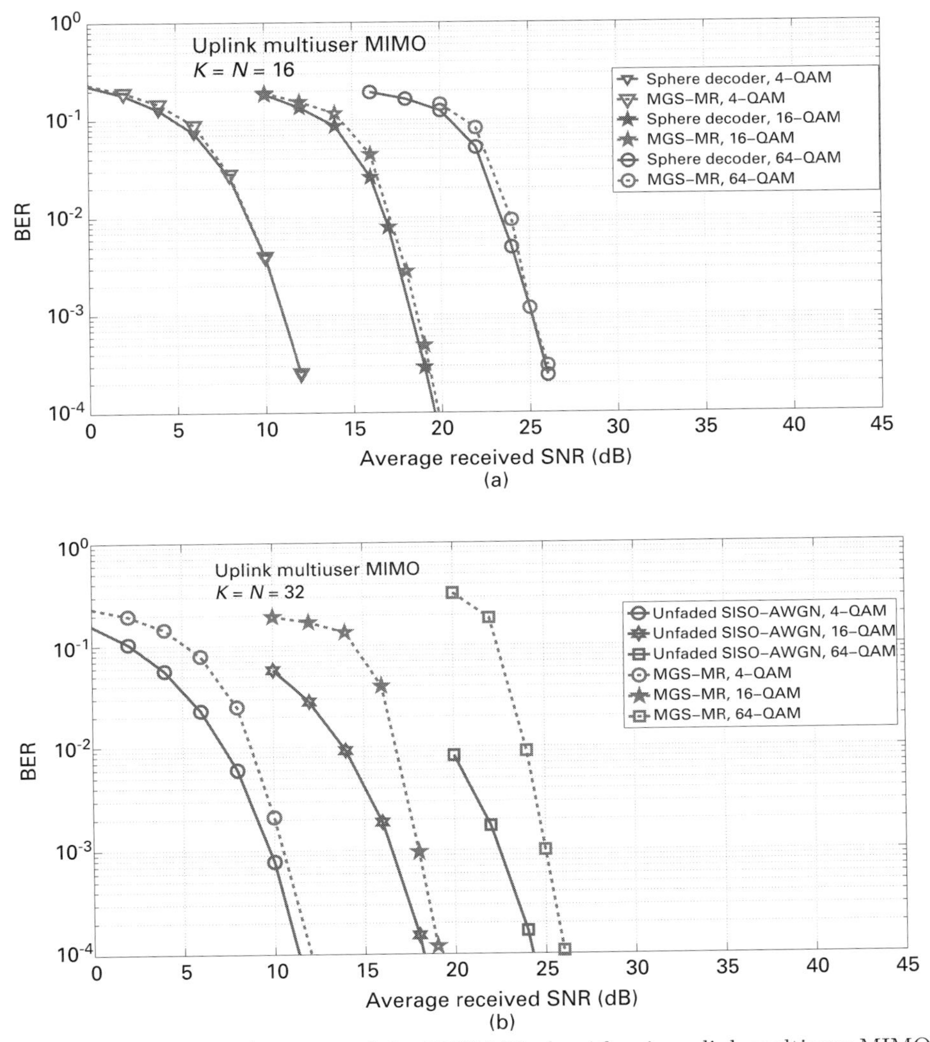

Figure 8.10 BER performance of the MGS-MR algorithm in uplink multiuser MIMO with 4-/16-/64-QAM: (a) $K = N = 16$; (b) $K = N = 32$.

close to unfaded SISO-AWGN performance, which is a lower bound on true ML performance.

8.4.13 Performance of the MGS-MR as a function of loading factor

Figure 8.11 shows BER and complexity plots as a function of the loading factor $\tau = K/N$, the ratio between the number of uplink users K and the number of BS antennas N, for 16-QAM. The BER and complexity plots for the MGS-MR detector and linear detectors, like MF, ZF, and MMSE detectors, are

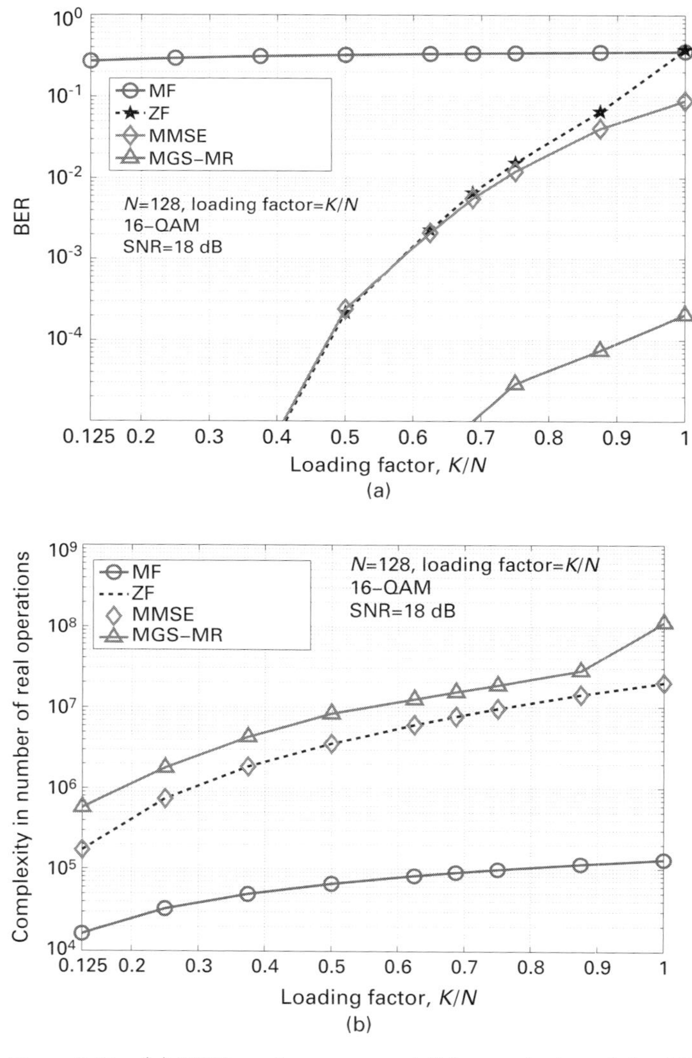

Figure 8.11 (a) BER performance and (b) complexity of the MGS-MR detector in comparison with those of linear (MF, ZF, MMSE) detectors as a function of loading factor $\tau = K/N$. $N = 128$, 16-QAM.

presented and compared. The number of BS antennas N is fixed at 128, and the number of uplink users K is varied from small values up to 128. From Fig. 8.11, it is observed that the MGS-MR detector performs better than the MF, ZF, and MMSE detectors: it is moderately better under low loading factors and significantly better (about 1–2 orders of improved BER) under medium to high loading factors. It is also seen that the complexity increase in MGS-MR detection compared to ZF and MMSE detection is nominal (not orders higher).

References

[1] W. R. Gilks, S. Richardson, and D. J. Spiegelhalter, *Markov Chain Monte Carlo in Practice*. London: Chapman & Hall, 1996.

[2] N. Metropolis and S. Ulam, "The Monte Carlo method," *J. Amer. Statist. Assoc.*, no. 44, pp. 335–341, 1949.

[3] N. Metropolis, A. W. Rosenbluth, M. N. Rosenbluth, A. Teller, and H. Teller, "Equations of state calculations by fast computing machines," *Journal of Chemical Physics*, no. 21, pp. 1087–1091, 1953.

[4] W. K. Hastings, "Monte Carlo sampling methods using Markov chains and their applications," *Biometrika*, no. 57, pp. 97–109, 1970.

[5] S. Chib and E. Greenberg, "Understanding the Metropolis–Hastings algorithm," *American Statistician*, no. 49, pp. 327–335, 1995.

[6] J. Geweke, "Evaluating the accuracy of sampling-based approaches to the calculation of posterior moments," in *Bayesian Statistics*, J. M. Bernardo, J. O. Berger, A. P. Dawid, and A. F. M. Smith, Eds. Oxford, UK: Oxford University Press, 1992, ch. 4, pp. 169–193.

[7] A. E. Raftery and S. Lewis, "How many iterations in the Gibbs sampler?" in *Bayesian Statistics*, J. M. Bernardo, J. O. Berger, A. P. Dawid, and A. F. M. Smith, Eds. Oxford, UK: Oxford University Press, 1992, ch. 4, pp. 763–773.

[8] S. Geman and D. Geman, "Stochastic relaxation, Gibbs distribution and Bayesian restoration of images," *IEEE Trans. Pattern Analysis and Machine Intelligence*, vol. PAMI-6, no. 6, pp. 721–741, Nov. 1984.

[9] R. Chen and T.-H. Li, "Blind restoration of linearly degraded discrete signals by Gibbs sampler," *IEEE Trans. Signal Process.*, vol. 43, no. 9, pp. 2410–2413, Sep. 1995.

[10] R. Chen, J. S. Liu, and X. Wang, "Convergence analyses and comparisons of Markov chain Monte Carlo algorithms in digital communications," *IEEE Trans. Signal Process.*, vol. 50, no. 2, pp. 255–270, Feb. 2002.

[11] B. Farhang-Boroujeny, H. Zhu, and Z. Shi, "Markov chain Monte Carlo algorithms for CDMA and MIMO communication systems," *IEEE Trans. Signal Process.*, vol. 54, no. 5, pp. 1896–1909, May 2006.

[12] S. Henriksen, B. Ninness, and S. R. Weller, "Convergence of Markov-Chain Monte-Carlo approaches to multiuser and MIMO detection," *IEEE J. Sel. Areas in Commun.*, vol. 26, no. 3, pp. 497–505, Apr. 2008.

[13] R. Peng, R.-R. Chen, and B. Farhang-Boroujeny, "Markov chain Monte Carlo detectors for channels with intersymbol interference," *IEEE Trans. Signal Process.*, vol. 58, no. 4, pp. 2206–2217, Apr. 2010.

[14] R. R. Chen, R. Peng, A. Ashikhmin, and B. Farhang-Boroujeny, "Approaching MIMO capacity using bitwise Markov chain Monte Carlo detection," *IEEE Trans. Commun.*, vol. 58, no. 2, pp. 423–428, Nov. 2010.

[15] D. J. C. MacKay, *Information Theory, Inference and Learning Algorithms*. Cambridge, UK: Cambridge University Press, 2003.

[16] M. K. Hanawal and R. Sundaresan, "Randomised attacks on passwords," in *Technical Report TR-PME-2010-11, DRDO-IISc Programme on Advanced Research in Mathematical Engineering*, IISc, Bangalore, 12 February 2010.

Online: http://www.pal.ece.iisc.ernet.in/PAM/docs/techreports/tech_rep10/TR-PME-2010-11.pdf.

[17] ——, "Guessing revisited: A large deviations approach," *IEEE Trans. Inform. Theory*, vol. 57, no. 1, pp. 70–78, Jan. 2011.

[18] E. Arikan, "An inequality on guessing and its application to sequential decoding," *IEEE Trans. Inform. Theory*, vol. 42, no. 1, pp. 99–105, Jan. 1996.

[19] M. Hansen, B. Hassibi, A. G. Dimakis, and W. Xu, "Near-optimal detection in MIMO systems using Gibbs sampling," in *IEEE GLOBECOM'2009*, Hondulu Nov.–Dec. 2009, pp. 1–6.

[20] T. Datta, N. A. Kumar, A. Chockalingam, and B. S. Rajan, "A novel Monte Carlo sampling based receiver for large-scale uplink multiuser MIMO systems," *IEEE Trans. Veh. Tech.*, vol. 62, no. 7, pp. 3019–3038, Sep. 2013.

[21] A. Kumar, S. Chandrasekaran, A. Chockalingam, and B. S. Rajan, "Near-optimal large-MIMO detection using randomized MCMC and randomized search algorithms," in *IEEE ICC'2011*, Kyoto, Jun. 2011, pp. 1–5.

[22] T. Datta, N. A. Kumar, A. Chockalingam, and B. S. Rajan, "A novel MCMC algorithm for near-optimal detection in large-scale uplink mulituser MIMO systems," in *ITA'2012*, San Diego, CA, Feb. 2012, pp. 69–77.

[23] X. Mao, P. Amini, and B. Farhang-Boroujeny, "Markov chain Monte Carlo MIMO detection methods for high signal-to-noise ratio regimes," in *IEEE GLOBECOM'2007*, Washington, DC, Nov. 2008, pp. 3979–3983.

[24] S. Akoum, R. Peng, R.-R. Chen, and B. Farhang-Boroujeny, "Markov chain Monte Carlo MIMO detection methods for high SNR regimes," in *IEEE ICC'2009*, Glasgow, Jun. 2009, pp. 1–5.

9 Channel estimation in large MIMO systems

In the previous chapters, large MIMO detection algorithms were presented under the assumption of perfect knowledge of channel gains at the receiver. However, in practice, these gains are *estimated* at the receiver, either blindly/semi-blindly or through pilot transmissions (training). In FDD systems, channel gains estimated at the receiver are fed back to the transmitter (e.g., for precoding purposes). In TDD systems, where channel reciprocity holds, the transmitter can estimate the channel and use it for precoding. Due to noise and the finite number of pilot symbols used for channel estimation, the channel estimates are not perfect, i.e., there are estimation errors. This has an influence on the achieved capacity of the MIMO channel and the error performance of detection and precoding algorithms. This chapter addresses the effect of imperfect CSI on MIMO capacity, how much training is needed for MIMO channel estimation, and channel estimation algorithms and their performance on the uplink in large-scale multiuser TDD MIMO systems.

9.1 MIMO capacity with imperfect CSI

The capacity of MIMO channels can be degraded if the CSI is not perfect. Gaussian input distribution, which is the capacity achieving distribution in the perfect CSI case, is suboptimal when CSI is imperfect [1],[2]. Lower and upper bounds on the mutual information for iid frequency-flat Rayleigh fading point-to-point MIMO channels have been derived for the imperfect CSI case in [3] assuming Gaussian input, where the MMSE channel estimate is assumed at the receiver and the same channel estimate is assumed to be available at the transmitter. Some key results on MIMO capacity with this imperfect CSI model are summarized below [3].

First, while the Gaussian mutual information saturates with increasing SNR with imperfect CSI, it still increases linearly with the smaller of the number of transmit and receive antennas.

Second, in the perfect CSI case, the capacity gain in knowing the channel at the transmitter decreases with increasing SNR. This is because the optimal input covariance matrix approaches the identity matrix for increasing SNRs, which is also the optimal covariance matrix when the channel is not known at the transmitter. This capacity gain trend, however, changes with imperfect CSI. The capacity gain due to exploitation of CSIT becomes significant with increasing estimation error and does not reduce much at high SNRs. This is because the estimation error causes the effective SNR to saturate, and thereby eliminates the high SNR capacity region where transmitter channel knowledge becomes unimportant.

Third, in terms of optimal power allocation strategies to exploit CSIT, for ergodic capacity, the optimal strategy is modified waterfilling over the spatial and temporal domain. For outage capacity, it is spatial waterfilling and temporally truncated channel inversion. The improvement in ergodic and outage capacities due to spatial power allocation becomes significant with channel estimation errors. Spatial power allocation with imperfect CSI helps even at high SNRs. Performing temporal power adaptation in addition to spatial power allocation enhances the outage capacity significantly but gives only negligible gains in terms of ergodic capacity.

Other key references for the effect of training sequence based channel estimation on the achievable rate and outage capacity are [4]–[8]. In particular, they address the question of how much training is needed in point-to-point frequency-flat [4]–[6] and frequency-selective [7] MIMO wireless links, and in multiuser MIMO links [8]. More on the results in [6], [8] and their relevance to channel estimation and performance in the context of large MIMO systems is presented in the next section.

The effect of imperfect channel knowledge on the achievable rates is also analyzed in [9], where a lower bound on capacity is expressed as a function of the Cramer–Rao bound (CRB). In several works (e.g., [10], [11]), the performance of channel estimation methods is investigated by deriving expressions of the CRB for different pilot symbol/placement designs. The relation between achievable rate and channel CRB established in [9] therefore enables the comparison of achievable rates under different pilot design and placement alternatives. The effect of MMSE and ML channel estimates on the decoding performance of space-time codes is studied in [12].

9.2 How much training is required?

Training based channel estimation and synchronization is a widely adopted approach in MIMO systems, where training/pilot signals are embedded in transmitted data streams (see [13] and references therein). In this approach, often, transmission is divided into a training (or pilot) phase and a data phase. In the training phase, a pilot signal known to the transmitter and the receiver is transmitted in order to get an estimate of the channel at the receiver. The receiver obtains an estimate of the channel, for example, using the ML or MMSE criterion [12]. The estimated channel thus obtained in the training phase is used for detection in the data phase. The estimate obtained in the training phase can be further refined using detected data in the data phase in an iterative manner.

A key question of interest in training based channel estimation in MIMO systems in general, and in large MIMO systems in particular, is "how much time should be spent in training, for a given number of transmit antennas (n_t), number of receive antennas (n_r), length of the channel coherence time (T, in number of channel uses), and average received SNR (γ)." This question is addressed in [6] for point-to-point MIMO wireless links, and in [8] for multiuser TDD MIMO systems.

9.2.1 Point-to-point MIMO training

Too little training leads to inadequate learning and inaccurate estimates of the channel, though the throughput loss due to pilot would be less. On the other hand, too much training means less time for data transmission before the channel changes (i.e., high throughput loss), though the quality of the channel estimate will be good. This tradeoff is captured in the analysis in [6] by computing a lower bound on the capacity of a channel that is learned by training, and maximizing the bound as a function of the received SNR (γ), channel coherence time (T), and number of transmit and receive antennas (n_t, n_r). It has been shown that, when the transmit powers for pilot and data are allowed to vary, the optimal number of training channel uses is equal to the number of transmit antennas. This number is also the smallest length of the training interval that guarantees meaningful estimates of the channel. On the other hand, if the training and data powers are to be made equal, then the optimal number of pilot channel uses can be larger than the number of antennas. Let us elaborate on this a little more.

Assume that the channel remains constant over one coherence interval of T channel uses, and varies iid from one coherent interval to the other (the block fading or quasi-static assumption). Assume that channel estimation (using pilots) and data transmission are to be done within the coherence interval T, and that this procedure repeats in subsequent intervals, each of length T. This can be viewed as transmission in frames, where each frame is of length $T = T_p + T_d$, and T_p, T_d are the duration of the pilot part and the data part in a frame, respectively. Also, the transmit power is split between pilot and data parts so that $\gamma T = \gamma_p T_p + \gamma_d T_d$, where γ_p and γ_d are the pilot and data SNRs, respectively. Define $\beta_p = \gamma_p/\gamma$ and $\beta_d = \gamma_d/\gamma$. Assume that the receiver obtains an estimate of the channel during the pilot phase using MMSE estimation. Let $\hat{\mathbf{H}}$ denote the MMSE estimate of the $n_r \times n_t$ channel matrix $\mathbf{H}$ and $\sigma^2_{\hat{\mathbf{H}}}$ denote the variance of the estimate. For this system model, a lower bound on the ergodic capacity is given by [6]:

$$C \geq \mathbb{E}\left[\frac{T - T_p}{T} \operatorname{logdet}\left(\mathbf{I}_{n_t} + \frac{\gamma^2 \beta_d \beta_p T_p}{n_t(1+\gamma\beta_d) + \gamma\beta_p T_p} \frac{\hat{\mathbf{H}}\hat{\mathbf{H}}^H}{n_t \sigma^2_{\hat{\mathbf{H}}}}\right)\right]. \tag{9.1}$$

The optimal length of the training interval T_p is n_t for all SNRs γ and coherence times T, if β_p and β_d are allowed to vary, whereas it can be more than n_t if β_p

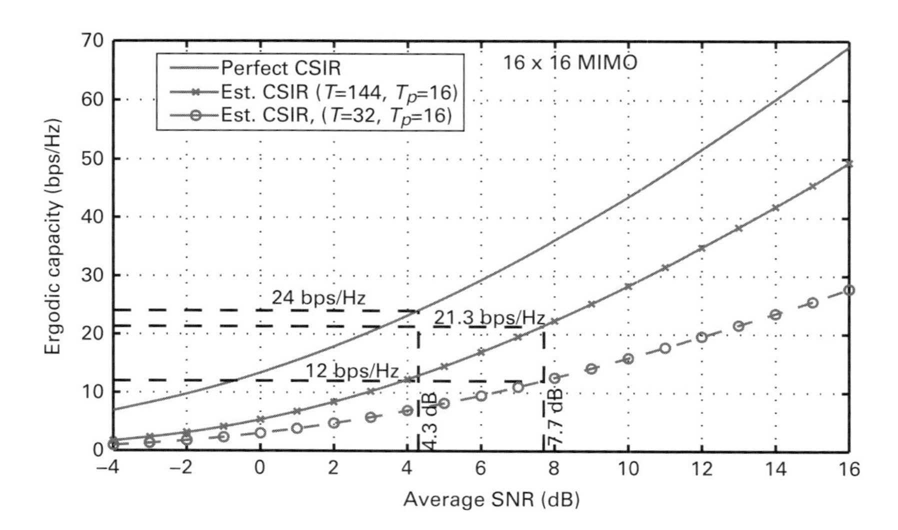

Figure 9.1 Lower bound on the ergodic capacity of 16×16 MIMO channel with (i) estimated CSIR, $T = 32, T_p = 16, \beta_p = \beta_d = 1$, ($ii$) estimated CSIR, $T = 144$, $T_p = 16$, $\beta_p = \beta_d = 1$, and (iii) perfect CSIR.

and β_d are made equal (Fig. 3 in [6]). In the latter case, the trend is that the optimal training length for a given n_t, n_r increases with increasing T and decreasing γ, such that as $\gamma \to 0$ the length increases until it reaches $T/2$.

Figure 9.1 shows the lower bound on the ergodic capacity with estimated CSI (9.1) evaluated for a 16×16 MIMO channel with (i) $T = 144, T_p = 16$ (large coherence time; a slowly fading channel) and (ii) $T = 32, T_p = 16$ (short coherence time; a relatively fast fading channel). Compared to the perfect CSI case, the capacity degradation in case (ii) is more than in case (i), i.e., for a given n_t, the larger the value of T the smaller the capacity/throughput loss will be compared to perfect CSI capacity. This implies that large MIMO systems with large n_t benefit from large coherence times (e.g., in slow fading as witnessed in no mobility/low mobility scenarios). Even after accounting for the throughput loss due to training overhead, the spectral efficiencies achieved with large n_t are in the tens of bps/Hz range, which are significantly higher than the spectral efficiencies achieved in current wireless systems.

Optimum n_t for a given T, n_r, γ

For a given SNR γ, coherence time T, and number of receive antennas n_r, there is an optimum number of transmit antennas n_t that maximizes the capacity. For small n_t, capacity is small because of fewer antennas. For large n_t (e.g., closer to T), the capacity is again small because most of the coherence time is spent for training. Figure 9.2 illustrates this tradeoff, where the variation of the capacity lower bound as a function of n_t is shown for a fixed $T = 100$, $\gamma = 18$ dB, and

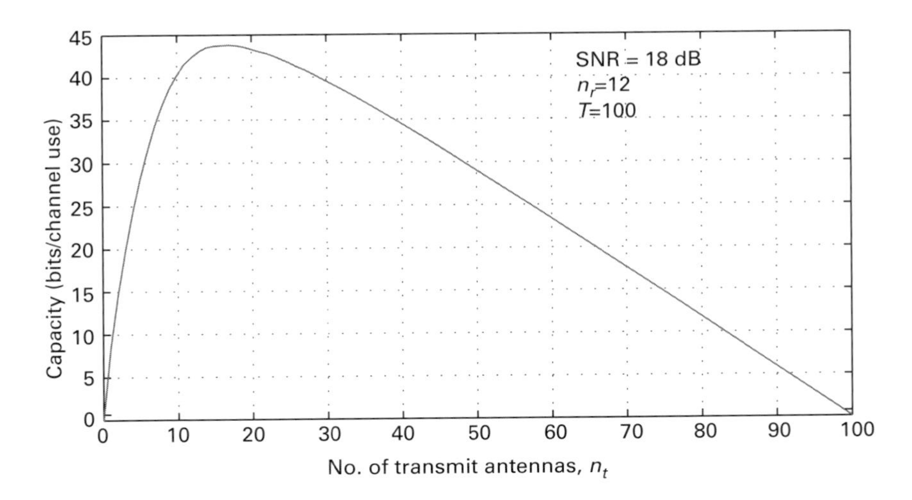

Figure 9.2 Variation of capacity lower bound as a function of n_t for a given $n_r = 12$, $T = 100$, and $\gamma = 18$ dB.

$n_r = 12$. It is seen that the capacity increases initially for increasing n_t, but starts diminishing beyond $n_t = 15$ and reaches zero when $n_t = T = 100$ (i.e., when the entire coherence time is used for training). Such a behavior, namely, fewer transmit antennas being optimum, has also been captured in [14] through simulation of practical MIMO system designs. In [14], it has been shown that a MIMO system with $n_t = 12$ achieves a higher spectral efficiency and better coded BER performance than a system with $n_t = 16$ for $T = 48$, $n_r = 16$ and training based channel estimation.

9.2.2 Multiuser MIMO training

A natural large MIMO system architecture in a multiuser setting is to have a BS with tens to hundreds of antennas, and a similar (or smaller) number of users each having one (or more) antenna. If such a system with a large number of BS antennas and users employs FDD, the feedback load to pass on the channel vectors estimated by all the users to the BS is very high. On the other hand, if TDD is employed, and if channel reciprocity holds,[1] then the channel vectors can be estimated at the BS itself through the pilots transmitted by the users, thereby eliminating the need for feedback to acquire CSIT. Such a TDD system is considered in [8], and the question of how much training is required in such a large multiuser MIMO system is addressed. For a given coherence time (T), number of BS antennas (N), uplink and downlink SINRs, and linear precoding at the BS, the optimum number of single-antenna users to serve simultaneously and the optimum number of uplink pilot symbols are obtained by maximizing

[1] In reciprocal MIMO channels, the downlink channel matrix will be the transpose of the uplink channel matrix.

a lower bound on the net sum-rate on the downlink. It has been shown that, given a large number of antennas at the BS ($N > 16$), even with short coherence intervals ($T = 10$) and low SINRs (0 dB on the downlink and -10 dB on the uplink), it is both possible as well as advantageous to learn the channel (with the pilot length equal to the number of users) and serve several users simultaneously.

In summary, a key observation to make here in the context of large MIMO systems is that, although the potential for capacity increase with an increasing number of antennas is diminished by training, the spectral efficiency achieved with the optimum number of antennas is still high (e.g., about 45 bps/Hz with $n_t = 15$ in Fig. 9.2 and about 20 bps/Hz with $N = 16, K = 1$ in Fig. 5 of [8]). Considering that the spectral efficiencies in current systems are much less than 10 bps/Hz, the large MIMO system approach is an attractive and viable approach to achieving a quantum jump in the efficiency of spectrum usage in future systems and standards.

9.3 Large multiuser MIMO systems

In this section, channel estimation schemes and their performance in frequency-flat and frequency-selective fading in large multiuser TDD MIMO systems are presented.

9.3.1 System model

Consider a large-scale multiuser MIMO system on the uplink consisting of a BS with N receive antennas and K uplink users, each having one transmit antenna, $K \leq N$ (as shown in Fig. 8.1). N and K are in the range of tens to hundreds. All users transmit symbols from a modulation alphabet $\mathbb{B}$. It is assumed that synchronization has been carried out, and that the sampled baseband signals are available at the BS receiver.

9.3.2 Iterative channel estimation/detection in frequency-flat fading

Let $x_k \in \mathbb{B}$ denote the symbol transmitted from user k. Let $\mathbf{x}_c = [x_1, x_2, \ldots, x_K]^T$ denote the vector comprising the symbols transmitted simultaneously by all users in one channel use. Let $\mathbf{H}_c \in \mathbb{C}^{N \times K}$, given by $\mathbf{H}_c = [\mathbf{h}_1, \mathbf{h}_2, \ldots, \mathbf{h}_K]$, denote the channel gain matrix, where $\mathbf{h}_k = [h_{1k}, h_{2k}, \ldots, h_{Nk}]^T$ is the channel gain vector from user k to the BS, and h_{jk} denotes the channel gain from the kth user to the jth receive antenna at the BS. Assuming rich scattering and adequate spatial separation between the BS antenna elements, $h_{jk}, \forall j$ are assumed to be independent Gaussian with zero mean and σ_k^2 variance such that $\sum_k \sigma_k^2 = K$. The imbalance in received powers from different users is modeled by σ_k^2 , and $\sigma_k^2 = 1$ corresponds to the perfect power control scenario. The complex-valued system model is given by

$$\mathbf{y}_c = \mathbf{H}_c\mathbf{x}_c + \mathbf{n}_c. \tag{9.2}$$

The real-valued system model corresponding to (9.2) is given by

$$\mathbf{y} = \mathbf{H}\mathbf{x} + \mathbf{n}, \tag{9.3}$$

as defined in Section 8.4.1.

Frame structure

In order to detect the transmitted data vector $\mathbf{x}$, knowledge of the channel matrix $\mathbf{H}$ is needed. The channel matrix is estimated based on a pilot based channel estimation scheme, where transmission is carried out in frames, with each frame consisting of several blocks as shown in Fig. 9.3. A slow fading channel (typical with no/low mobility users) is assumed, where the channel is assumed to be constant over one frame duration. Each frame consists of a pilot block for the purpose of initial channel estimation, followed by Q data blocks. The pilot block consists of K channel uses in which a K-length pilot symbol vector comprising pilot symbols transmitted from K users (one pilot symbol per user) is received by N receive antennas at the BS. Each data block consists of K channel uses, where K information symbol vectors, each of length K (one data symbol from each user) are transmitted. Taking both pilot and data channel uses into account, the total number of channel uses per frame is $(Q+1)K$. Data blocks are detected using any of the known large MIMO detection algorithms (e.g., the MGS-MR algorithm presented in Chapter 8) using an initial channel estimate. The detected data blocks are then iteratively used to refine the channel estimates during the data phase employing a Gibbs sampling based channel estimation algorithm described below.

Initial channel estimate during pilot phase

Let $\mathbf{x}_P^k = [x_P^k(0), x_P^k(1), \ldots, x_P^k(K-1)]$ denote the the pilot symbol vector transmitted from user k in K channel uses in a frame. Let $\mathbf{X}_P = [(\mathbf{x}_P^1)^T, (\mathbf{x}_P^2)^T, \ldots, (\mathbf{x}_P^K)^T]^T$ denote the $K \times K$ pilot matrix formed by the pilot symbol vectors transmitted by all the users in the pilot phase. The received signal matrix at the BS, $\mathbf{Y}_P$, of size $N \times K$ is given by

$$\mathbf{Y}_P = \mathbf{H}_c\mathbf{X}_P + \mathbf{N}_P, \tag{9.4}$$

where $\mathbf{N}_P$ is the $N \times K$ noise matrix at the BS. The following pilot sequence is used:

$$\mathbf{x}_P^k = [\mathbf{0}_{(k-1)\times 1} \quad p \quad \mathbf{0}_{(K-k)\times 1}], \tag{9.5}$$

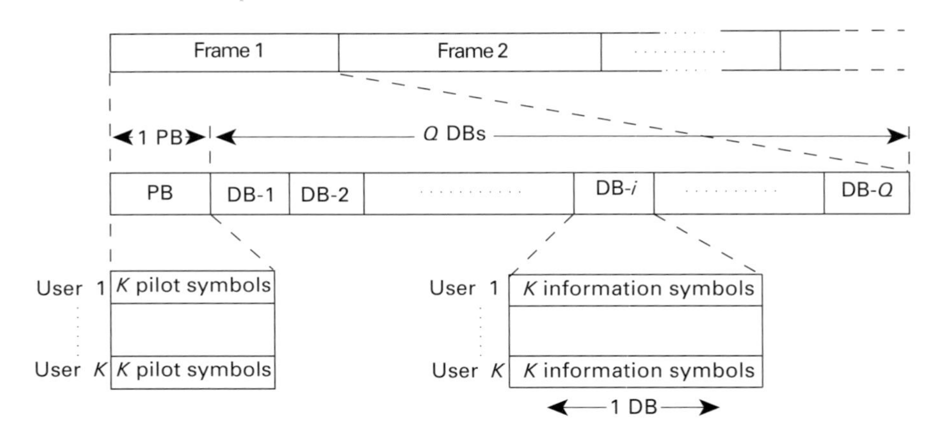

Figure 9.3 Frame structure for uplink multiuser MIMO system in frequency-flat fading (PB: pilot block; DB: data block).

where p is chosen to be $p = \sqrt{KE_s}$, and E_s is the average symbol energy. Using the scaled identity nature of $\mathbf{x}_P$, an initial channel estimate $\widehat{\mathbf{H}}_c$ is obtained as

$$\widehat{\mathbf{H}}_c = \mathbf{Y}_P/p. \tag{9.6}$$

Data detection using initial channel estimate

Let $\mathbf{x}_i^k = [x_i^k(0), x_i^k(1), \ldots, x_i^k(K-1)]$ denote the data symbol vector transmitted from user k in K channel uses in the ith data block in a frame. Let $\mathbf{X}_i = [(\mathbf{x}_i^1)^T, (\mathbf{x}_i^2)^T, \ldots, (\mathbf{x}_i^K)^T]^T$ denote the $K \times K$ data matrix formed by the data symbol vectors transmitted by the users in the ith data block during the data phase, $i = 1, 2, \ldots, Q$. The received signal matrix at the BS in the ith data block, $\mathbf{Y}_i$ of size $N \times K$, is given by

$$\mathbf{Y}_i = \mathbf{H}_c\mathbf{X}_i + \mathbf{N}_i, \tag{9.7}$$

where $\mathbf{N}_i$ is the $N \times K$ noise matrix at the BS during the ith data block. Detection is performed on a vector by vector basis using the independence of data symbols transmitted by the users. Let $\mathbf{y}_i^{(t)}$ denote the tth column of $\mathbf{Y}_i$, $t = 0, 2, \ldots, K-1$. Denoting the tth column of $\mathbf{X}_i$ as $\mathbf{x}_i^{(t)} = [x_i^1(t), x_i^2(t), \ldots, x_i^K(t)]^T$, the system equation (9.4) can be rewritten as

$$\mathbf{y}_i^{(t)} = \mathbf{H}_c\mathbf{x}_i^{(t)} + \mathbf{n}_i^{(t)}, \tag{9.8}$$

where $\mathbf{n}_i^{(t)}$ is the tth column of $\mathbf{N}_i$. The initial channel estimate $\widehat{\mathbf{H}}_c$ obtained from (9.6) is used to detect the transmitted data vectors using, say, the MGS-MR algorithm presented in Chapter 8.

From (9.4) and (9.6), it is observed that $\widehat{\mathbf{H}}_c = \mathbf{H}_c + \mathbf{N}_P/p$. This knowledge about imperfection of channel estimates is used to calculate the statistics of the error-free ML cost required in the MGS-MR algorithm. In the case of perfect channel knowledge, the error-free ML cost is nothing but $\|\mathbf{n}^2\|$. In the case of imperfect channel knowledge at the receiver, at channel use t,

$$\|\mathbf{y}_i^{(t)} - \widehat{\mathbf{H}}_c\mathbf{x}_i^{(t)}\|^2 = \|\mathbf{n}_i^{(t)} - \mathbf{N}_P\mathbf{x}_i^{(t)}/p\|^2.$$

Each entry of the vector $\mathbf{n}_i^{(t)} - \mathbf{N}_P\mathbf{x}_i^{(t)}/p$ has mean zero and variance $2\sigma^2$. Using this knowledge at the receiver, the transmitted data are detected using the MGS-MR algorithm and $\widehat{\mathbf{x}}_i^{(t)}$ is obtained. Let the detected data matrix in data block i be denoted $\widehat{\mathbf{X}}_i = [\widehat{\mathbf{x}}_i^{(0)}, \widehat{\mathbf{x}}_i^{(1)}, \ldots, \widehat{\mathbf{x}}_i^{(K-1)}]$.

Channel estimation using Gibbs sampling in data phase

Let $\mathbf{Y}_{tot} = [\mathbf{Y}_P\,\mathbf{Y}_1 \cdots \mathbf{Y}_Q]$, $\mathbf{X}_{tot} = [\mathbf{X}_P\,\mathbf{X}_1 \cdots \mathbf{X}_Q]$, $\mathbf{N}_{tot} = [\mathbf{N}_P\,\mathbf{N}_1 \cdots \mathbf{N}_Q]$ denote the matrices corresponding to one full frame. $\mathbf{Y}_{tot}$ can be expressed as

$$\mathbf{Y}_{tot} = \mathbf{H}_c\mathbf{X}_{tot} + \mathbf{N}_{tot}. \tag{9.9}$$

This system model corresponding to the full frame is converted into a real-valued system model. That is, (9.9) can be written in the form

$$\mathbf{Y} = \mathbf{H}\mathbf{X} + \mathbf{N}, \tag{9.10}$$

where

$$\mathbf{Y} = \begin{bmatrix} \Re(\mathbf{Y}_{tot}) & -\Im(\mathbf{Y}_{tot}) \\ \Im(\mathbf{Y}_{tot}) & \Re(\mathbf{Y}_{tot}) \end{bmatrix}, \quad \mathbf{H} = \begin{bmatrix} \Re(\mathbf{H}_c) & -\Im(\mathbf{H}_c) \\ \Im(\mathbf{H}_c) & \Re(\mathbf{H}_c) \end{bmatrix},$$

$$\mathbf{X} = \begin{bmatrix} \Re(\mathbf{X}_{tot}) & -\Im(\mathbf{X}_{tot}) \\ \Im(\mathbf{X}_{tot}) & \Re(\mathbf{X}_{tot}) \end{bmatrix}, \quad \mathbf{N} = \begin{bmatrix} \Re(\mathbf{N}_{tot}) & -\Im(\mathbf{N}_{tot}) \\ \Im(\mathbf{N}_{tot}) & \Re(\mathbf{N}_{tot}) \end{bmatrix}.$$

Equation (9.10) can be written as

$$\mathbf{Y}^T = \mathbf{X}^T\mathbf{H}^T + \mathbf{N}^T. \tag{9.11}$$

Vectorizing the matrices $\mathbf{Y}^T$, $\mathbf{H}^T$, and $\mathbf{N}^T$, define

$$\mathbf{r} \stackrel{\triangle}{=} \text{vec}(\mathbf{Y}^T), \quad \mathbf{g} \stackrel{\triangle}{=} \text{vec}(\mathbf{H}^T), \quad \mathbf{z} \stackrel{\triangle}{=} \text{vec}(\mathbf{N}^T).$$

With the above definitions, (9.11) can be written in vector form as

$$\mathbf{r} = \underbrace{\mathbf{I}_{2N} \otimes \mathbf{X}^T}_{\stackrel{\triangle}{=} \mathbf{S}}\, \mathbf{g} + \mathbf{z}. \tag{9.12}$$

Now, the goal is to estimate $\mathbf{g}$ knowing $\mathbf{r}$, the estimate of $\mathbf{S}$, and the statistics of $\mathbf{z}$ using Gibbs sampling. The estimate of $\mathbf{S}$ is obtained as

$$\widehat{\mathbf{S}} = \mathbf{I}_{2N} \otimes \widehat{\mathbf{X}}^T,$$

where

$$\widehat{\mathbf{X}} = \begin{bmatrix} \Re(\widehat{\mathbf{X}}_{tot}) & -\Im(\widehat{\mathbf{X}}_{tot}) \\ \Im(\widehat{\mathbf{X}}_{tot}) & \Re(\widehat{\mathbf{X}}_{tot}) \end{bmatrix} \text{ and } \widehat{\mathbf{X}}_{tot} = [\mathbf{X}_{\mathrm{P}}\, \widehat{\mathbf{X}}_1\, \widehat{\mathbf{X}}_2 \ldots \widehat{\mathbf{X}}_Q].$$

The initial vector for the algorithm is obtained as

$$\widehat{\mathbf{g}}^{(0)} = \mathrm{vec}(\widehat{\mathbf{H}}^T), \tag{9.13}$$

where

$$\widehat{\mathbf{H}} = \begin{bmatrix} \Re(\widehat{\mathbf{H}}_c) & -\Im(\widehat{\mathbf{H}}_c) \\ \Im(\widehat{\mathbf{H}}_c) & \Re(\widehat{\mathbf{H}}_c) \end{bmatrix}. \tag{9.14}$$

Gibbs sampling based estimation

The vector $\mathbf{g}$ is of length $4KN \times 1$. To estimate $\mathbf{g}$, the algorithm starts with an initial estimate, takes samples from the conditional distribution of each coordinate in $\mathbf{g}$, and updates the estimate. This is carried out for a certain number of iterations. At the end of the iterations, a weighted average of the previous and current estimates is given as the output.

Let the ith coordinate in $\mathbf{g}$ be denoted by g_i, and let $\mathbf{g}_{-i}$ denote all elements in $\mathbf{g}$ other than the ith element. Let $\widehat{\mathbf{s}}_q$ denote the qth column of $\widehat{\mathbf{S}}$. The conditional probability distribution for the ith coordinate is given by

$$p\left(g_i|\mathbf{r}, \widehat{\mathbf{S}}, \mathbf{g}_{-i}\right) \propto p(g_i).\, p\left(\mathbf{r}|g_i, \widehat{\mathbf{S}}, \mathbf{g}_{-i}\right) \tag{9.15}$$

$$\propto \exp\left(-|g_i|^2\right) \exp\left(-\frac{||\mathbf{r} - \sum_{q=1,q\neq i}^{4KN} g_q\widehat{\mathbf{s}}_q - g_i\widehat{\mathbf{s}}_i||^2}{\sigma^2}\right) \tag{9.16}$$

$$= \exp\left(-|g_i|^2 - \frac{||\widetilde{\mathbf{r}}^{(i)} - g_i\widehat{\mathbf{s}}_i||^2}{\sigma^2}\right) \tag{9.17}$$

$$= \exp\left(-\frac{||\bar{\mathbf{r}}^{(i)} - g_i\bar{\mathbf{s}}_i||^2}{\sigma^2}\right), \tag{9.18}$$

where $\widetilde{\mathbf{r}}^{(i)} = \mathbf{r} - \sum_{q=1,q\neq i}^{4KN} g_q\widehat{\mathbf{s}}_q$, $\bar{\mathbf{r}}^{(i)} = [\widetilde{\mathbf{r}}^{(i)}, 0]^T$, and $\bar{\mathbf{s}}_i = [\widehat{\mathbf{s}}_i, \sigma]^T$. The quantity $||\bar{\mathbf{r}}^{(i)} - g_i\bar{\mathbf{s}}_i||^2$ in (9.18) is minimized for $g_i = \left(\bar{\mathbf{r}}^{(i)}\right)^T \bar{\mathbf{s}}_i/||\bar{\mathbf{s}}_i||^2$. Therefore,

$$||\bar{\mathbf{r}}^{(i)} - g_i\bar{\mathbf{s}}_i||^2 = ||\bar{\mathbf{r}}^{(i)} - \left(\frac{\left(\bar{\mathbf{r}}^{(i)}\right)^T \bar{\mathbf{s}}_i}{||\bar{\mathbf{s}}_i||^2} + g_i - \frac{\left(\bar{\mathbf{r}}^{(i)}\right)^T \bar{\mathbf{s}}_i}{||\bar{\mathbf{s}}_i||^2}\right)\bar{\mathbf{s}}_i||^2$$

$$= ||\bar{\mathbf{r}}^{(i)} - \frac{\left(\bar{\mathbf{r}}^{(i)}\right)^T \bar{\mathbf{s}}_i}{||\bar{\mathbf{s}}_i||^2}\bar{\mathbf{s}}_i||^2 + \left(g_i - \frac{\left(\bar{\mathbf{r}}^{(i)}\right)^T \bar{\mathbf{s}}_i}{||\bar{\mathbf{s}}_i||^2}\right)^2 ||\bar{\mathbf{s}}_i||^2. \tag{9.19}$$

Hence,

$$p\left(g_i|\mathbf{r}, \widehat{\mathbf{S}}, \mathbf{g}_{-i}\right) \propto \exp\left(-\frac{\left(g_i - \frac{\left(\bar{\mathbf{r}}^{(i)}\right)^T \bar{\mathbf{s}}_i}{||\bar{\mathbf{s}}_i||^2}\right)^2}{\frac{\sigma^2}{||\bar{\mathbf{s}}_i||^2}}\right), \tag{9.20}$$

which is Gaussian with mean $\mu_{g_i} = \left(\bar{\mathbf{r}}^{(i)}\right)^T \bar{\mathbf{s}}_i / \|\bar{\mathbf{s}}_i\|^2$, and variance $\sigma^2_{g_i} = \sigma^2/2\|\mathbf{s}_i\|^2$. Let MAX denote the number of iterations. In each iteration, for each coordinate, the probability distribution specified by its mean and variance has to be calculated to draw samples. Let the mean and variance in the rth iteration and the ith coordinate be denoted as $\mu^{(r)}_{g_i}$ and $\sigma^{2\ (r)}_{g_i}$, respectively, where $r = 1, 2, \ldots, MAX$ and $i = 1, 2, \ldots, 4KN$. Use $\widehat{\mathbf{g}}^{(0)}$in (9.13), which is the estimate from the pilot phase, as the initial estimate. In the rth iteration, $\widehat{\mathbf{g}}^{(r)}$ is obtained from $\widehat{\mathbf{g}}^{(r-1)}$ as follows:

- Take $\widehat{\mathbf{g}}^{(r)} = \widehat{\mathbf{g}}^{(r-1)}$.
- Update the ith coordinate of $\widehat{\mathbf{g}}^{(r)}$ by sampling from $\mathcal{N}\left(\mu^{(r)}_{g_i}, \sigma^{2\ (r)}_{g_i}\right)$ for all i. Let $\widehat{g}^{(r)}_i$ denote the updated ith coordinate of $\mathbf{g}^{(r)}$.
- Compute weights $\alpha^{(r)}_i = \exp\left(-\left(\widehat{g}^{(r)}_i - \mu^{(r)}_{g_i}\right)^2 / \, 2\sigma^{2\ (r)}_{g_i}\right)$ for all i. This gives more weight to samples closer to the mean.

After MAX iterations, compute the final estimate of the ith coordinate, denoted by g^*_i, to be the following weighted sum of the estimates from previous and current iterations:

$$g^*_i = \frac{\sum\limits_{r=1}^{MAX} \alpha^{(r)}_i \widehat{g}^{(r)}_i}{\sum\limits_{r=1}^{MAX} \alpha^{(r)}_i}. \tag{9.21}$$

Finally, the updated $2N \times 2K$ channel estimate $\widehat{\mathbf{H}}$ is obtained by restructuring $\mathbf{g}^* = [g^*_1, g^*_2, \ldots, g^*_{4KN}]^T$ as

$$\widehat{\mathbf{H}}(p, q) = g^*_n, \quad p = 1, 2, \ldots, 2N, \quad q = 1, 2, \ldots, 2K, \tag{9.22}$$

where $n = 2N(p-1)+q$ and $\widehat{\mathbf{H}}(p, q)$ denotes the element in the pth row and qth column of $\widehat{\mathbf{H}}$. A listing of the above Gibbs sampling based channel estimation algorithm is given in Algorithm 3.

The matrix $\widehat{\mathbf{H}}$ obtained thus is used for data detection using the MGS-MR algorithm. This ends one iteration between channel estimation and detection. The detected data matrix is fed back for channel estimation in the next iteration, whose output is then used to detect the data matrix again. This iterative channel estimation and detection procedure is carried out for a certain number of iterations.

Performance results

In Fig. 9.4(a), the mean square error (MSE) performance of the iterative channel estimation/detection scheme using Gibbs sampling based channel estimation and MGR-MR based detection with 4-QAM for $K = N = 128$ and $Q = 9$ is shown. In the simulations, the MGS-MR algorithm parameter values used are the same as in Section 8.4.12. For the channel estimation algorithm, the value of MAX used is 2. The MSEs of the initial channel estimate, and the channel

Algorithm 3. Channel estimation using Gibbs sampling

1. **input:** $\mathbf{r}$, $\widehat{\mathbf{S}}$, σ^2, $\widehat{\mathbf{g}}^{(0)}$: initial vector $\in \mathbb{R}^{4KN}$; MAX: max. # iterations;
2. $r = 1$; $g^*(0) = \widehat{\mathbf{g}}^{(0)}$; $\alpha_i^{(0)} = 0, \forall i = 1, 2, \ldots, 4KN$;
3. **while** $r < MAX$ **do**
4. $\quad \widehat{\mathbf{g}}^{(r)} = \widehat{\mathbf{g}}^{(r-1)}$;
5. $\quad \widetilde{\mathbf{r}}^* = \mathbf{r} - \widehat{\mathbf{S}}\widehat{\mathbf{g}}^{(r)}$;
6. $\quad$ **for** $i = 1$ to $4KN$ **do**
7. $\quad\quad$ Compute $\widetilde{\mathbf{r}}^{(i)} = \widetilde{\mathbf{r}}^* + \widehat{g}_i^{(r)}\widehat{\mathbf{s}}_i$, $\bar{\mathbf{r}}^{(i)} = [\widetilde{\mathbf{r}}^{(i)}, 0]^T$, and $\bar{\mathbf{s}}_i = [\widehat{\mathbf{s}}_i, \sigma]^T$;
8. $\quad\quad$ Compute $\mu_{g_i}^{(r)} = \left(\bar{\mathbf{r}}^{(i)}\right)^T \bar{\mathbf{s}}_i \,/\, \|\bar{\mathbf{s}}_i\|^2$ and $\sigma_{g_i}^{2\,(r)} = \sigma^2 \,/\, 2\|\mathbf{s}_i\|^2$;
9. $\quad\quad$ Sample $\widehat{g}_i^{(r)} \sim \mathcal{N}\left(\mu_{g_i}^{(r)}, \sigma_{g_i}^{2\,(r)}\right)$;
10. $\quad\quad \widetilde{\mathbf{r}}^* = \widetilde{\mathbf{r}}^{(i)} - \widehat{g}_i^{(r)}\widehat{\mathbf{s}}_i$;
11. $\quad\quad$ Compute $\alpha_i^{(r)} = \exp\left(-\left(\widehat{g}_i^{(r)} - \mu_{g_i}^{(r)}\right)^2 \,/\, 2\sigma_{g_i}^{2\,(r)}\right)$;
12. $\quad\quad g_i^*(r) = \dfrac{\alpha_i^{(r)}\widehat{g}_i^{(r)} + \left(\sum_{z=o}^{r-1}\alpha_i^{(z)}\right) g_i^*(r-1)}{\sum_{z=0}^{r}\alpha_i^{(z)}}$;
13. $\quad$ **end for**
14. $\quad r = r + 1$;
15. **end while**
16. **output:** $\mathbf{g}^* = \mathbf{g}^*(MAX)$. $\quad\mathbf{g}^*$: output solution vector

estimates after one and two iterations between channel estimation and detection are shown. For comparison, the Cramer–Rao lower bound (CRLB) for this system is also plotted. It can be seen that the channel estimation/detection scheme results in good MSE performance with improved MSE for an increased number of iterations between channel estimation and detection. For the same set of system and algorithm parameters as in Fig. 9.4(a), the BER performance curves are plotted in Fig. 9.4(b). For comparison, the BER performance with perfect channel knowledge is also plotted. It can be seen that with two iterations between channel estimation and detection the channel estimation/detection scheme can achieve 10^{-3} BER within about 1 dB of the performance with perfect channel knowledge.

9.3.3 Iterative channel estimation/equalization in ISI channels

In the previous subsection, channel estimation and detection procedures are considered for large multiuser MIMO channels with frequency-flat fading. In several practical scenarios, channels can be frequency-selective, causing ISI. Channel estimation and equalization in large MIMO-ISI channels is considered next.

MIMO-OFDM

One popular way to deal with ISI channels is to use multicarrier techniques like OFDM which can transform a frequency-selective channel into several narrow-

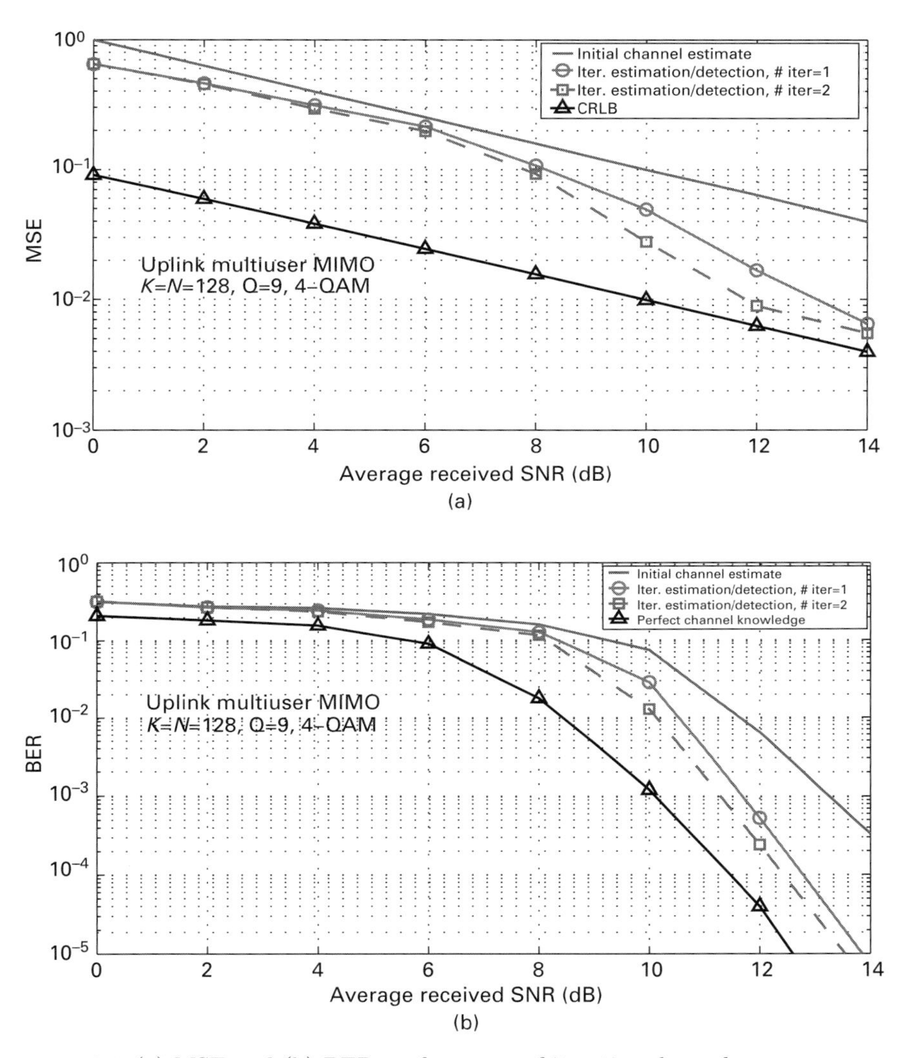

Figure 9.4 (a) MSE and (b) BER performance of iterative channel estimation/detection using Gibbs sampling based channel estimation and MGS-MR based detection in uplink multiuser MIMO system with $K = N = 128, Q = 9$, 4-QAM.

band frequency-flat subchannels. In OFDM, use of an inverse FFT (IFFT) at the transmitter and a FFT at the receiver (Fig. 9.5(a)) converts an ISI channel into parallel ISI-free subchannels with gains equal to the channel's frequency response values on the FFT grid. Inserting a cyclic prefix (CP) of length equal to or more than the channel delay spread at the transmitter and dropping it at the receiver eliminates inter-frame interference. In addition, use of a CP converts linear channel convolution into circular convolution, which facilitates diag-

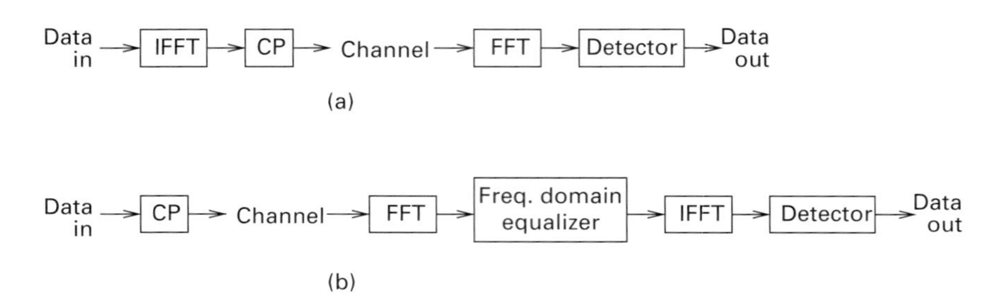

Figure 9.5 (a) OFDM and (b) single-carrier schemes.

onalization of the associated channel matrix. Channel equalization in OFDM systems thus takes the form of a simple multiplier bank at the FFT output in the receiver. Zero padding (ZP) can be used in place of a CP. ZP has the advantage of guaranteed symbol recovery even in the presence of channel nulls and hence improved performance. This improved performance comes at the cost of some increased receiver complexity; the single FFT required by cyclic prefixed OFDM is replaced by FIR filtering in zero-padded OFDM [15].

In large MIMO-OFDM systems, the channel estimation/detection procedure outlined in the previous subsection for frequency-flat channels can be employed on the resulting narrowband subchannels. Other simplified channel estimation schemes, e.g., as in [16],[17], can also be used in MIMO-OFDM settings.

Single-carrier communication

Instead of adopting a multicarrier approach, one can resort to a single-carrier block transmission approach and perform equalization at the receiver. The preference for single-carrier communication over OFDM communication is motivated by the peak-to-average power ratio (PAPR) problem encountered in the multicarrier approach [18]. In OFDM systems, the PAPR of the transmitted signal is large. This results in non-linear distortion in the power amplifier. Unless PAPR-reduction techniques are incorporated to control the non-linear distortion, power backoff in the amplifier becomes necessary. Several PAPR-reduction algorithms have been reported in the literature [19]–[21]. However, the resulting PAPRs are still (at least a few dB) larger than those of single-carrier block transmissions. Therefore, single-carrier schemes are considered to be good alternatives to address the PAPR issue that arises in multicarrier systems [18]–[25]. Single-carrier schemes alleviate the PAPR problem by discarding the IFFT at the transmitter (Fig. 9.5(b)). In addition, they also retain the FFT at the receiver which facilitates low complexity equalization in the frequency domain. As in OFDM, a CP or ZP can be used in single-carrier schemes [25]. In the following, system model development and the channel estimation/equalization approach for large multiuser MIMO-CPSC (MIMO cyclic prefixed single carrier) systems are presented. Likewise, the channel estimation/equalization approach can be developed for MIMO-ZPSC (MIMO zero-padded single carrier) systems using an overlap-and-add (OLA) technique [15], [26].

Channel estimation/equalization in large MIMO-CPSC

Consider CPSC signaling, where the overall channel includes an FFT operation at the receiver so that the transmitted symbols are estimated from the received frequency-domain signal [22]–[24]. The optimal training sequence that minimizes the channel estimation MSE of the linear channel estimator is shown to be of length KL per transmit antenna in [27]. Blind/semi-blind channel estimation methods can be considered, but they require long data samples and the complexity is high [28]–[30]. Here, channel estimation using uplink pilots and iterations between channel estimation and equalization in multiuser MIMO-CPSC systems is considered.

Multiuser MIMO-CPSC system model

Consider the uplink multiuser MIMO system shown in Fig. 8.1. The channel between each user transmit antenna–BS receive antenna pair is assumed to be frequency-selective with L multipath components. Let $h^{(j,k)}(l)$ denote the channel gain between the kth user and the jth receive antenna at the BS on the lth path, which is modeled as $\mathcal{CN}(0, \Omega_l^2)$. Perfect synchronization is assumed.

Transmission is carried out in frames, where each frame consists of several blocks as shown in Fig. 9.6. The channel is assumed to be constant over one frame duration. Each frame consists of a pilot block for the purpose of initial channel estimation, followed by Q data blocks. The pilot block consists of $(L-1)+KL$ channel uses. In the first $L-1$ channel uses in the pilot block, padding of $L-1$ zeros is used to avoid inter-frame interference. In each of the remaining KL channel uses, a K-length pilot symbol vector comprising pilot symbols transmitted from K users (one pilot symbol per user) is received by N receive antennas at the BS. Each data block consists of $I+L-1$ channel uses, where I information symbol vectors, each of length K (one data symbol from each user), preceded by a $(L-1)$-length cyclic prefix from each user (to avoid inter-block interference) are transmitted. With Q data blocks in a frame, the number of channel uses in the data part of the frame is $(I+L-1)Q$. Taking both pilot and data channel uses into account, the total number of channel uses per frame is $(L+1)K+(I+L-1)Q-1$.

Data blocks are detected with the MGS-MR algorithm (presented in Chapter 8) using an initial channel estimate. The detected data blocks are then iteratively used to refine the channel estimates during the data phase. The padding of $L-1$ zeros at the beginning of the pilot block makes the transmitters silent during the first $L-1$ channel uses in a frame. The channel output in these channel uses is ignored at the receiver. Accordingly, the zeroth channel use in a frame at

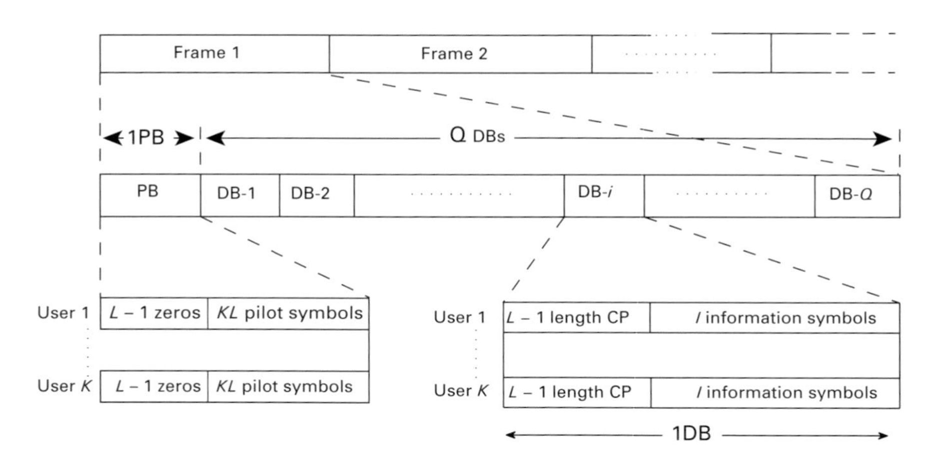

Figure 9.6 Frame structure for multiuser MIMO-CPSC system in ISI channels (PB: pilot block; DB: data block).

the receiver is taken to be the channel use in which the first pilot symbol in the frame is sent.

Initial channel estimate during pilot phase

Let $\mathbf{b}^k = [b^k(0), b^k(1), \ldots, b^k(KL-1)]$ denote the pilot symbol vector transmitted from user k in KL channel uses in a frame. The signal received by the jth receive antenna at the BS during the pilot phase in the nth channel use is given by

$$y_{\rm P}^j(n) = \sum_{k=1}^{K}\sum_{l=0}^{L-1} h^{(j,k)}(l)b^k(n-l) + q_{\rm P}^j(n), \tag{9.23}$$

$j = 1, 2, \ldots, N$, $n = 0, 1, \ldots, KL-1$, where the subscript P in $y_{\rm P}^j(n)$ and $q_{\rm P}^j(n)$ denotes pilot phase. $\{q_{\rm P}^j(n)\}$ are noise samples modeled as iid $\mathcal{CN}(0, \sigma^2)$. The following training sequence is used:

$$\mathbf{b}^k = [\mathbf{0}_{(k-1)L\times 1} \quad b \quad \mathbf{0}_{(K-(k-1))L-1)\times 1}], \tag{9.24}$$

where b is taken to be $\sqrt{KE_s}\left(\sum_{l=0}^{L-1}\Omega_l^2\right)$ so that the same average receive SNR is maintained in the pilot phase and the data phase. Writing (9.23) in matrix notation after substituting (9.24),

$$\mathbf{y}_{\rm P}^j = \mathbf{B}_{\rm P}\mathbf{h}^j + \mathbf{q}_{\rm P}^j, \quad j = 1, 2, \ldots, N, \tag{9.25}$$

where

$$\mathbf{y}_{\rm P}^j = [y_{\rm P}^j(0), y_{\rm P}^j(1), \ldots, y_{\rm P}^j(KL-1)]^T,\ \mathbf{h}^j = [(\mathbf{h}^{(j,1)})^T, \ldots, (\mathbf{h}^{(j,k)})^T, \ldots, (\mathbf{h}^{(j,K)})^T]^T,$$

$$\mathbf{h}^{(j,k)} = [h^{(j,k)}(0), h^{(j,k)}(1), \ldots, h^{(j,k)}(L-1)]^T, \ \mathbf{q}_{\mathrm{P}}^j = [q_{\mathrm{P}}^j(0), q_{\mathrm{P}}^j(1), \ldots, q_{\mathrm{P}}^j(KL-1)]^T,$$

$$\mathbf{B}_P = [\mathbf{B}_{\mathrm{P}_1} \mathbf{B}_{P_2} \ldots \mathbf{B}_{P_K}], \ \mathbf{B}_{P_k} = [\mathbf{0}_{L\times(k-1)L} \quad b\mathbf{I}_L \quad \mathbf{0}_{L\times(K-k)L}]^T.$$

From the signal observed at the jth receive antenna from time 0 to $KL-1$ during the pilot phase, an initial estimate of the channel vector $\mathbf{h}^j$ is obtained using the scaled identity nature of $\mathbf{B}_P$, as

$$\widehat{\mathbf{h}}^j = \mathbf{y}_P^j / b, \quad j = 1, 2, \ldots, N. \tag{9.26}$$

These initial channel estimates are used for equalization and detection of data vectors in the data phase.

9.3.4 Equalization using initial channel estimates

In the data phase, let $\mathbf{a}_i^k = [a_i^k(0), a_i^k(1), \ldots, a_i^k(I+L-2)]^T$ denote the data vector of size $(I+L-1)\times 1$, which includes $(L-1)$ CP symbols and I information symbols transmitted from the kth user during the ith data block, $i = 1, 2, \ldots, Q$. The signal received at the jth receive antenna at the nth channel use of the ith data block is given by

$$y_i^j(n) = \sum_{k=1}^{K}\sum_{l=0}^{L-1} h^{(j,k)}(l) a_i^k(n-l) + q_i^j(n), \tag{9.27}$$

$j = 1, 2, \ldots, N$, $n = 0, 1, \ldots, I+L-2$, where $q_i^j(n)$ is the noise sample modeled as iid $\mathcal{CN}(0, \sigma^2)$.

Define the following vectors and matrices: $\mathbf{y}_i^j \triangleq [y_i^j(L-1), y_i^j(L), \ldots, y_i^j(I+L-2)]^T$, $\mathbf{q}_i^j \triangleq [q_i^j(L-1), q_i^j(L), \ldots, q_i^j(I+L-2)]^T$, $\mathbf{x}_i^k \triangleq [a_i^k(L-1), a_i^k(L), \ldots, a_i^k(I+L-2)]^T$, and $\mathbf{H}^{j,k}$ as a $(I+L)\times I$ circulant matrix with $[h^{(j,k)}(0), h^{(j,k)}(1), \ldots, h^{(j,k)}(L-1), 0, \ldots, 0]^T$ as the first column. With these definitions, (9.27) can be written as

$$\mathbf{y}_i^j = \sum_{k=1}^{K} \mathbf{H}^{j,k}\mathbf{x}_i^k + \mathbf{q}_i^j, \quad j = 1, 2, \ldots, N. \tag{9.28}$$

The above set of equations can be written in the form

$$\mathbf{y}_i = \mathbf{H}\mathbf{x}_i + \mathbf{q}_i, \quad i = 1, 2, \ldots, Q, \tag{9.29}$$

where $\mathbf{y}_i = [(\mathbf{y}_i^1)^T, (\mathbf{y}_i^2)^T, \ldots, (\mathbf{y}_i^N)^T]^T$, $\mathbf{x}_i = [(\mathbf{x}_i^1)^T, (\mathbf{x}_i^2)^T, \ldots, (\mathbf{x}_i^K)^T]^T$, $\mathbf{q}_i = [(\mathbf{q}_i^1)^T, (\mathbf{q}_i^2)^T, \ldots, (\mathbf{q}_i^N)^T]^T$, and

$$\mathbf{H} = \begin{bmatrix} \mathbf{H}^{1,1} & \mathbf{H}^{1,2}\ldots & \mathbf{H}^{1,K} \\ \mathbf{H}^{2,1} & \mathbf{H}^{2,2}\ldots & \mathbf{H}^{2,K} \\ \vdots & \vdots & \vdots \\ \mathbf{H}^{N,1} & \mathbf{H}^{N,2}\ldots & \mathbf{H}^{N,K} \end{bmatrix}.$$

9.3.5 Equalization using the MGS-MR algorithm

The MGS-MR algorithm presented in Chapter 8 can be employed in the frequency domain using FFT based processing for equalization. The circulant matrix $\mathbf{H}^{j,k}$ can be decomposed as

$$\mathbf{H}^{j,k} = \mathbf{F}_I^H \mathbf{D}^{j,k} \mathbf{F}_I, \tag{9.30}$$

where $\mathbf{F}_I$ is an $I \times I$ DFT matrix, and $\mathbf{D}^{j,k}$ is a diagonal matrix in which diagonal elements are the DFT of the vector $[h^{(j,k)}(0), h^{(j,k)}(1), \ldots, h^{(j,k)}(L-1), 0, \ldots, 0]^T$. Taking the DFT of $\mathbf{y}_i^j$ in (9.28),

$$\mathbf{z}_i^j = \mathbf{F}_I \mathbf{y}_i^j = \sum_{k=1}^{K} \mathbf{D}^{j,k} \mathbf{b}_i^k + \mathbf{w}_i^j, \quad j = 1, 2, \ldots, N, \tag{9.31}$$

where $\mathbf{z}_i^j = [z_i^j(0), z_i^j(1), \ldots, z_i^j(I-1)]^T$, $\mathbf{b}_i^k \triangleq \mathbf{F}_I \mathbf{x}_i^k = [b_i^k(0), b_i^k(1), \ldots, b_i^k(I-1)]^T$, and $\mathbf{w}_i^j \triangleq \mathbf{F}_I \mathbf{q}_i^j = [w_i^j(0), w_i^j(1), \ldots, w_i^j(I-1)]^T$. Writing (9.31) in matrix form,

$$\mathbf{z}_i = \mathbf{D}\mathbf{b}_i + \mathbf{w}_i, \quad i = 1, 2, \ldots, Q, \tag{9.32}$$

where $\mathbf{z}_i = [(\mathbf{z}_i^1)^T, (\mathbf{z}_i^2)^T, \ldots, (\mathbf{z}_i^N)^T]^T$, $\mathbf{b}_i = [(\mathbf{b}_i^1)^T, (\mathbf{b}_i^2)^T, \ldots, (\mathbf{b}_i^K)^T]^T$, $\mathbf{w}_i = [(\mathbf{w}_i^1)^T, (\mathbf{w}_i^2)^T, \ldots, (\mathbf{w}_i^N)^T]^T$, and

$$\mathbf{D} = \begin{bmatrix} \mathbf{D}^{1,1} & \mathbf{D}^{1,2} & \ldots & \mathbf{D}^{1,K} \\ \mathbf{D}^{2,1} & \mathbf{D}^{2,2} & \ldots & \mathbf{D}^{2,K} \\ \vdots & \vdots & & \vdots \\ \mathbf{D}^{N,1} & \mathbf{D}^{N,2} & \ldots & \mathbf{D}^{N,K} \end{bmatrix}.$$

Rearranging the terms, (9.32) can be written as

$$\bar{\mathbf{z}}_i = \bar{\mathbf{D}}\bar{\mathbf{b}}_i + \bar{\mathbf{w}}_i, \quad i = 1, 2, \ldots, Q, \tag{9.33}$$

where

$$\bar{\mathbf{z}}_i = \begin{bmatrix} \bar{\mathbf{z}}_i(0) \\ \bar{\mathbf{z}}_i(1) \\ \vdots \\ \bar{\mathbf{z}}_i(I-1) \end{bmatrix}, \ \bar{\mathbf{b}}_i = \begin{bmatrix} \bar{\mathbf{b}}_i(0) \\ \bar{\mathbf{b}}_i(1) \\ \vdots \\ \bar{\mathbf{b}}_i(I-1) \end{bmatrix}, \ \bar{\mathbf{w}}_i = \begin{bmatrix} \bar{\mathbf{w}}_i(0) \\ \bar{\mathbf{w}}_i(1) \\ \vdots \\ \bar{\mathbf{w}}_i(I-1) \end{bmatrix},$$

$$\bar{\mathbf{D}} = \begin{bmatrix} \bar{\mathbf{D}}(0) & \ldots & 0 \\ \vdots & \ddots & \vdots \\ 0 & \ldots & \bar{\mathbf{D}}(I-1) \end{bmatrix}, \ \bar{\mathbf{z}}_i(m) = [z_i^1(m), \ldots, z_i^N(m)]^T,$$

$$\bar{\mathbf{b}}_i(m) = [b_i^1(m), \ldots, b_i^N(m)]^T, \ \bar{\mathbf{w}}_i(m) = [w_i^1(m), \ldots, w_i^N(m)]^T,$$

and

$$\bar{\mathbf{D}}(m) = \begin{bmatrix} \mathbf{D}^{1,1}(m) & \mathbf{D}^{1,2}(m) \ldots & \mathbf{D}^{1,K}(m) \\ \mathbf{D}^{2,1}(m) & \mathbf{D}^{2,2}(m) \ldots & \mathbf{D}^{2,K}(m) \\ \vdots & \vdots & \vdots \\ \mathbf{D}^{N,1}(m) & \mathbf{D}^{N,2}(m) \ldots & \mathbf{D}^{N,K}(m) \end{bmatrix}.$$

$\mathbf{D}^{j,k}(m)$ is the mth diagonal element of the matrix $\mathbf{D}^{j,k}$. Also, $\bar{\mathbf{b}}_i = \bar{\mathbf{F}}\bar{\mathbf{x}}_i$, where $\bar{\mathbf{F}} \triangleq \mathbf{F}_I \otimes \mathbf{I}_K$, $\bar{\mathbf{x}}_i = [a_i^1(L-1)\ldots a_i^K(L-1), a_i^1(L)\ldots a_i^K(L), \ldots, a_i^1(I+L-2)\ldots a_i^K(I+L-2)]^T$. Now, (9.33) can written as

$$\bar{\mathbf{z}}_i = \bar{\mathbf{D}}\bar{\mathbf{F}}\bar{\mathbf{x}}_i + \bar{\mathbf{w}}_i \quad = \quad \bar{\mathbf{H}}\bar{\mathbf{x}}_i + \bar{\mathbf{w}}_i, \quad i = 1, 2, \ldots, Q, \tag{9.34}$$

where $\bar{\mathbf{H}} \triangleq \bar{\mathbf{D}}\bar{\mathbf{F}}$.

For each i in (9.34), run the MGS-MR detection algorithm and detect the information symbols in the ith block. In the first iteration of data detection, the channel estimates from (9.26) are used to calculate $\widehat{\bar{\mathbf{H}}}$, an estimate of $\bar{\mathbf{H}}$. Each coordinate of the vector $(\bar{\mathbf{z}}_i - \widehat{\bar{\mathbf{H}}}\bar{\mathbf{x}}_i)$ has zero mean and $2\sigma^2$ variance. Using this knowledge, the statistics of the ML cost of error-free vectors are recalculated in the MGS-MR algorithm. MGS-MR detector outputs are denoted $\hat{\mathbf{x}}_i^k$, $k = 1, \ldots, K$, $i = 1, \ldots, Q$. These output vectors are then used to improve the channel estimates through iterations between equalization and channel estimation. The channel estimation in these iterations is based on a Gibbs sampling approach which is presented next. Other approaches like MMSE channel estimation can also be used.

Gibbs sampling based channel estimation in data phase

Consider (9.31), which can be rewritten as

$$\mathbf{z}_i^j = \sum_{k=1}^{K} \mathbf{B}_i^k \mathbf{d}^{j,k} + \mathbf{w}_i^j, \quad j = 1, 2, \ldots, N, \tag{9.35}$$

where $\mathbf{B}_i^k = \text{diag}(\mathbf{b}_i^k)$ and $\mathbf{d}^{j,k}$ is a vector consisting of the diagonal elements of matrix $\mathbf{D}^{j,k}$, which is the I-point DFT of $\mathbf{h}^{(j,k)}$ (zero padded to length I), i.e., $\mathbf{d}^{j,k} = \tilde{\mathbf{F}}_{I\times L}\mathbf{h}^{(j,k)}$, where $\tilde{\mathbf{F}}_{I\times L}$ is the matrix with the first L columns of $\mathbf{F}_I$. Now, (9.35) can be written as

$$\mathbf{z}_i^j = \sum_{k=1}^{K} \mathbf{B}_i^k \tilde{\mathbf{F}}_{I\times L}\mathbf{h}^{(j,k)} + \mathbf{w}_i^j. \tag{9.36}$$

Defining $\mathbf{A}_i^k \triangleq \mathbf{B}_i^k \tilde{\mathbf{F}}_{I\times L}$, (9.36) can be written as

$$\mathbf{z}_i^j = \mathbf{A}_i \mathbf{h}^j + \mathbf{w}_i^j, \quad i = 1, \ldots, Q, \tag{9.37}$$

where $\mathbf{A}_i = [\mathbf{A}_i^1\ \mathbf{A}_i^2\ \ldots\ \mathbf{A}_i^K]$. Now, (9.37) can be written as

$$\mathbf{z}^j = \mathbf{A}\mathbf{h}^j + \mathbf{w}^j, \tag{9.38}$$

where

$$\mathbf{z}^j = \begin{bmatrix} \mathbf{z}_1^j \\ \mathbf{z}_2^j \\ \vdots \\ \mathbf{z}_Q^j \end{bmatrix}, \ \mathbf{A} = \begin{bmatrix} \mathbf{A}_1 \\ \mathbf{A}_2 \\ \vdots \\ \mathbf{A}_Q \end{bmatrix}, \ \mathbf{w}^j = \begin{bmatrix} \mathbf{w}_1^j \\ \mathbf{w}_2^j \\ \vdots \\ \mathbf{w}_Q^j \end{bmatrix}.$$

Using the signal received at antenna j from blocks 1–Q in a frame (i.e., using $\mathbf{z}^j$) and the matrix $\hat{\mathbf{A}}$ which is formed by replacing the information symbols $\{\mathbf{x}_i^k\}$ in $\mathbf{A}$ by the detected information symbols $\{\hat{\mathbf{x}}_i^k\}$, the channel coefficients $\{\mathbf{h}^j\}$

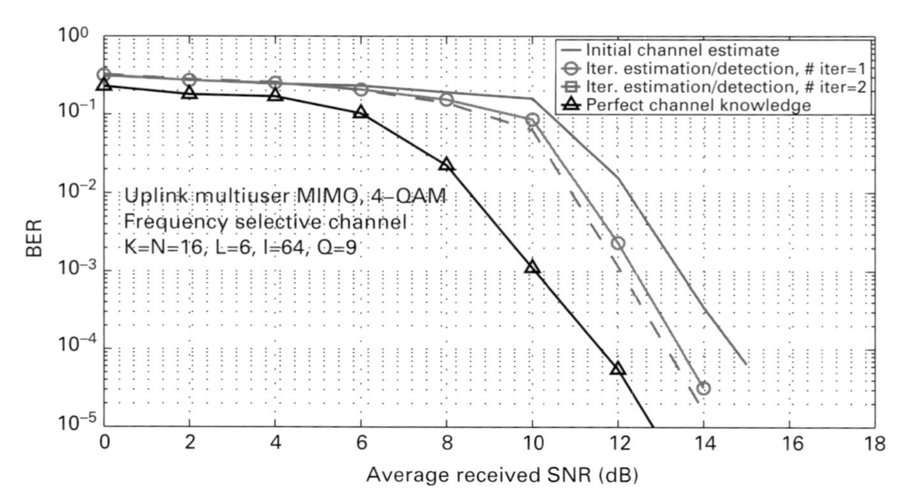

Figure 9.7 BER performance of iterative channel estimation/equalization on ISI channel. $K = N = 16, L = 6, I = 64, Q = 9$, 4-QAM.

are estimated using the Gibbs sampling based estimation technique presented in Section 9.3.2. This ends one iteration between channel estimation and detection. The detected data matrix is fed back for channel estimation in the next iteration, whose output is then used to detect the data matrix again. This iterative channel estimation/equalization procedure is carried out for a certain number of iterations.

Performance results

Figure 9.7 presents the BER performance of the iterative channel estimation/equalization scheme described above in an uplink multiuser MIMO system on frequency-selective fading with $K = N = 16, L = 6, I = 64, Q = 9$ and 4-QAM. For the same settings, the BER performance with perfect channel knowledge is also plotted. It can be seen that the BER improves as the number of iterations between channel estimation and detection increases. It can also be seen that with estimated channel knowledge, a performance close to that with perfect channel knowledge is achieved.

References

[1] M. Medard, "The effect upon channel capacity in wireless communications of perfect and imperfect knowledge of the channel," *IEEE Trans. Inform. Theory*, vol. 46, no. 3, pp. 933–946, May 2000.

[2] A. Lapidoth and S. Shamai (Shitz), "Fading channels: how perfect need "perfect side information" be?" *IEEE Trans. Inform. Theory*, vol. 48, no. 5, pp. 1118–1134, May 2002.

[3] T. Yoo and A. Goldsmith, "Capacity and power allocation for fading MIMO channels with channel estimation error," *IEEE Trans. Inform. Theory*, vol. 52, no. 5, pp. 2203–2214, May 2006.

[4] T. L. Marzetta, "BLAST training: estimating channel characteristics for high capacity space-time wireless," in *37th Annual Allerton Conf. on Commun. Contr. and Comput.*, Monticello, IL, Sep. 1999, pp. 958–966.

[5] J. Baltersee, G. Fock, and H. Meyr, "Achievable rate of MIMO channels with data-aided channel estimation and perfect interleaving," *IEEE J. Sel. Areas Commun.*, vol. 19, no. 12, pp. 2358–2368, Dec. 2001.

[6] B. Hassibi and B. M. Hochwald, "How much training is needed in multiple-antenna wireless links?" *IEEE Trans. Inform. Theory*, vol. 49, no. 4, pp. 951–963, Apr. 2003.

[7] X. Ma, L. Yang, and G. B. Giannakis, "Optimal training for MIMO frequency-selective fading channels," *IEEE Trans. Wireless Commun.*, vol. 4, no. 2, pp. 453–456, Mar. 2005.

[8] T. L. Marzetta, "How much training is required for multiuser MIMO?" in *Proc. 40th Asilomar Conf. on Signals, Systems and Computers*, Pacific Grove, CA, Oct.–Nov. 2006, pp. 359–363.

[9] L. Berriche, K. Abed-Meraim, and J.-C. Belfiore, "Investigation of the channel estimation error on MIMO system performance," in *European Signal Process. Conf.*, Antalya, Sep. 2005.

[10] M. Dong and L. Tong, "Optimal design and placement of pilot symbols for channel estimation," *IEEE Trans. Signal Process.*, vol. 50, no. 12, pp. 3055–3069, Dec. 2002.

[11] L. Berriche, K. Abed-Meraim, and J.-C. Belfiore, "Cramer-Rao bounds for MIMO channel estimation," in *IEEE ICASSP'2004*, Montreal, vol. 4, May 2004, pp. 397–400.

[12] G. Tarrico and E. Biglieri, "Space-time coding with imperfect channel estimation," *IEEE Trans. Wireless Commun.*, vol. 4, no. 4, pp. 1874–1888, Apr. 2005.

[13] Y. Sung, T. E. Sung, B. M. Sadler, and L. Tong, "Training for MIMO wireless communications," in *Space-time Wireless Systems: From Array Processing to MIMO Communications*, H. Bolcskei, D. Gesbert, C. B. Papadias, and A.-J. van der Veen, Eds. Cambridge, UK: Cambridge University Press, 2006, ch. 17.

[14] S. K. Mohammed, A. Zaki, A. Chockalingam, and B. S. Rajan, "High-rate space-time coded large-MIMO systems: low-complexity detection and channel estimation," *IEEE J. Sel. Topics in Signal Process.*, vol. 3, no. 6, pp. 958–974, Dec. 2009.

[15] B. Muquet, Z. Wang, G. B. Giannakis, M. de Courville, and P. Duhamel, "Cyclic prefixing or zero padding for wireless multicarrier transmissions?" *IEEE Trans. Commun.*, vol. 50, no. 12, pp. 2136–2148, Dec. 2002.

[16] Y. Li, "Simplified channel estimation for OFDM systems with multiple transmit antennas," *IEEE Trans. Wireless Commun.*, vol. 1, no. 1, pp. 67–75, Jan. 2002.

[17] K. Higuchi, H. Kawai, N. Maeda, H. Taoka, and M. Sawahashi, "Experiments on real-time 1-Gb/s packet transmission using MLD-based signal detection in MIMO-OFDM broadband radio access," *IEEE J. Sel. Areas Commun.*, vol. 24, no. 6, pp. 1141–1153, Jun. 2006.

[18] Z. Wang, X. Ma, and G. B. Giannakis, "OFDM or single-carrier block transmissions," *IEEE Trans. Commun.*, vol. 52, no. 3, pp. 380–394, Mar. 2004.

[19] J. L. Cimini, Jr. and N. R. Sollenberger, "Peak-to-average power ratio reduction of an OFDM signal using partial transmit sequences," *IEEE Commun. Lett.*, vol. 4, no. 3, pp. 86–88, Mar. 2000.

[20] S. Han and J. Lee, "An overview of peak-to-average power ratio reduction techniques for multicarrier transmission," *IEEE Wireless Commun.*, vol. 12, no. 2, pp. 56–65, Apr. 2005.

[21] T. T. Nguyen and L. Lampe, "On partial transmit sequences for PAR reduction in OFDM systems," *IEEE Trans. Wireless Commun.*, vol. 7, no. 2, pp. 74–755, Feb. 2008.

[22] H. Sari, G. Karam, and I. Jeanclaude, "Transmission techniques for digital terrestrial TV broadcasting," *IEEE Commun. Mag.*, vol. 33, no. 2, pp. 100–109, Feb. 1995.

[23] D. Falconer, S. L. Ariyavisitakul, A. Benyamin-Seeyar, and B. Eidson, "Frequency domain equalization for single-carrier broadband wireless systems," *IEEE Commun. Mag.*, vol. 40, no. 4, pp. 58–66, Apr. 2002.

[24] B. Devillers, J. Louveaux, and L. Vandendorpe, "About the diversity in cyclic prefixed single-carrier systems," *Physical Commun.*, vol. 1, no. 4, pp. 266–276, Dec. 2008.

[25] S. Ohno, "Performance of single-carrier block transmissions over multipath fading channels with linear equalization," *IEEE Trans. Signal Process.*, vol. 54, no. 10, pp. 3678–3687, Oct. 2006.

[26] H. Eshwaraiah and A. Chockalingam, "Cooperative particle swarm optimization based receiver for large-dimension MIMO-ZPSC systems," in *IEEE WCNC'2012*, Apr. 2012, pp. 336–341.

[27] X. Ma, L. Yang, and G. B. Giannakis, "Optimal training for MIMO frequency-selective fading channels," *IEEE Trans. Wireless Commun.*, vol. 4, no. 2, pp. 453–466, Mar. 2005.

[28] D. Slock and A. Medles, "Blind and semiblind MIMO channel estimation," in *Space-time Wireless Systems: From Array Processing to MIMO Communications*, H. Bolcskei, D. Gesbert, C. B. Papadias, and A.J. van der Veen, Eds. Cambridge, UK: Cambridge University Press, 2006, ch. 14.

[29] Y. S. Chen and C. A. Lin, "Blind-channel identification for MIMO single-carrier zero-padding block-transmission systems," *IEEE Trans. Circuits and Systems*, vol. 55, no. 6, pp. 1571–1579, Jul. 2008.

[30] Y. S. Chen, "Semiblind channel estimation for MIMO single carrier with frequency-domain equalization systems," *IEEE Trans. Veh. Tech.*, vol. 59, no. 1, pp. 53–62, Jan. 2010.

10 Precoding in large MIMO systems

Channel state information at the transmitter (CSIT) can be exploited to improve performance in MIMO wireless systems through a precoding operation at the transmitter. Precoding techniques use CSIT to encode the information symbols into transmit vectors. Typically, an information symbol vector $\mathbf{u}$ is encoded into a transmit vector $\mathbf{x}$ using the transformation $\mathbf{x} = \mathbf{T}\mathbf{u}$, where $\mathbf{T}$ is referred to as the precoding matrix. The precoding matrix $\mathbf{T}$ is chosen based on the available CSIT. Precoding on point-to-point MIMO links and multiuser MIMO links is common. In addition, multiuser MIMO precoding in multicell scenarios is of interest. In this chapter, precoding schemes for large MIMO systems are considered.

10.1 Precoding in point-to-point MIMO

In point-to-point MIMO links, precoding techniques can achieve improved performance in terms of enhanced communication reliability, which is typically quantified in terms of the diversity gain/order achieved by the precoding scheme. In addition to performance, precoding/decoding complexities are also of interest. Often, one encounters a tradeoff between performance (diversity gain) and precoding/decoding complexity. Well-known precoders for point-to-point MIMO are presented in the following subsections.

System model

Consider an $n_t \times n_r$ point-to-point MIMO system ($n_r \leq n_t$), where n_t and n_r denote the number of transmit and receive antennas, respectively. Assume CSI to be known perfectly at both the transmitter and the receiver. Let $\mathbf{x} = (x_1, \ldots, x_{n_t})^T$ be the vector of symbols transmitted by the n_t transmit antennas in one channel use. Let $\mathbf{H} = \{h_{ij}\}$, $i = 1, \ldots, n_r$, $j = 1, \ldots, n_t$, be the $n_r \times n_t$ channel coefficient matrix, where h_{ij} is the complex channel gain between the jth transmit antenna and the ith receive antenna and h_{ij}s are modeled as iid $\mathcal{CN}(0,1)$. The $n_r \times 1$ received vector is given by

$$\mathbf{y} = \mathbf{H}\mathbf{x} + \mathbf{n}, \tag{10.1}$$

where $\mathbf{n}$ is a spatially uncorrelated Gaussian noise vector such that $\mathbb{E}[\mathbf{n}\mathbf{n}^H] = N_0\mathbf{I}_{n_r}$. Let the number of transmitted information symbols per channel use be n_s ($n_s \leq n_r$). In each channel use, b information bits are first mapped to the information symbol vector $\mathbf{u} = (u_1, \ldots, u_{n_s})^T \in \mathbb{C}^{n_s}$, which is then mapped to the data symbol vector $\mathbf{z} = (z_1, \ldots, z_{n_s})^T \in \mathbb{C}^{n_s}$ using an $n_s \times n_s$ encoding matrix $\mathbf{G}$ as

$$\mathbf{z} = \mathbf{G}\mathbf{u} + \mathbf{u}^0, \tag{10.2}$$

where $\mathbf{u}^0 \in \mathbb{C}^{n_s}$ is a displacement vector used to reduce the average transmitted power. Let $\mathbf{T}$ be the $n_t \times n_s$ precoding matrix which is applied to the data symbol vector to yield the transmitted vector

$$\mathbf{x} = \mathbf{T}\mathbf{z}. \tag{10.3}$$

In general, $\mathbf{T}$, $\mathbf{G}$, and $\mathbf{u}_0$ are derived from the knowledge of $\mathbf{H}$ at the transmitter and they are crucial to the system performance and complexity. The transmit power constraint is given by

$$\mathbb{E}[\|\mathbf{x}\|^2] = P_T, \tag{10.4}$$

where P_T is the total transmit power, and SNR is defined as $\gamma \triangleq P_T/N_0$.

In slow fading MIMO channels, where transmissions are subject to block fading, diversity gain/order is a relevant performance metric. In fast fading MIMO channels, ergodic capacity is the relevant metric. The rate and diversity order for the precoding schemes are defined as follows. The rate R is defined as the number of information bits transmitted in each channel use (bits per channel use or bpcu). Since b bits are transmitted in each channel use, $R = b$ bpcu. To define the achieved diversity order d_{ord}, let $P(\mathbf{H}, \gamma)$ be the word error probability of $\mathbf{u}$ for a given channel realization $\mathbf{H}$ and SNR γ. The average word error probability, averaged over the channel fading statistics, is $P(\gamma) = \mathbb{E}_{\mathbf{H}}[P(\mathbf{H}, \gamma)]$. The diversity order is then defined as

$$d_{ord} \triangleq \lim_{\gamma \to \infty} \frac{-\log P(\gamma)}{\log(\gamma)}. \tag{10.5}$$

10.1.1 SVD precoding

A well-known precoding scheme in point-to-point MIMO is based upon the singular value decomposition (SVD) of the channel matrix, which transforms the MIMO channel into parallel subchannels. Because of this decomposition, ML decoding of the transmitted information symbol vector at the receiver reduces to separate ML decoding for the information symbols transmitted on each subchannel, thereby resulting in low ML detection complexity. The SVD of the channel matrix $\mathbf{H}$ is given by [1]

$$\mathbf{H} = \mathbf{U}\mathbf{\Lambda}\mathbf{V}, \tag{10.6}$$

where $\mathbf{U} \in \mathbb{C}^{n_r \times n_r}$, $\mathbf{\Lambda} \in \mathbb{C}^{n_r \times n_r}$, $\mathbf{V} \in \mathbb{C}^{n_r \times n_t}$, such that $\mathbf{U}\mathbf{U}^H = \mathbf{V}\mathbf{V}^H = \mathbf{I}_{n_r}$, and $\mathbf{\Lambda} = \text{diag}(\lambda_1, \ldots, \lambda_{n_r})$ is the diagonal matrix of singular values with $\lambda_1 \geq \lambda_2 \geq \cdots \geq \lambda_{n_r} \geq 0$. Let $\tilde{\mathbf{V}} \in \mathbb{C}^{n_s \times n_t}$ be the submatrix with the first n_s rows of $\mathbf{V}$. The SVD precoder uses

$$\mathbf{T} = \tilde{\mathbf{V}}^H, \quad \mathbf{G} = \mathbf{I}_{n_s}, \quad \mathbf{u}^0 = \mathbf{0}. \tag{10.7}$$

The received vector is then given by

$$\mathbf{y} = \mathbf{H}\mathbf{T}\mathbf{u} + \mathbf{n}. \tag{10.8}$$

Let $\tilde{\mathbf{U}} \in \mathbb{C}^{n_r \times n_s}$ be the submatrix with the first n_s columns of $\mathbf{U}$. The receiver computes

$$\mathbf{r} = \tilde{\mathbf{U}}^H \mathbf{y} \;=\; \tilde{\mathbf{\Lambda}}\mathbf{u} + \mathbf{w}, \tag{10.9}$$

where $\mathbf{w} \in \mathbb{C}^{n_s}$ is still an uncorrelated Gaussian noise vector with $\mathbb{E}[\mathbf{w}\mathbf{w}^H] = N_0\mathbf{I}_{n_s}$, $\tilde{\mathbf{\Lambda}} \triangleq \text{diag}(\lambda_1, \ldots, \lambda_{n_s})$, and $\mathbf{r} = (r_1, \ldots, r_{n_s})^T$. Hence, SVD precoding transforms the channel into n_s parallel subchannels

$$r_i = \lambda_i u_i + w_i, \quad i = 1, \ldots, n_s, \tag{10.10}$$

with non-negative fade coefficients λ_i, where the channel gain of the kth subchannel is λ_k, the kth singular value of the channel matrix. The diversity order achieved by the kth stream (i.e., the asymptotic slope of the average error probability for the information symbol u_k wrt γ) depends on how the pdf of λ_k behaves around $\lambda_k = 0$ [2],[3]. For iid Rayleigh fading, the pdf of the kth singular value around $\lambda_k = 0$ is [2]

$$p(\lambda_k) = c_k \lambda_k^{(n_r-k+1)(n_t-k+1)-1} + o\big(\lambda_k^{(n_r-k+1)(n_t-k+1)-1}\big). \tag{10.11}$$

So, the diversity order of the kth stream is given by $(n_r - k + 1)(n_t - k + 1)$. The lowest diversity order is achieved by the n_sth stream, i.e., the overall error performance is dominated by the minimum singular value λ_{n_s}. When $n_s = n_r = n_t$, the resulting diversity order of SVD precoding is only 1.

10.1.2 Pairing of good and bad subchannels

Pairing of subchannels, which refers to joint coding of information symbols across two subchannels, is a low complexity technique to improve the overall diversity order of SVD precoding. Significant improvement in achievable diversity gain is possible by jointly coding over pairs of subchannels as long as the pairs are appropriately chosen. In particular, subchannels with high diversity gain when paired with those having low diversity gain can provide improved overall diversity order [4]–[6]. The motivation for the pairing of subchannels arises from the idea of rotation coding [7], which is described below.

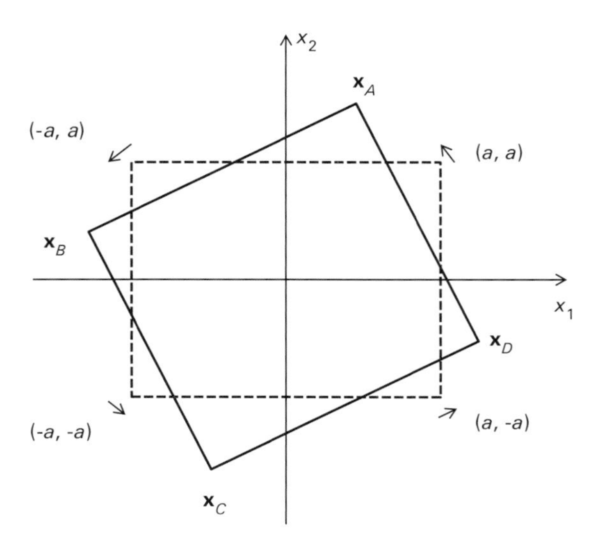

Figure 10.1 Original and rotated constellations.

Rotation coding

Consider an SISO fading channel and signaling in two channel uses. Let $\mathbf{u} = [u_1 \ u_2]^T$, $u_i \in \pm a$, denote the information vector. The transmit vector $\mathbf{x}$ is obtained as $\mathbf{x} = \mathbf{Tu}$, where the rotation matrix $\mathbf{T}$ is parameterized by a single angle θ, which is given by

$$\mathbf{T} = \begin{bmatrix} \cos\theta & -\sin\theta \\ \sin\theta & \cos\theta \end{bmatrix}, \quad \theta \in (0, 2\pi). \tag{10.12}$$

Then the following four codewords are possible: $\mathbf{x}_A = \mathbf{T}[a \ \ a]^T$, $\mathbf{x}_B = \mathbf{T}[-a \ \ a]^T$, $\mathbf{x}_C = \mathbf{T}[-a \ \ -a]^T$, $\mathbf{x}_D = \mathbf{T}[a \ \ -a]^T$. Figure 10.1 shows the original constellation and the rotated constellation. The received signal in the kth channel use is

$$y_k = h_k x_k + w_k, \quad k = 1, 2, \tag{10.13}$$

where h_1 and h_2 are iid fade coefficients in channel uses 1 and 2, respectively, and w_k is the additive noise component. For this system model, the pairwise symbol error probability is given by [7]

$$P(\mathbf{x}_A \to \mathbf{x}_B) \leq (1 + \gamma|d_1|^2/4)^{-1}(1 + \gamma|d_2|^2/4)^{-1}, \tag{10.14}$$

where $\gamma = a^2/N_0$ and $\mathbf{d} = [d_1 \ d_2]^T = (1/a)(\mathbf{x}_A - \mathbf{x}_B) = [2\cos\theta \ \ 2\sin\theta]^T$. If $d_1, d_2 \neq 0$, at high SNR,

$$P(\mathbf{x}_A \to \mathbf{x}_B) \leq \frac{16}{|d_1 d_2|^2}\gamma^{-2}. \tag{10.15}$$

The squared product distance between $\mathbf{x}_A$ and $\mathbf{x}_B$ is $\beta_{AB} = |d_1 d_2|^2$, and the overall error probability is given by

$$P_e \leq \frac{48}{\min\limits_{j=B,C,D} \beta_{Aj}} \gamma^{-2}. \tag{10.16}$$

As long as $\beta_{ij} > 0$ for all i, j, a diversity gain of 2 is achieved. The θ that maximizes the minimum squared product distance is $\theta^* = (1/2)\tan^{-1} 2$, and $\min \beta_{ij} = 16/5$. The bound on P_e then becomes $P_e \leq 15\gamma^{-2}$. This shows that the rotation coding allows one to achieve second order diversity even in SISO fading channels.

The idea of rotation coding can be applied individually on pairs of parallel subchannels to achieve improved diversity gains [4]–[6]. The idea is to pair subchannels (good with bad) followed by SVD. Each pair is encoded by a 2×2 complex [4] or real [5] rotation matrix. The choice of the optimal pairing and the optimal rotation angle for each pair is important.

X-precoding

Without loss of generality, consider even n_r and $n_s = n_r$. Precoding using real rotation matrices, known as X-codes [5], is performed on pairs of subchannels. The X-precoding scheme is shown in Fig. 10.2. A linear encoder defined by matrix $\mathbf{X} \in \mathbb{C}^{n_r \times n_r}$ pairs the subchannels so that the precoder matrix $\mathbf{T}$ and transmit vector $\mathbf{x}$ are given by

$$\mathbf{T} = \mathbf{V}^H \mathbf{X}, \qquad \mathbf{x} = \mathbf{V}^H \mathbf{X} \mathbf{u}.$$

The $\mathbf{X}$ matrix is determined by the list of subchannel pairings $\{(i_k, j_k),\ k = 1, 2, \ldots, n_r/2,\ i_k < j_k\}$ and the 2×2 encoder matrix $\mathbf{A}_k$ for the kth pair. On the kth pair consisting of subchannels i_k and j_k, the symbols u_{i_k} and u_{j_k} are jointly coded using a 2×2 matrix

$$\mathbf{A}_k = \begin{bmatrix} \cos\theta_k & \sin\theta_k \\ -\sin\theta_k & \cos\theta_k \end{bmatrix}, \qquad k = 1, \ldots, \frac{n_r}{2}. \tag{10.17}$$

Each A_k is used as a 2×2 submatrix of the $\mathbf{X}$ matrix so that

$$X_{i_k,i_k} = \cos\theta_k, \quad X_{i_k,j_k} = \sin\theta_k, \quad X_{j_k,i_k} = -\sin\theta_k, \quad X_{j_k,j_k} = \cos\theta_k.$$

The optimal pairing in terms of achieving the best diversity order is the one in which the kth subchannel is paired with the $(n_r - k + 1)$th subchannel [4], i.e., the optimal pairing is given by $i_k = k,\ \ j_k = (n_r - k + 1)$, $k = 1, 2, \ldots, n_r/2$. This ordering achieves diversity order $d_{ord} \geq (n_r/2 + 1)(n_t - n_r/2 + 1)$. This is a significantly improved diversity order compared to that achieved in the case of no pairing. For example, with $n_r = n_t = n_s$, the overall diversity order in the scheme with pairing is $(n_r/2 + 1)^2$, whereas the diversity order is 1 for the scheme without pairing. If only n_s (n_s even) out of the n_r subchannels are used for transmission, the lower bound on the overall achievable diversity order

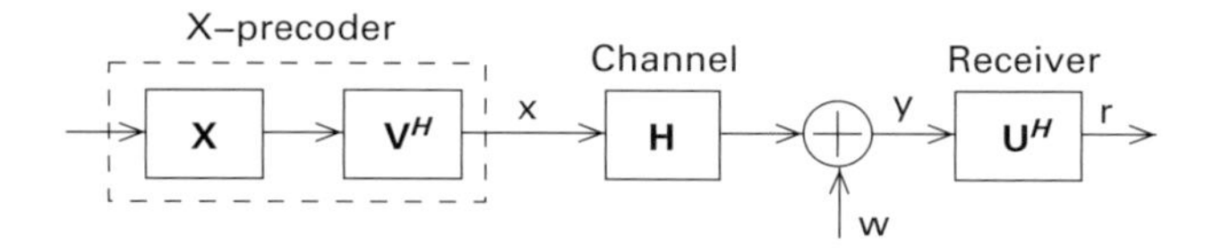

Figure 10.2 X-precoding scheme.

is $(n_r - n_s/2 + 1)\,(n_t - n_s/2 + 1)$. The structure of the $\mathbf{X}$ matrix thus obtained by the above pairing for $n_t = n_r = 6$ is shown below:

$$\mathbf{X} = \begin{bmatrix} \cos\theta_1 & & & & & \sin\theta_1 \\ & \cos\theta_2 & & & \sin\theta_2 & \\ & & \cos\theta_3 & \sin\theta_3 & & \\ & & -\sin\theta_3 & \cos\theta_3 & & \\ & -\sin\theta_2 & & & \cos\theta_2 & \\ -\sin\theta_1 & & & & & \cos\theta_1 \end{bmatrix}. \tag{10.18}$$

The best rotation angles have to be found for each realization of $\mathbf{H}$. It is tedious to compute them due to the lack of an exact expression for the word error probability $P_k(\mathbf{H})$. Alternatively, optimal angles that maximize a generalized minimum distance [4], independent of $\mathbf{H}$, can be computed. For M^2-QAM, the best rotation angle θ_k^* for the kth pair can be computed based on such a maximization as

$$\theta_k^* = \underset{[0,\pi/4]}{\operatorname{argmax}}\ d_{\min}(\theta_k), \tag{10.19}$$

where $d_{\min}(\theta_k) = \min_{(p,q)\in\mathbb{S}_M}(p^2+q^2)\cos^2(\theta_k - \tan^{-1}(q/p))$, $\mathbb{S}_M \triangleq \{(p,q) \neq (0,0)\,|\,0 \leq p \leq (M-1),\ 0 \leq q \leq (M-1)\}$, and $\gamma = P_T/N_0$. The maximization in (10.19) can be done numerically. It can be done offline, since these angles can be fixed a priori. The performance of X-codes with these precomputed fixed angles is found to be good (see the BER plot for 16×16 MIMO with 16-QAM in Fig. 10.4).

Decoding at the receiver is carried out as follows. The receiver computes

$$\mathbf{r} = \mathbf{U}^H\mathbf{y} = \mathbf{\Lambda}\mathbf{X}\mathbf{u} + \mathbf{w} = \mathbf{M}\mathbf{u} + \mathbf{w}. \tag{10.20}$$

This is equivalent to

$$\mathbf{r}_k = \mathbf{M}_k\mathbf{u}_k + \mathbf{w}_k, \quad k = 1, 2, \ldots, n_r/2, \tag{10.21}$$

where

$$\mathbf{r}_k = \begin{bmatrix} r_{i_k} \\ r_{j_k} \end{bmatrix},\ \mathbf{u}_k = \begin{bmatrix} u_{i_k} \\ u_{j_k} \end{bmatrix},\ \mathbf{w}_k = \begin{bmatrix} w_{i_k} \\ w_{j_k} \end{bmatrix},\ \mathbf{M}_k = \begin{bmatrix} \lambda_{i_k}\cos\theta_k & \lambda_{i_k}\sin\theta_k \\ -\lambda_{j_k}\sin\theta_k & \lambda_{j_k}\cos\theta_k \end{bmatrix}.$$

Therefore, ML decoding of $\mathbf{u}$ reduces to independent ML decoding of the k pairs. Also, ML decoding for each pair is separable into independent ML decoding of the real and imaginary components of $\mathbf{u}_k$. Hence, the overall complexity is n_r two-dimensional real ML decoders (e.g., SDs). This low complexity allows

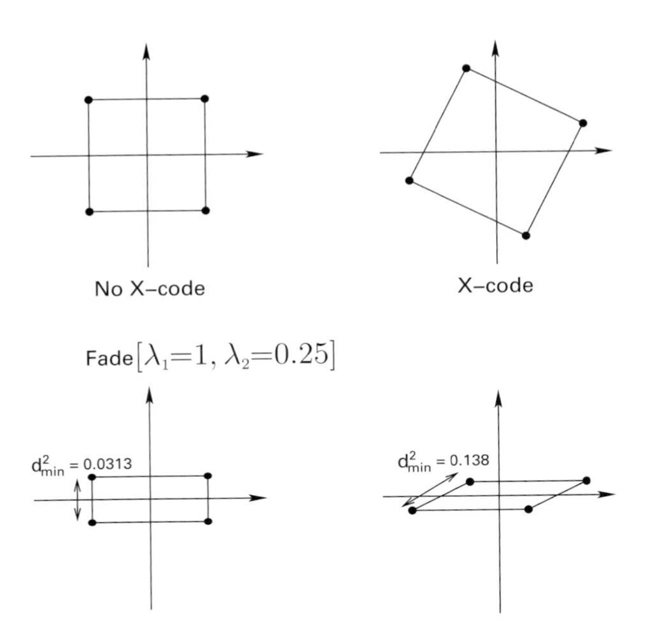

Figure 10.3 Signal space of the transmit and received two-dimensional codewords.

the X-precoding scheme to scale well for a large number of antennas. The reason why X-precoding achieves good performance can be explained as follows. In the case of no coding across two subchannels, a deep fade along any one subchannel can result in an arbitrarily small minimum distance between the received codewords, and this would increase the word error probability. However, with rotation using the X-code, the minimum distance between the received codewords of the rotated constellation is larger and not vanishing even when there is a deep fade along one of the component subchannels as illustrated in Fig. 10.3.

Y-precoding

When a pair of subchannels is well conditioned (i.e., λ_1/λ_2 is close to 1), performance of X-codes is good. However, their performance degrades when the pair of subchannels is ill conditioned (i.e., $\lambda_1/\lambda_2 \gg 1$). To improve performance in ill-conditioned subchannels, Y-precoding [5] can be used. The idea behind Y-precoding is as follows. In SVD precoding, the subchannel gains are the ordered singular values of the SVD of the channel matrix. By pairing these subchannels, one of the subchannels in a pair is stronger than the other. So, it is intuitive that the codewords be chosen so that the minimum Euclidean distance between the received code words along the stronger subchannel component is larger than that along the weaker subchannel component. By doing so, the code design can make use of the total constrained transmit power to achieve a minimum received codeword Euclidean distance greater than that achieved with

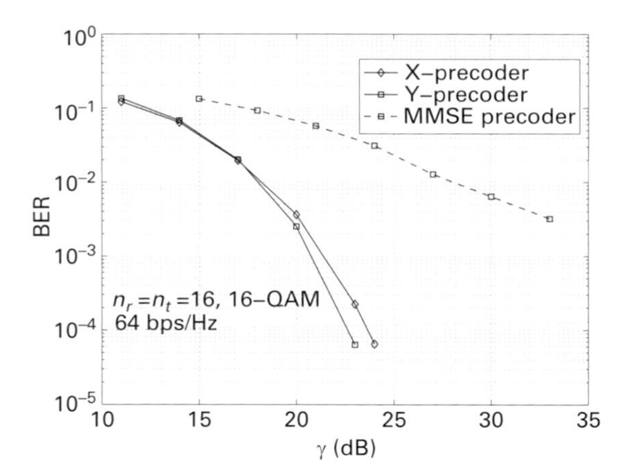

Figure 10.4 BER performance of the X-precoder, the Y-precoder, and the MMSE precoder for a 16×16 MIMO system with 16-QAM.

rotated constellations used in X-codes. Y-codes are designed based on this intuition, and the codewords form a subset of a two-dimensional real skewed lattice. Y-codes are parameterized with two parameters a_k and b_k related to power allocated to the two subchannels so that the $\mathbf{A}_k$ matrix in Y-codes is of the form

$$\mathbf{A}_k = \begin{bmatrix} a_k & 2a_k \\ 2b_k & 0 \end{bmatrix}, \quad a_k, b_k \in \mathbb{R}^+, \tag{10.22}$$

and a_k, b_k are computed so as to minimize the average error probability. The overall structure of Y-codes using these parameters for three pairs of subchannels for $n_t = n_r = 6$ is given by

$$\mathbf{G} = \begin{bmatrix} a_{1,1,1} & & & & & a_{1,1,2} \\ & a_{2,1,1} & & & a_{2,1,2} & \\ & & a_{3,1,1} & a_{3,1,2} & & \\ & & a_{3,2,1} & & & \\ & a_{2,2,1} & & & & \\ a_{1,2,1} & & & & & \end{bmatrix}. \tag{10.23}$$

As in X-codes, the optimum parameters for Y-codes can be computed independent of $\mathbf{H}$ whose performance is found to be good [5]. In addition, Y-codes have the advantage of lower detection complexity than X-codes. This is because ML detection of X-codes requires a two-dimensional search whereas ML detection of Y-codes needs only one-dimensional search.

10.1.3 Performance of X-codes and Y-codes

Figure 10.4 shows the BER performance of X-codes and Y-codes for a 16×16 MIMO system with 16-QAM and 64 bps/Hz spectral efficiency (i.e., $n_s = n_t =$

$n_r = 16$). The performance of the MMSE precoder is also plotted for comparison. It can be seen that both X- and Y-codes perform significantly better than the MMSE precoder. Also, Y-codes perform better than X-codes.

10.2 Precoding in a multiuser MIMO downlink

MIMO techniques applied to multiuser systems are of practical interest. A multiuser MIMO system on the downlink consists of a BS transmitter equipped with multiple transmit antennas communicating with multiple users each equipped with one or more receive antennas. In a single-user MIMO system with n_t transmit antennas and n_r receive antennas, the capacity grows linearly with $\min(n_t, n_r)$. It has been shown that the same capacity scaling applies in a multiuser scenario, where a transmitter with n_t transmit antennas communicates with n_u users [8]. In a multiuser MIMO downlink, the knowledge of CSIT can almost always be utilized to increase the system performance, whereas single-user MIMO systems benefit from having CSIT only when $n_t > n_r$ or at low SNRs [9]. The achievable rate in multiuser MIMO can be significantly larger than in single-user MIMO [10]–[12].

The optimal transmission scheme is based on DPC, where the transmitter jointly encodes the data symbols for all users using perfect knowledge of CSI [13]–[16]. Since the transmitter has perfect knowledge of the CSI and the data symbols for all users, it can compute the interference and subtract it prior to transmission. The symbols can be jointly encoded such that the achievable data rate is the same as if the interference from other user data streams was not present. Though optimal, the DPC based transmission scheme is prohibitively complex for practical implementation. Therefore, transmit-side preprocessing techniques suited for practical implementation are important. Such techniques fall into two types of precoding schemes, namely, linear precoding and non-linear precoding.

10.2.1 Linear precoding

Linear precoding involves a linear transformation of the data symbols meant for the users on the downlink using a precoding matrix. A block diagram of a multiuser MIMO system on the downlink is shown in Fig. 10.5. The BS equipped with n_t transmit antennas transmits simultaneously to n_u users, where the ith user is equipped with n_{r_i} receive antennas. The data for the ith user $\mathbf{u}_i \in \mathbb{C}^{m_i}$, where m_i is the number of data streams for the ith user, are transformed by the precoding matrix $\mathbf{B}_i$. The signal vector transmitted by the BS can be represented as

$$\mathbf{x} = \sum_{i=1}^{n_u} \mathbf{B}_i \mathbf{u}_i = \mathbf{B}\mathbf{u}, \tag{10.24}$$

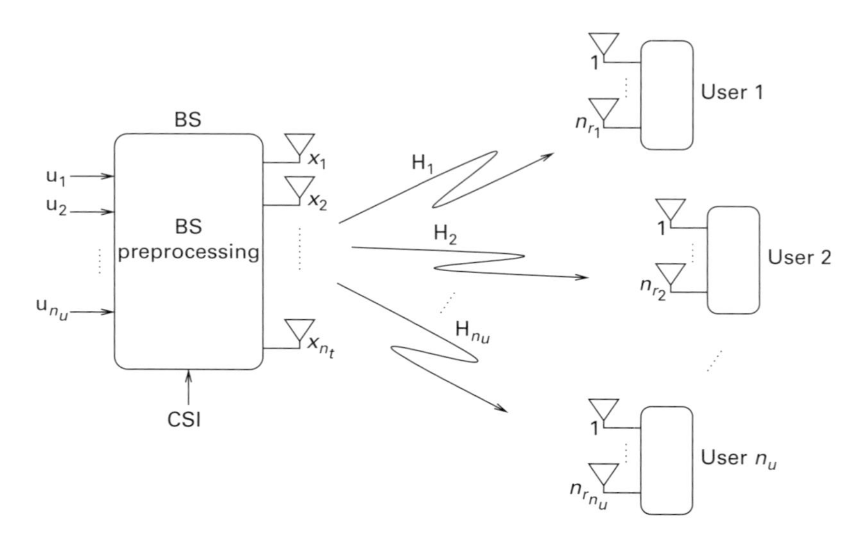

Figure 10.5 Multiuser MIMO system on the downlink.

where $\mathbf{B} = [\mathbf{B}_1\, \mathbf{B}_2 \cdots \mathbf{B}_{n_u}]$ is the global precoding matrix, and $\mathbf{u} = [\mathbf{u}_1^T\, \mathbf{u}_2^T \cdots \mathbf{u}_{n_u}^T]^T$ is the global data vector. The matrix of channel gains for the ith user is represented by $\mathbf{H}_i$, and the global channel matrix is given by $\mathbf{H} = [\mathbf{H}_1^T\, \mathbf{H}_2^T \cdots \mathbf{H}_{n_u}^T]^T$. The signal vector received at the ith user is given by

$$\mathbf{y}_i = \mathbf{H}_i \mathbf{B}_i \mathbf{u}_i + \mathbf{H}_i \sum_{j=1, j\neq i}^{n_u} \mathbf{B}_j \mathbf{u}_j + \mathbf{n}_i, \tag{10.25}$$

where $\mathbf{n}_i \in \mathbb{C}^{n_{r_i}}$ is the noise vector at the ith user. The global received signal vector can be written as

$$\mathbf{y} = \mathbf{HBu} + \mathbf{n}, \tag{10.26}$$

where $\mathbf{y} = [\mathbf{y}_1^T\, \mathbf{y}_2^T \cdots \mathbf{u}_{n_u}^T]^T$ and $\mathbf{n} = [\mathbf{n}_1^T\, \mathbf{n}_2^T \cdots \mathbf{u}_{n_u}^T]^T$.

Linear precoder designs based on different performance criteria are known in the literature [9],[10],[17]. For receivers with a single antenna, a simple approach is to design the precoding matrix such that the combined channel **HB** results in interference-free reception. This approach, called ZF precoding, results in the matrix given by

$$\mathbf{B} = \mathbf{H}^H (\mathbf{H}\mathbf{H}^H)^{-1}. \tag{10.27}$$

This method does not lead to the linear capacity growth with $\min(n_t, n_u)$ that is possible in the multiuser channel. This is because, with a power constraint, an ill-conditioned channel matrix when inverted needs a large normalization factor that dramatically reduces the SNR at the receivers. Generalization of the ZF method to the case of users with more than one antenna can be done [18]. At the receiver, if the ith user performs the receive processing with a receiver filter $\mathbf{C}_i$, then the estimate of its data symbol is expressed as

$$\widehat{\mathbf{u}}_i = \mathbf{C}_i\mathbf{H}_i\mathbf{B}_i\mathbf{u}_i + \mathbf{C}_i\mathbf{H}_i \sum_{j=1,j\neq i}^{n_u} \mathbf{B}_j\mathbf{u}_j + \mathbf{C}_i\mathbf{n}_i. \tag{10.28}$$

The generalized ZF approach, known as block-diagonalization, is to jointly design $\{\mathbf{B}_i\}_{i=1}^{n_u}$ and $\{\mathbf{C}_i\}_{i=1}^{n_u}$ such that $\mathbf{C}_i\mathbf{H}_i\mathbf{B}_j = \mathbf{0}\delta_{ij}$, where δ_{ij} is the Kronecker delta, i.e., $\delta_{ij} = 1$ if $i = j$, and zero otherwise. From (10.28), it can be observed that the above constraint ensures interference-free reception at each user. As it is difficult to obtain a closed-form solution, several iterative solutions have been proposed, e.g., [19],[20]. In such approaches, the transmitter generally computes a new effective channel for each user i employing the initial receive combining vector. Using this new effective channel, the transmitter recomputes the transmit filter $\mathbf{B}_i$ to enforce a zero interference condition and the receive filter $\mathbf{C}_i$, for all i. This process is repeated until a convergence criterion is satisfied. To extend this approach to multiple data streams for each user, the matrix of right singular vectors based on the number of data streams is used to calculate the effective channel matrix [19]–[21]. To avoid the use of extra feedback between the users and the BS, the computation of all the transmit and receive filters takes place at the BS. After this computation, either the users must acquire the effective combined channel or the information about the filters must be sent [20].

The disadvantages of the ZF precoder result from the stringent requirement of complete cancelation of interference. Relaxation of this requirement leads to a larger set of solutions that can potentially give higher capacity for a given transmit power, or require a lower transmit power for a given rate point. This behavior is seen in the solutions that maximize the sum capacity; they allow some level of multiuser interference at each receiver. Precoders based on the sum of mean square errors (SMSE) at the users do not suffer from the defects of the ZF precoder [22]–[26]. The algorithms reported in [22],[23] are iterative algorithms that minimize SMSE with a constraint on total transmit power, where the minimization is done alternately between the transmit precoder and receive filter. These algorithms are not guaranteed to converge to the global minimum. Minimum SMSE precoder and receiver designs reported in [24],[25] are based on uplink–downlink duality, and these algorithms are guaranteed to converge to the global minimum. Linear precoder designs have also been developed based on other performance criteria like SINR and information rate [27].

10.2.2 Non-linear precoding

Though linear precoders are of low complexity, they lag in performance compared to non-linear precoders, particularly when the set of active users is small [10]. Tomlinson–Harashima precoding (THP) is a popular non-linear precoding scheme. THP was originally proposed for time-domain equalization [28],[29]. This technique has been applied in multiuser MIMO systems for spatial equalization

[24],[30],[31]. Assuming perfect CSI, for a given user ordering, it is possible to perfectly presubtract the interference such that the ith user can estimate its signal as

$$\widehat{\mathbf{u}}_i = \mathbf{C}_i\mathbf{H}_i\mathbf{B}_i\mathbf{u}_i + \mathbf{C}_i\mathbf{H}_i\sum_{j\geq i}^{n_u}\mathbf{B}_j\mathbf{u}_j + \mathbf{C}_i\mathbf{n}_i. \tag{10.29}$$

From (10.28) and (10.29), it can be observed that the interference to the ith user, $i > 1$, is less in THP than in linear precoding. Here again, joint optimization of the precoder and user receive filters can improve performance. The additional operations involved in non-linear precoding like interference presubtraction and optimal user ordering result in improved performance of non-linear precoders compared to linear precoders, but also increase the complexity which is not desired for large multiuser systems.

10.2.3 Precoding in large multiuser MISO systems

This section considers low complexity precoding for large multiuser MISO systems having a large number of transmit antennas at the BS and a large number of downlink users, each user having one receive antenna. High sum-rates are possible in such large multiuser systems.

System model

Consider a multiuser MISO system, where a BS communicates with n_u users on the downlink (Fig. 10.6). The BS employs n_t transmit antennas and each user is equipped with one receive antenna (i.e., $n_r = 1$). Let $\mathbf{u}_c \in \mathbb{C}^{n_u}$ denote the complex information symbol vector, where the ith symbol in $\mathbf{u}_c$ is meant for the ith user, $i = 1, \ldots, n_u$. Precoding on the information symbol vector $\mathbf{u}_c$ is carried out to obtain the precoded symbol vector $\mathbf{x}_c \in \mathbb{C}^{n_t}$, which is transmitted using n_t transmit antennas such that the jth symbol of $\mathbf{x}_c$ is transmitted on the jth transmit antenna, $j = 1, \ldots, n_t$.

Let y_i denote the received complex signal at user i, and $\mathbf{y}_c = [y_1 y_2 \cdots y_{n_u}]^T$. Let $\mathbf{H}_c \in \mathbb{C}^{n_u \times n_t}$ denote the channel matrix such that its (i, j)th entry $h_{i,j}$ is the complex channel gain from the jth transmit antenna to the ith user's receive antenna. Assuming rich scattering, the entries of $\mathbf{H}_c$ are modeled as iid and $\mathcal{CN}(0,1)$. Let n_i denote the noise at the ith user, and $\mathbf{n}_c = [n_1 n_2 \cdots n_{n_u}]^T$. The elements of $\mathbf{n}_c$ are modeled as iid and $\mathcal{CN}(0, \sigma^2)$. Therefore, $\mathbf{y}_c$ can be expressed in terms of $\mathbf{H}_c$, $\mathbf{x}_c$, and $\mathbf{n}_c$ as

$$\mathbf{y}_c = \mathbf{H}_c\mathbf{x}_c + \mathbf{n}_c. \tag{10.30}$$

The complex-valued system model in (10.30) is converted into the real-valued system model given by

$$\mathbf{y} = \mathbf{H}\mathbf{x} + \mathbf{n}, \tag{10.31}$$

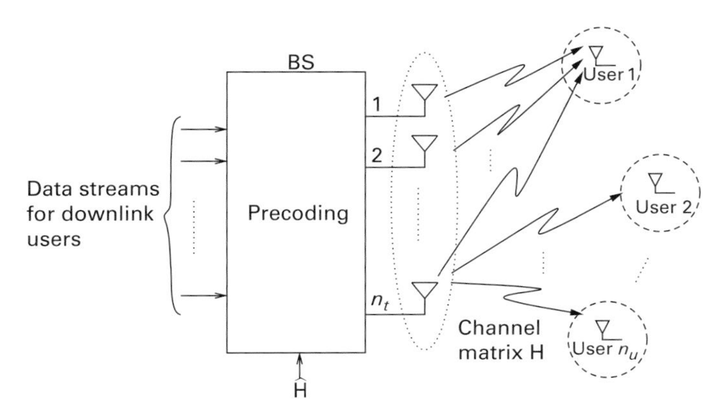

Figure 10.6 Precoding in multiuser MISO system on the downlink.

where

$$\mathbf{u} = [\mathbf{u}_I^T \ \mathbf{u}_Q^T]^T \in \mathbb{R}^{2n_u}, \ \mathbf{H} = \begin{pmatrix} \mathbf{H}_I & -\mathbf{H}_Q \\ \mathbf{H}_Q & \mathbf{H}_I \end{pmatrix} \in \mathbb{R}^{2n_u \times 2n_t},$$

$$\mathbf{x} = [\mathbf{x}_I^T \ \mathbf{x}_Q^T]^T \in \mathbb{R}^{2n_t}, \ \mathbf{y} = [\mathbf{y}_I^T \ \mathbf{y}_Q^T]^T, \in \mathbb{R}^{2n_u}, \ \mathbf{n} = [\mathbf{n}_I^T \ \mathbf{n}_Q^T]^T \in \mathbb{R}^{2n_u},$$

and

$$\mathbf{u}_c = \mathbf{u}_I + \mathrm{j}\mathbf{u}_Q, \ \mathbf{x}_c = \mathbf{x}_I + \mathrm{j}\mathbf{x}_Q, \ \mathbf{y}_c = \mathbf{y}_I + \mathrm{j}\mathbf{y}_Q, \ \mathbf{H}_c = \mathbf{H}_I + \mathrm{j}\mathbf{H}_Q, \ \mathbf{n}_c = \mathbf{n}_I + \mathrm{j}\mathbf{n}_Q.$$

With the above real-valued system model, the real part of the original complex information symbols (i.e., symbols in $\mathbf{u}_c$) will be mapped to $[u_1, \ldots, u_{n_u}]$ and the imaginary part of these symbols will be mapped to $[u_{n_u+1}, \ldots, u_{2n_u}]$. For M-PAM modulation, $[u_{n_u+1}, \ldots, u_{2n_u}]$ will be zeros since M-PAM symbols take only real values. In the case of M-QAM, $[u_1, \ldots, u_{n_u}]$ can be viewed to be from an underlying M-PAM signal set, and so is $[u_{n_u+1}, \ldots, u_{2n_u}]$.

Vector perturbation

With the above system model, let $\mathbf{G} \in \mathbb{R}^{2n_t \times 2n_u}$ denote the precoding matrix. Therefore, the unit-norm transmitted symbol vector $\mathbf{x}$ can be written as

$$\mathbf{x} = \frac{\mathbf{Gu}}{\|\mathbf{Gu}\|}. \tag{10.32}$$

For the ZF precoder [32] with $n_t \geq n_u$, the precoding matrix is given by

$$\mathbf{G} \ = \ \mathbf{G}_{ZF} = \mathbf{H}^T(\mathbf{HH}^T)^{-1}, \tag{10.33}$$

and the corresponding received signal vector $\mathbf{y}$ is given by

$$\mathbf{y} = \frac{\mathbf{u}}{\|\mathbf{Gu}\|} + \mathbf{n}. \tag{10.34}$$

From (10.34), it is seen that $\|\mathbf{Gu}\|$ has a scaling effect on the instantaneous received SNR at the users, and for poorly conditioned channels this results in a significant loss in SNR. It is assumed that $\|\mathbf{Gu}\|$ is known at the receiver so that the received signal is scaled by $\|\mathbf{Gu}\|$ prior to detection. Simulations show that using $\mathbb{E}\{\|\mathbf{Gu}\|\}$ instead of the instantaneous value of $\|\mathbf{Gu}\|$ results in almost the same performance. Hence, in order to improve performance, $\|\mathbf{Gu}\|$ needs to be minimized. One technique suggested in the literature is to perturb the information symbol vector $\mathbf{u}$ in such a way that the perturbed vector $\tilde{\mathbf{u}}$ is another point in the lattice, but $\|\mathbf{G}\tilde{\mathbf{u}}\|$ is much less than $\|\mathbf{Gu}\|$ [33]. Specifically, $\tilde{\mathbf{u}}$ can be defined as

$$\tilde{\mathbf{u}} = \mathbf{u} + \tau\,\mathbf{p}, \tag{10.35}$$

where $\mathbf{p} \in \mathbb{Z}^{2n_u}$ is the perturbation vector and τ is a positive real number. The optimal value of $\tilde{\mathbf{u}}$, denoted by $\tilde{\mathbf{u}}_{opt}$, is given by

$$\tilde{\mathbf{u}}_{opt} = \mathbf{u} + \tau\,\mathbf{p}_{opt}, \tag{10.36}$$

where

$$\mathbf{p}_{opt} = \underset{\mathbf{p}\in\mathbb{Z}^{2n_u}}{\operatorname{argmin}}\ \|\mathbf{G}(\mathbf{u} + \tau\mathbf{p})\|^2. \tag{10.37}$$

Exact solution of (10.37) requires exponential complexity in n_u. Approximate methods (with polynomial complexity) have been proposed in the literature to solve the problem in (10.37) [34]. Even these polynomial complexity precoders are prohibitively complex for large MISO systems with tens to hundreds of BS transmit antennas and users.

Detection at the receiver

In terms of detection at the receiver, let $\tilde{\mathbf{p}}$ be an approximate solution to (10.37). Then, the received signal vector (after scaling by $\|\mathbf{Gu} + \tau\tilde{\mathbf{p}}\|$) is given by

$$\tilde{\mathbf{y}} = (\mathbf{u} + \tau\,\tilde{\mathbf{p}}) + \tilde{\mathbf{n}}, \tag{10.38}$$

where

$$\tilde{\mathbf{n}} = \|\mathbf{G}(\mathbf{u} + \tau\,\tilde{\mathbf{p}})\|\mathbf{n}. \tag{10.39}$$

The detected symbol vector at the receiver is given by

$$\hat{\mathbf{u}} = \tilde{\mathbf{y}} - \tau \left\lfloor \frac{\tilde{\mathbf{y}} + \frac{\tau}{2}}{\tau} \right\rfloor. \tag{10.40}$$

In (10.40), the operation is defined on each entry of the vector since each user gets only one entry of the vector $\mathbf{y}$. The value of the positive real scalar τ is fixed. The choice of the value of τ affects the overall performance. Too high a value is good as far as mitigating the effect of receiver noise is concerned (since the constellation replicas are placed far apart, and there is little probability that noise may push a point from one replica to another), but on the other hand a high value of τ results in a high value of $\|\mathbf{G}(\mathbf{u} + \tau\,\tilde{\mathbf{p}})\|$. It has been empirically observed that a good choice of τ is given by [33]

$$\tau = 2|c_{max}| + d, \tag{10.41}$$

where $|c_{max}|$ is the maximum value of either the real or imaginary component of the constellation symbols, and d is the spacing between the constellation symbols.

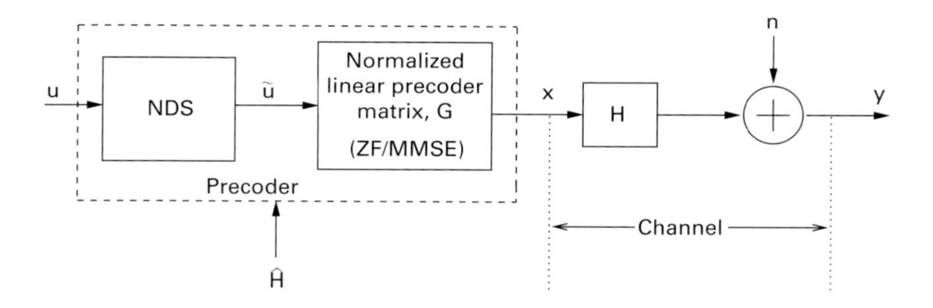

Figure 10.7 Low complexity precoder using NDS.

For example, 16-QAM is effectively two 4-PAM constellations in quadrature (taking values of -3, -1, 1, 3 on the real and imaginary axes). Therefore, for 16-QAM, $|c_{max}|$ is 3, d is 2, and so τ is 8. Similarly, τ is 4 for 4-QAM.

10.2.4 Precoder based on norm descent search (NDS)

Since vector perturbation precoding based on the solution of (10.37) does not scale well for large systems, low complexity alternatives are desired. Precoder design based on an approximate solution of (10.37) is presented in this section. The approximate solution is obtained through an iterative algorithm using NDS [35]. Figure 10.7 shows the block diagram of the precoder based on NDS.

The NDS algorithm works as follows. Let $\tilde{\mathbf{u}}^{(k)}$ be the perturbed information symbol vector after the kth iteration. The algorithm starts with $\tilde{\mathbf{u}}^{(0)} = \mathbf{u}$, where $\mathbf{u}$ is the unperturbed information symbol vector. Perturb $\tilde{\mathbf{u}}^{(k)}$ to get $\tilde{\mathbf{u}}^{(k+1)}$ as

$$\tilde{\mathbf{u}}^{(k+1)} = \tilde{\mathbf{u}}^{(k)} + \tau\,\mathbf{p}^{(k)}, \tag{10.42}$$

where $\mathbf{p}^{(k)} \in \mathbb{Z}^{2n_u}$ is the perturbation vector for the $(k+1)$th iteration. To reduce computational complexity, constrain $\mathbf{p}^{(k)}$ to have only one non-zero entry. This can be viewed to be similar to the one-symbol neighborhood definition in the 1-LAS algorithm for detection in Chapter 5.

Let $\mathbf{F} \triangleq \mathbf{G}^T\mathbf{G}$, where $\mathbf{G} \in \mathbb{R}^{2n_t \times 2n_u}$ is the precoding matrix, which can be one of the linear precoders (e.g., ZF or MMSE). Let $q^{(k)}$ be the power (squared-norm) of the precoded symbol vector after the kth iteration. Therefore, $q^{(k)}$ is given by

$$q^{(k)} = \|\mathbf{G}\tilde{\mathbf{u}}^{(k)}\|^2 \quad = \quad \tilde{\mathbf{u}}^{(k)^T}\mathbf{F}\,\tilde{\mathbf{u}}^{(k)}. \tag{10.43}$$

In the $(k+1)$th iteration, the algorithm finds a constrained integer vector $\mathbf{p}^{(k)}$ such that $q^{(k+1)} \leq q^{(k)}$. Let

$$\Delta q^{(k+1)} \triangleq q^{(k+1)} - q^{(k)}. \tag{10.44}$$

Let $\mathbf{e}_i$ denote a $2n_u$-dimensional vector of which only the ith entry is one, and all the other entries are zeros. Since only one non-zero entry in $\mathbf{p}^{(k)}$ is allowed, $\mathbf{p}^{(k)}$ can be expressed as a scaled integer multiple of some $\mathbf{e}_i$, $i = 1, \ldots, 2n_u$. Because $\Delta q^{(k+1)}$ can be negative for more than one choice of i, an appropriate i has to be chosen. Let $\Delta q_i^{(k+1)}$ denote the value of $\Delta q^{(k+1)}$ when $\mathbf{p}^{(k)}$ is a scaled integer multiple of $\mathbf{e}_i$. For each i, there exists a scaling integer for $\mathbf{e}_i$, $\lambda_i^{(k)}$, which minimizes $\Delta q_i^{(k+1)}$. Let this minimum value of $\Delta q_i^{(k+1)}$ be denoted $\Delta q_{i,opt}^{(k+1)}$. Therefore $\Delta q_{i,opt}^{(k+1)}$ can be expressed as

$$\Delta q_{i,opt}^{(k+1)} = {\lambda_i^{(k)}}^2 \tau^2 \, \mathbf{F}_{i,i} + 2\lambda_i^{(k)} \tau \, {z_i}^{(k)}, \tag{10.45}$$

where $\mathbf{F}_{i,i}$ is the ith diagonal entry of $\mathbf{F}$, ${z_i}^{(k)}$ is the ith entry of the vector

$$\mathbf{z}^{(k)} \triangleq \mathbf{F}\tilde{\mathbf{u}}^{(k)}, \tag{10.46}$$

and

$$\begin{aligned} \lambda_i^{(k)} &= \underset{\lambda \in \mathbb{Z}}{\operatorname{argmin}} \ \Delta q_i^{(k+1)} \\ &= \underset{\lambda \in \mathbb{Z}}{\operatorname{argmin}} \ \|\mathbf{G}(\tilde{\mathbf{u}}^{(k)} + \lambda\tau\mathbf{e}_i)\|^2 - \|\mathbf{G}\tilde{\mathbf{u}}^{(k)}\|^2 \\ &= \underset{\lambda \in \mathbb{Z}}{\operatorname{argmin}} \ \lambda^2 \mathbf{F}_{i,i} + \frac{2\lambda}{\tau} {\tilde{\mathbf{u}}^{(k)}}^T \mathbf{F} \, \mathbf{e}_i \\ &= \underset{\lambda \in \mathbb{Z}}{\operatorname{argmin}} \ \lambda^2 \mathbf{F}_{i,i} + \frac{2\lambda}{\tau} {z_i}^{(k)}. \end{aligned} \tag{10.47}$$

It can be shown that the exact solution to the minimization problem in (10.47) is given by

$$\lambda_i^{(k)} = -\text{sgn}({z_i}^{(k)}) \left\lfloor \frac{|{z_i}^{(k)}|}{\tau \mathbf{F}_{i,i}} \right\rceil. \tag{10.48}$$

Though (10.48) gives a closed-form solution for $\lambda_i^{(k)}$, it has been observed in the simulations that in cases in which $\lambda_i^{(k)}$ is large, the algorithm tends to become trapped in some poor local minima. In order to alleviate this phenomenon, the value of $\lambda_i^{(k)}$ is constrained to be within a set $\mathbb{S} = \{-s_{max}, -(s_{max} - 1), \ldots, (s_{max} - 1), s_{max}\}$, which is a finite subset of $\mathbb{Z}$, and where s_{max} denotes the maximum absolute value in $\mathbb{S}$. For example, the appropriate set $\mathbb{S}$ for 4-QAM is found (through simulations) to be $\mathbb{S} = \{-1, 0, 1\}$. If $|\lambda_i^{(k)}| > s_{max}$, then $\lambda_i^{(k)}$ is set to 0, and so is $\Delta q_{i,opt}^{(k+1)}$. If $|\lambda_i^{(k)}| \leq s_{max}$, then $\Delta q_{i,opt}^{(k+1)}$ is computed as per (10.45). We refer to this correction in $\lambda_i^{(k)}$ as λ-adjustment. Therefore, in the $(k+1)$th iteration, $\Delta q_{i,opt}^{(k+1)}$ for $i = 1, \ldots, 2n_u$ can be calculated. Given these values of $\lambda_i^{(k)}$, $i = 1, \ldots, 2n_u$, $\tilde{\mathbf{u}}^{(k)}$ is updated as follows:

$$\tilde{\mathbf{u}}^{(k+1)} = \tilde{\mathbf{u}}^{(k)} + \tau \, \lambda_j^{(k)} \mathbf{e}_j, \tag{10.49}$$

where

$$j = \underset{i}{\operatorname{argmin}} \ \Delta q_{i,opt}^{(k+1)}. \tag{10.50}$$

Table 10.1. *Listing of the precoding algorithm based on NDS*

1. Choose the set $\mathbb{S}$; $s_{\max} = \max_{s \in \mathbb{S}} s$
2. $\tilde{\mathbf{u}}^{(0)} = \mathbf{u}$; $\mathbf{F} = \mathbf{G}^T\mathbf{G}$; $k = 0$ (k: iteration index)
3. $\mathbf{z}^{(0)} = \mathbf{F}\tilde{\mathbf{u}}^{(0)}$; $\tau = 2|c_{max}| + d$
4. $nsymb = 2n_u$; ($nsymb$: $2n_u$ for QAM and n_u for PAM)
5. for $i = 1, 2, \ldots, nsymb$
6. $\quad \lambda_i^{(k)} = -\mathrm{sgn}({z_i}^{(k)}) \left\lfloor \dfrac{|{z_i}^{(k)}|}{\tau \mathbf{F}_{(i,i)}} \right\rceil$
7. $\quad$ if $(|\lambda_i^{(k)}| > s_{max})$ $\lambda_i^{(k)} = 0$
8. $\quad \Delta q_{i,opt}^{(k+1)} = {\lambda_i^{(k)}}^2 \tau^2 \mathbf{F}_{i,i} + 2\lambda_i^{(k)} \tau {z_i}^{(k)}$
9. end; (end of the for loop starting in Step 5)
10. $\Delta q_{min} = \min\limits_i \Delta q_{i,opt}^{(k+1)}$
11. if $(\Delta q_{min} \geq 0)$ goto Step 16
12. $j = \operatorname*{argmin}\limits_i \ \Delta q_{i,opt}^{(k+1)}$
13. $\tilde{\mathbf{u}}^{(k+1)} = \tilde{\mathbf{u}}^{(k)} + \tau \lambda_j^{(k)} \mathbf{e}_j$
14. $\mathbf{z}^{(k+1)} = \mathbf{z}^{(k)} + \tau \lambda_j^{(k)} \mathbf{f}_j$
15. $k = k + 1$, goto Step 5
16. Terminate

The values of $\lambda_j^{(k)}$ used in (10.49) are those after the λ-adjustment described above. From (10.46), $\mathbf{z}^{(k+1)}$ can be written as

$$\mathbf{z}^{(k+1)} - \mathbf{z}^{(k)} = \mathbf{F}\left(\tilde{\mathbf{u}}^{(k+1)} - \tilde{\mathbf{u}}^{(k)}\right), \tag{10.51}$$

which can be rewritten, using (10.49), as

$$\mathbf{z}^{(k+1)} = \mathbf{z}^{(k)} + \tau \lambda_j^{(k)} \mathbf{f}_j, \tag{10.52}$$

where $\mathbf{f}_j$ refers to the jth column of $\mathbf{F}$. The algorithm terminates after the nth iteration if

$$\min_i \Delta q_{i,opt}^{(n+1)} \geq 0. \tag{10.53}$$

It is easy to see that the algorithm guarantees a monotonic descent in $\|\mathbf{G}\tilde{\mathbf{u}}^{(k)}\|^2$ with every iteration until a local minimum is reached. Since $\lambda_i^{(k)}$ can take values only from a finite integer valued set $\mathbb{S}$ and $\|\mathbf{G}\tilde{\mathbf{u}}^{(k)}\|^2$ has a global minimum for perturbations with $\lambda_i^{(k)} \in \mathbb{S}$, it can be seen that the algorithm will terminate in a finite number of iterations. The listing of the precoding algorithm based on NDS is presented in Table 10.1.

10.2.5 Complexity and performance

In Table 10.1 the per-symbol computation complexities of $\mathbf{G}^T\mathbf{G}$ in Step 2 and $\mathbf{z}^{(0)}$ in Step 3 are $O(n_u n_t)$ and $O(n_u)$, respectively. Steps 5–15 constitute one basic iteration of the NDS algorithm, whose per-symbol complexity is constant, i.e., $O(1)$. The mean number of iterations till the algorithm terminates (obtained through simulations) has been found to be proportional to n_u. Combining the above individual complexities, the overall per-symbol complexity of the NDS precoder is $O(n_t n_u)$, which, as with linear precoders, can scale well for large systems.

The BER performance of the NDS precoder with the $\mathbf{G}$ matrix taken to be either the ZF or the MMSE precoding matrix is obtained. The ZF precoding matrix $\mathbf{G}_{ZF}$ is given by

$$\mathbf{G}_{ZF} = \mathbf{H}^T(\mathbf{H}\mathbf{H}^T)^{-1}, \tag{10.54}$$

and the MMSE precoding matrix is given by

$$\mathbf{G}_{MMSE} = \mathbf{H}^T(\mathbf{H}\mathbf{H}^T + \sigma^2 n_t \mathbf{I}_{n_u})^{-1}. \tag{10.55}$$

The NDS precoder is referred to as the NDS-MMSE precoder when $\mathbf{G}_{MMSE}$ is used as the precoding matrix, and as the NDS-ZF precoder when $\mathbf{G}_{ZF}$ is used.

Figure 10.8 shows the BER performances of NDS-MMSE and NDS-ZF precoders for $n_t = n_u = 50, 200$, $n_r = 1$, and 4-QAM. Perfect knowledge of $\mathbf{H}$ is assumed. It is noted that the NDS-MMSE precoder performs better than the NDS-ZF precoder. Also, increasing $n_t = n_u$ improves the performance of the NDS-MMSE precoder: the performance for $n_t = n_u = 200$ is better than the performance for $n_t = n_r = 50$. This shows that precoding based on NDS not only scales well for large systems with tens to hundreds of transmit antennas and users, but can achieve increasingly better performance for increasing $n_t = n_u$.

Figure 10.9 presents a comparison between the BER performances of the NDS-MMSE precoder, the MMSE precoder (without NDS), and the VP precoder with sphere encoding (VP-SE) in [33] for $n_u = 8$, $n_r = 1$, and 4-QAM. The VP-SE precoder employs the VP scheme in [33] which uses sphere encoding to solve (10.37). The performances of these three precoders for $n_t = 8, 16$, are shown. The following observations can be made from Fig. 10.9. First, the system with $n_t = 16, n_u = 8$ performs significantly better than the system with $n_t = n_u = 8$, which is expected because of the availability of $n_t - n_u$ additional dimensions for the precoder to exploit. Second, comparing the performances of the MMSE precoder (without NDS) and the NDS-MMSE precoders, it is seen that carrying out NDS prior to MMSE precoding leads to much better diversity order than just MMSE precoding. Given that the NDS operation itself is of low complexity, i.e., $O(n_u)$ per-symbol complexity, compared to the $O(n_t n_u)$ per-symbol complexity of $\mathbf{G}^T\mathbf{G}$ and MMSE operations, this improvement due to the NDS operation is quite appealing. Third, comparing the performances of NDS-MMSE and VP-

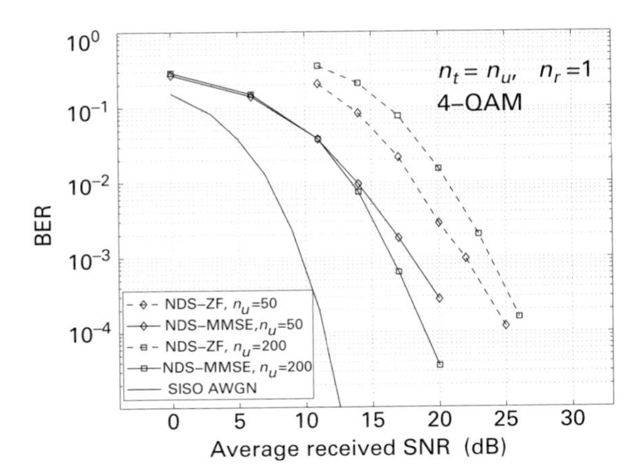

Figure 10.8 BER performances of the NDS-ZF and NDS-MMSE precoders for $n_t = n_u = 50, 200$, $n_r = 1$, and 4-QAM.

SE precoders, it is seen that the VP-SE precoder gives the better performance at moderate to high SNRs. However, the performance of the NDS-MMSE precoder is quite close to that of the VP-SE precoder at these SNRs. For example, for $n_t = 16$, the SNR gap between the VP-SE and NDS-MMSE performances at 10^{-3} BER is just about 0.4 dB. It is important to note that the NDS-MMSE precoder achieves this good performance at a much reduced complexity compared to the exponential complexity of the VP-SE precoder in solving (10.37). This low complexity advantage of the NDS-MMSE over the VP-SE precoder is illustrated in Fig. 10.10. It can be seen that the VP-SE precoder has exponential complexity in $n_t = n_u$, whereas the NDS-MMSE precoder has similar complexity to the MMSE precoder.

In the above context, it is of interest to note that a new type of cellular structure that comprises inexpensive single-antenna terminals working with BSs with 50 or 100 antennas, each driven by its own tower-top amplifier, of power no greater than a typical cell-phone power amplifier, is envisioned in [36]. Precoders like the NDS precoder can address the need for low complexity near-optimal precoding algorithms in such large multiuser MISO systems.

10.2.6 Closeness to sum capacity

The coded BER and the closeness to capacity performance of the NDS-MMSE precoder are of interest. A relevant metric is the ergodic sum capacity of the broadcast MISO channel.

The ergodic sum capacity of the system model in (10.30) is given by [33]

$$C_{sum} = \mathbb{E}\left[\sup_{\mathbf{D}\in\mathcal{A}} \log\det\left(\mathbf{I}_{n_t} + \rho\mathbf{H}_c^H\mathbf{D}\mathbf{H}_c\right)\right], \tag{10.56}$$

where $\mathcal{A}$ is the set of $n_u \times n_u$ diagonal matrices with non-negative elements that sum to 1 (i.e., $\text{tr}\{\mathbf{D}\} = 1$), and ρ is the average SNR defined as $1/\sigma^2$.

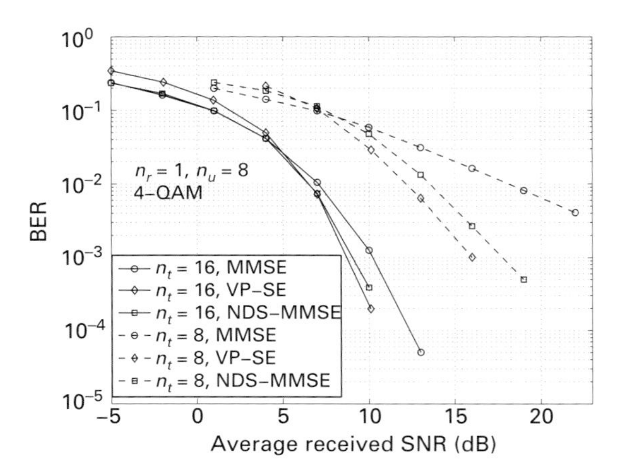

Figure 10.9 BER performance comparison between MMSE, NDS-MMSE, and VP-SE precoders. $n_t = 8, 16$, $n_u = 8$, $n_r = 1$ and 4-QAM.

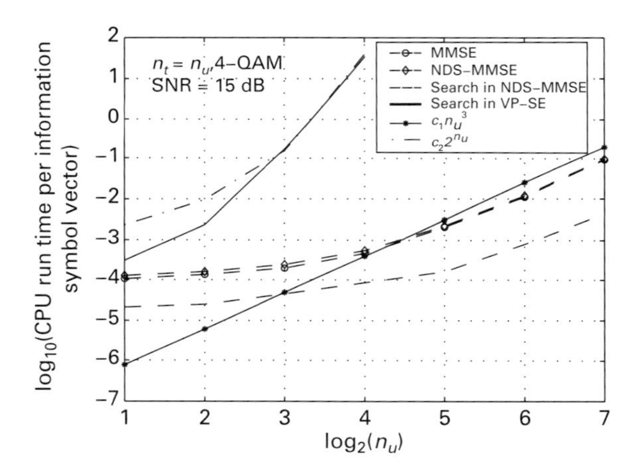

Figure 10.10 Complexity comparison between MMSE, NDS-MMSE, and VP-SE precoders. $n_t = n_u$, $n_r = 1$, 4-QAM, SNR = 15 dB.

Since there is no closed-form expression for the optimization in (10.56), it has to be evaluated through Monte Carlo simulations, which are prohibitive for large systems. So upper and lower bounds to the sum capacity can be considered. Note that $\mathbf{D} = \mathbf{D}_{CSIR} \triangleq (1/n_t)\mathbf{I}_{n_t}$ satisfies the trace constraint, and therefore

$$C_{CSIR} \triangleq \mathbb{E}\Big[\log\det\Big(\mathbf{I}_{n_t} + \frac{\rho}{n_t}\mathbf{H}_c^H\mathbf{H}_c\Big)\Big] \tag{10.57}$$

is a lower bound for C_{sum}, i.e., $C_{sum} \geq C_{CSIR}$. Also note that C_{CSIR} is the ergodic capacity of a point-to-point single-user MIMO system with n_t receive antennas and n_u transmit antennas with CSIR only. On the other hand, receiver cooperation between the users increases the capacity, and therefore the sum capacity C_{sum} is upper bounded by the capacity of a point-to-point MIMO sys-

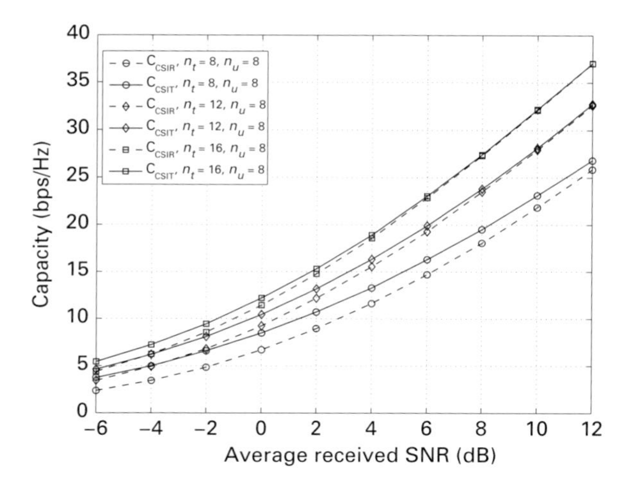

Figure 10.11 Upper and lower bounds for the ergodic sum capacity, C_{sum}.

tem with n_t transmit antennas and n_u receive antennas with CSIT and CSIR. We denote this upper bound as C_{CSIT}.

In Fig. 10.11, the upper and lower bounds (C_{CSIT} and C_{CSIR}) of the sum capacity C_{sum} are plotted. It is observed that the gap between the bounds diminishes with increasing SNR, and therefore any of these bounds is a good approximation at high SNR. However, at low SNRs, there is a gap between the bounds, which diminishes as the system becomes more asymmetrical. For example, with $n_u = 8$ users and a target spectral efficiency of 1.5 bps/Hz for each user, the gap between the upper and lower bounds is 0.5, 0.8, and 1.3 dB for $n_t = 16$, 12, and 8, respectively. The performance of the NDS-MMSE precoder with turbo coding and its closeness to the upper bound on C_{sum} is illustrated in Fig. 10.12.

Figure 10.12 shows the turbo coded BER performance of the NDS-MMSE and VP-SE precoders for $n_u = 8$, $n_t = 8, 12, 16$, 4-QAM, $n_r = 1$, rate-3/4 turbo code, and sum-rate $8 \times 2 \times 3/4 = 12$ bps/Hz. The minimum SNRs required to achieve a sum-rate of 12 bps/Hz obtained from the upper bound on the sum capacity curves in Fig. 10.11 are also shown. From Fig. 10.12, it can be seen that the VP-SE precoder achieves a vertical fall in coded BER at about 9.2, 7.8, and 7.2 dB away from the respective theoretical minimum SNRs required for $n_t = 8$, 12, and 16. It is further seen that the vertical fall for the NDS-MMSE precoder for $n_t = n_u = 8$ occurs at about 1.5 dB away from that of the VP-SE precoder. For the asymmetric cases of $n_t = 12$ and $n_t = 16$, the performance of the NDS-MMSE precoder is quite close to that of the VP-SE precoder.

10.3 Multicell precoding

Since multicell operation is typical in practical systems, multiuser precoding in multicell systems is of interest. Multicell precoding with and without BS

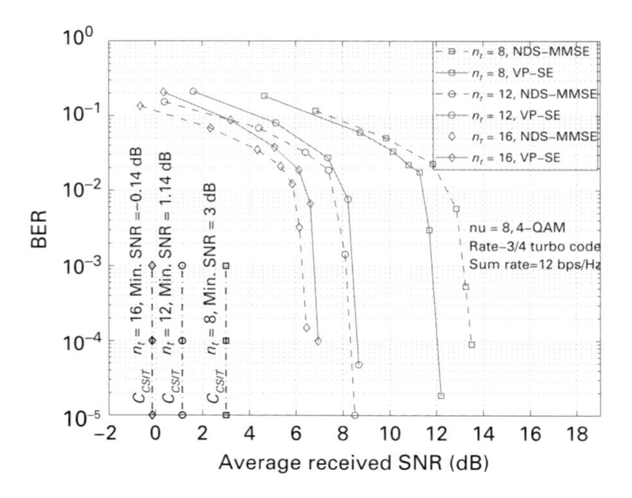

Figure 10.12 Turbo coded BER performance of NDS-MMSE and VP-SE precoders for $n_u = 8$, $n_t = 8, 12, 16$, $n_r = 1$, 4-QAM, rate-3/4 turbo code, sum-rate = 12 bps/Hz.

cooperation is considered in this section. Consider a multicell multiuser MIMO TDD system, where there are L cells and K users in each cell. If orthogonal pilots are used on the uplink, the length of orthogonal sequences is KL, which limits the throughput of the system in channels with smaller coherence times. Therefore, non-orthogonal pilots need to be used. Use of non-orthogonal pilots has a serious impact on the sum-rate because of the inter-cell interference caused by the non-orthogonal pilots, which is referred to as the 'pilot contamination problem' in the literature [37]. In [37]–[39], the impact of pilot contamination on the achievable rate is derived analytically. It has been shown that as the number of BS antennas grows to infinity, the achievable rate approaches a constant limit that depends on the path loss of the other users employing the same pilot sequence. This implies that if the path loss factors of the other users who employ the same pilot sequence are the same order as the path loss of the user of interest, the impact of pilot contamination will be very significant.

Some studies have focused on mitigating the pilot contamination problem [37],[40]. In [37], a linear precoding method called multicell MMSE precoding is derived to mitigate the pilot contamination problem. The precoding matrix for each cell is found based on minimizing a metric which constitutes the sum of the expected errors of all users and the interference caused by this cell to the other cells. In deriving the precoding matrix, the statistics of the CSI estimation error, which is a function of pilot sequences, has been used. This work assumes that BSs do not cooperate. However, cooperation among BSs using high-speed backhaul links is increasingly being adopted. BS cooperation can be exploited to mitigate the pilot contamination problem in multicell systems. This is because the inter-cell interference can be decreased effectively by precoding based on the information provided by the other BSs. Here, the design of the precoding matrices of all the BSs can be done jointly by minimizing the SMSE in the system. This design of joint precoding matrix requires the CSI to be estimated at all the BSs. So, the symbols transmitted at the antennas of each BS will be the linear transformation of all the information symbols in the system. Precoding matrices for all BSs with BS cooperation can be designed considering the minimization of the MSE of all users in the system.

10.3.1 System model

Consider a multicell TDD multiuser MIMO system where there are L cells and K users in each cell. Each user has one antenna and each BS has M antennas. The propagation factor between the mth BS antenna of the lth cell and the kth user of the jth cell is given by $\sqrt{\beta_{jlk}}h_{jlkm}$, as shown in Fig. 10.13. The non-negative real coefficients $\{\beta_{jlk}\}$ model the path loss and shadowing. It is assumed that these coefficients change slowly over long coherence times and are known perfectly at the BS. The complex coefficients $\{h_{jlkm}\}$ model the multipath fading and are assumed to be iid zero mean, circularly symmetric complex Gaussian $\mathcal{CN}(0,1)$ random variables. These coefficients are estimated at the BS through uplink pilots for every coherence interval. We assume that channel reciprocity, i.e., the propagation factor $\sqrt{\beta_{jlk}}h_{jlkm}$ is the same on both the uplink and the downlink. The fading coefficients between the BS of the lth cell and the users in the jth cell are represented by a matrix of size $K \times M$, given by

$$\mathbf{H}_{jl} = \begin{bmatrix} h_{jl11} & h_{jl12} & \cdots & h_{jl1M} \\ h_{jl21} & h_{jl22} & \cdots & h_{jl2M} \\ \vdots & \vdots & \ddots & \vdots \\ h_{jlK1} & h_{jlK2} & \cdots & h_{jlKM} \end{bmatrix}. \tag{10.58}$$

The coefficients that correspond to path loss and shadowing between the BS of the lth cell and the users in the jth cell are represented by a $K \times K$ diagonal matrix, given by

$$\mathbf{D}_{jl} = \mathrm{diag}\left\{\beta_{jl1} \quad \beta_{jl2} \quad \cdots \quad \beta_{jlK}\right\}. \tag{10.59}$$

There are two phases in the communication scheme: (i) uplink training, and (ii) downlink data transmission. The uplink training phase consists of users transmitting pilots to the corresponding BSs, and downlink data transmission consists of BSs transmitting data to their users.

Uplink training

In uplink training, each user employs a training sequence of length N_p. The training sequence used by the kth user in the jth cell is denoted as $\mathbf{s}_{jk}$. The

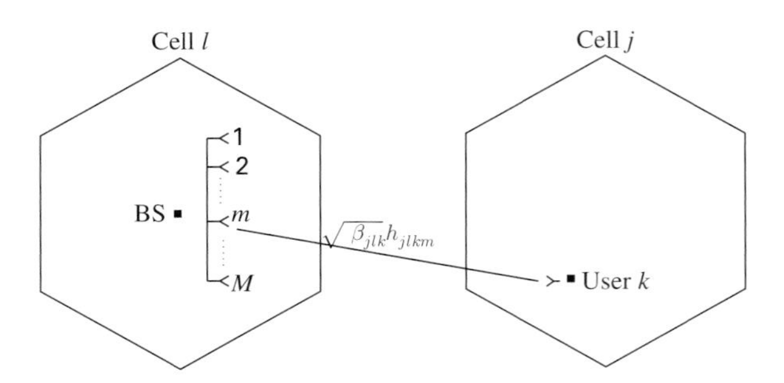

Figure 10.13 System model showing the BS in the lth cell and the kth user in the jth cell.

received $N_p \times M$ matrix at the lth BS is given by

$$\mathbf{Y}_l = \sqrt{P_u N_p} \sum_{j=1}^{L} \mathbf{S}_j \mathbf{D}_{jl}^{\frac{1}{2}} \mathbf{H}_{jl} + \mathbf{W}_l, \tag{10.60}$$

where $\mathbf{S}_j$ is a $N_p \times K$ matrix given by $\mathbf{S}_j = [\mathbf{s}_{j1} \ \mathbf{s}_{j2} \ \cdots \ \mathbf{s}_{jK}]$, P_u is the average transmit power at each user on the uplink, and $\mathbf{W}_l$ is the additive noise matrix of size $N_p \times M$ whose elements are $\mathcal{CN}(0,1)$ random variables. The MMSE estimate of $\mathbf{H}_{jl}$ is given by

$$\widehat{\mathbf{H}}_{jl} = \sqrt{P_u N_p} \mathbf{D}_{jl}^{\frac{1}{2}} \mathbf{S}_j^H \left(\mathbf{I} + P_u N_p \sum_{i=1}^{L} \mathbf{S}_i \mathbf{D}_{il} \mathbf{S}_i^H \right)^{-1} \mathbf{Y}_l. \tag{10.61}$$

The matrix $\mathbf{M}_{jl} =: \sqrt{P_u N_p} \mathbf{D}_{jl}^{\frac{1}{2}} \mathbf{S}_j^H \left(\mathbf{I} + P_u N_p \sum_{i=1}^{L} \mathbf{S}_i \mathbf{D}_{il} \mathbf{S}_i^H \right)^{-1}$ in the above equation is obtained by solving the following MMSE problem:

$$\underset{\mathbf{M}_{jl}}{\operatorname{argmin}} \ \mathbb{E}\left[\|\mathbf{H}_{jl} - \mathbf{M}_{jl} \mathbf{Y}_l\|_F^2 \right]. \tag{10.62}$$

At the end of the training phase, the lth BS has $\widehat{\mathbf{H}}_{jl}, j = 1, 2, \ldots, L$. The error in CSI estimate is denoted $\widetilde{\mathbf{H}}_{jl} = \mathbf{H}_{jl} - \widehat{\mathbf{H}}_{jl}$.

Downlink data transmission

Two types of data transmission schemes on the downlink are possible: one without BS cooperation [37], and another with BS cooperation.

System without BS cooperation

In the system without BS cooperation, the M-length symbol vector to be transmitted from the lth BS is a linear transformation of the information symbols to be transmitted to the K users in the same cell. The linear transformation is a multicell precoding matrix. Let $\mathbf{x}_l = [x_{l1} \ x_{l2} \ \cdots \ x_{lK}]^T$ denote the information symbol vector to be transmitted to the users in the lth cell. Let $\mathbf{A}_l$ denote the $M \times K$ precoding matrix used at the lth BS. $\mathbf{A}_l$ is a function of

$\widehat{\mathbf{H}}_{jl}, j = 1, 2, \ldots, L$. We assume that $\mathbb{E}[\mathbf{x}_l \mathbf{x}_l^H] = \mathbf{I}_K$ and $\mathrm{tr}\{\mathbf{A}_l^H \mathbf{A}_l\} = 1$. The $K \times 1$ received signal vector at the users in the jth cell is given by

$$\mathbf{r}_j = \sum_{l=1}^{L} \sqrt{P_d}\, \mathbf{D}_{jl}^{1/2}\, \mathbf{H}_{jl} \underbrace{\mathbf{A}_l \mathbf{x}_l}_{\text{Tx vector}} + \mathbf{z}_j, \tag{10.63}$$

where P_d is the average transmit power from each BS on the downlink, and $\mathbf{z}_j = [z_{j1}\ z_{j2}\ \cdots\ z_{jK}]^T$ is the $K \times 1$ additive noise vector whose elements are $\mathcal{CN}(0,1)$ random variables.

System with BS cooperation

In the system with BS cooperation, we assume that each BS knows the information symbols and CSI of all other BSs through a high-speed backbone network. Here, the M-length symbol vector to be transmitted from a BS is a linear transformation of the information symbols of all the users in the system. The precoding matrix used at each BS is a function of all the CSI matrices at all the BSs. Let $\mathbf{x} = [\mathbf{x}_1^T\ \mathbf{x}_2^T\ \cdots\ \mathbf{x}_L^T]^T$ denote the vector of information symbols of all the users in system. Let $\mathbf{B}_l$ denote the $M \times KL$ precoding matrix used at lth BS. The $K \times 1$ received signal vector at all the users in the jth cell is given by

$$\mathbf{r}_j = \sum_{l=1}^{L} \sqrt{P_d}\, \mathbf{D}_{jl}^{1/2}\, \mathbf{H}_{jl} \underbrace{\mathbf{B}_l \mathbf{x}}_{\text{Tx vector}} + \mathbf{z}_j. \tag{10.64}$$

Achievable rates

The received signal at the kth user of the jth cell can be expressed as

$$\begin{aligned} r_{jk} &= \sum_{l=1}^{M} \sum_{i=1}^{K} g_{jl}^{ki}\, x_{li} + z_{jk} \\ &= g_{jj}^{kk}\, x_{jk} + \sum_{(l,i) \neq (j,k)} g_{jl}^{ki} x_{li} + z_{jk}, \end{aligned} \tag{10.65}$$

where g_{jl}^{ki} depends on the system model. In the system without BS cooperation, g_{jl}^{ki} is given as the (k,i)th element of the matrix $\sqrt{P_d}\, \mathbf{D}_{jl}^{1/2}\, \mathbf{H}_{jl} \mathbf{A}_l$. In the system with BS cooperation, g_{jl}^{ki} is given as the $(k,(j-1)K)$th element of the matrix $\sum_{l=1}^{L} \sqrt{P_d}\, \mathbf{D}_{jl}^{1/2}\, \mathbf{H}_{jl} \mathbf{B}_l$. The complex term g_{jj}^{kk} is not known at the user. So, the received symbol expression is rewritten as

$$\begin{aligned} r_{jk} &= \mathbb{E}\left[g_{jj}^{kk}\right] x_{jk} + \left(g_{jj}^{kk} - \mathbb{E}\left[g_{jj}^{kk}\right]\right) x_{jk} + \sum_{(l,i) \neq (j,k)} g_{jl}^{ki}\, x_{li} + z_{jk}, \\ &= \mathbb{E}\left[g_{jj}^{kk}\right] x_{jk} + z_{jk}^{'}, \end{aligned} \tag{10.66}$$

where $z_{jk}^{'}$ is the effective additive noise term. The motivation for signal term $\mathbb{E}\left[g_{jj}^{kk}\right] x_{jk}$ in (10.66) comes from the fact that $\mathbb{E}\left[g_{jj}^{kk}\right]$ is known at the user as it depends only on the channel statistics, and not on the instantaneous channel. The variance of the effective additive noise is minimum with this rearrangement of terms. The variance of the effective noise term $z_{jk}^{'}$ is given by $\text{var}\{z_{jk}^{'}\} = \text{var}\{g_{jj}^{kk}\} + \sum_{(l,i)\neq(j,k)} \mathbb{E}\left[|g_{jl}^{ki}|^2\right] + 1$. It can be verified that $\mathbb{E}\left[x_{jk} z_{jk}^{'}\right] = 0$. So, the additive noise term is uncorrelated with the signal term. In [41], it has been proved that the worst case uncorrelated additive noise is independent Gaussian noise of the same variance. This implies that the rate achievable with independent Gaussian noise is achievable in the case of any uncorrelated additive noise whose distribution may not be Gaussian. So, the rate given by the following expression is achievable for the kth user of the jth cell:

$$R_{jk} = \log_2 \left(1 + \frac{\left|\mathbb{E}\left[g_{jj}^{kk}\right]\right|^2}{\text{var}\{g_{jj}^{kk}\} + \sum_{(l,i)\neq(j,k)} \mathbb{E}\left[|g_{jl}^{ki}|^2\right] + 1} \right). \tag{10.67}$$

10.3.2 Precoding without BS cooperation

The multicell MMSE precoding method developed in [37] without BS cooperation is presented in this section. The following minimization of an objective function for the lth cell is used:

$$\min_{\mathbf{A}_l, \alpha_l} \mathbb{E}_{\widetilde{\mathbf{F}}_{jl}, \mathbf{z}_l, \mathbf{x}_l} \left[\|\alpha_l \left(\mathbf{F}_{ll} \mathbf{A}_l \mathbf{x}_l + \mathbf{z}_l\right) - \mathbf{x}_l\|^2 + \sum_{j\neq l} \|\alpha_l \gamma \left(\mathbf{F}_{jl} \mathbf{A}_l \mathbf{x}_l\right)\|^2 \,\Big|\, \widehat{\mathbf{F}}_{jl} \right] \tag{10.68}$$

$$\text{subject to } \operatorname{tr}\{\mathbf{A}_l^H \mathbf{A}_l\} = 1,$$

where $\mathbf{F}_{jl} = \sqrt{P_d}\, \mathbf{D}_{jl}^{1/2}\, \mathbf{H}_{jl}$, $\widehat{\mathbf{F}}_{jl} = \sqrt{P_d}\, \mathbf{D}_{jl}^{1/2}\, \widehat{\mathbf{H}}_{jl}$, and $\widetilde{\mathbf{F}}_{jl} = \sqrt{P_d}\, \mathbf{D}_{jl}^{1/2}\, \widetilde{\mathbf{H}}_{jl}$, for all j, l. This objective function is quite intuitive. It consists of two parts: (i) the expected sum of squares of errors seen by the users in the lth cell, and (ii) the expected sum of squares of interference seen by the users in the other cells. The parameter γ controls the relative weights associated with these two parts. The real scalar parameter α_l is important as it corresponds to scaling that can be performed at the users. The following property of the CSI estimation error is used in solving the optimization problem in (10.68): $\mathbb{E}\left[\widetilde{\mathbf{F}}_{jl}^H \widetilde{\mathbf{F}}_{jl} \,\middle|\, \widehat{\mathbf{F}}_{jl}\right] = \eta_{jl} \mathbf{I}_M$, where

$$\eta_{jl} = P_d \operatorname{tr}\left\{ \mathbf{D}_{jl} \left(\mathbf{I}_K - P_u N_p \mathbf{D}_{jl}^{1/2} \mathbf{S}_j^H \Delta_l \mathbf{S}_j \mathbf{D}_{jl}^{1/2} \right) \mathbf{D}_{jl}^{1/2} \right\}, \tag{10.69}$$

and $\Delta_l = \left(\mathbf{I} + P_u N_p \sum_{i=1}^{L} \mathbf{S}_i \mathbf{D}_{il} \mathbf{S}_i^H\right)^{-1}$. It can be seen that the above matrix includes the training sequences of the users. The solution to the optimization problem in (10.68) in closed form is obtained as

$$\mathbf{A}_l^{opt} = \frac{1}{\alpha_l^{opt}} \left(\widehat{\mathbf{F}}_{ll}^H \widehat{\mathbf{F}}_{ll} + \gamma^2 \sum_{j\neq l} \widehat{\mathbf{F}}_{jl}^H \widehat{\mathbf{F}}_{jl} + \zeta \mathbf{I}_M \right)^{-1} \widehat{\mathbf{F}}_{ll}^H, \tag{10.70}$$

where $\zeta = \eta_{ll} + \gamma^2 \sum_{j\neq l} \eta_{jl} + K$, and α_l^{opt} satisfies $\operatorname{tr}\{(\mathbf{A}_l^{opt})^H \mathbf{A}_l^{opt}\} = 1$. This precoding method outperforms single-cell precoding methods as the optimization in this method considers the inter-cell interference and the statistics of the CSI estimation error. The effect of pilot contamination can be further reduced and increased sum-rates can be achieved through precoding with BS cooperation.

10.3.3 Precoding with BS cooperation

A precoding matrix design that exploits cooperation among the BSs is presented in this section. Define the following $KL \times ML$ matrices:

$$\widehat{\mathbf{F}} = \begin{bmatrix} \widehat{\mathbf{F}}_{11} & \widehat{\mathbf{F}}_{12} & \cdots & \widehat{\mathbf{F}}_{1L} \\ \widehat{\mathbf{F}}_{21} & \widehat{\mathbf{F}}_{22} & \cdots & \widehat{\mathbf{F}}_{2L} \\ \vdots & \vdots & \ddots & \vdots \\ \widehat{\mathbf{F}}_{L1} & \widehat{\mathbf{F}}_{L2} & \cdots & \widehat{\mathbf{F}}_{LL} \end{bmatrix}, \quad \widetilde{\mathbf{F}} = \begin{bmatrix} \widetilde{\mathbf{F}}_{11} & \widetilde{\mathbf{F}}_{12} & \cdots & \widetilde{\mathbf{F}}_{1L} \\ \widetilde{\mathbf{F}}_{21} & \widetilde{\mathbf{F}}_{22} & \cdots & \widetilde{\mathbf{F}}_{2L} \\ \vdots & \vdots & \ddots & \vdots \\ \widetilde{\mathbf{F}}_{L1} & \widetilde{\mathbf{F}}_{L2} & \cdots & \widetilde{\mathbf{F}}_{LL} \end{bmatrix},$$

where $\widehat{\mathbf{F}}_{jl}$ and $\widetilde{\mathbf{F}}_{jl}$ are as defined before. The optimization problem is defined as follows:

$$\min_{\alpha, \{\mathbf{B}_l\}_{l=1}^{L}} \mathbb{E}_{\widetilde{\mathbf{F}}, \mathbf{z}, \mathbf{x}} \sum_{l=1}^{L} \left[\left\| \alpha \left(\left(\sum_{j=1}^{L} \mathbf{F}_{lj} \mathbf{B}_j \right) \mathbf{x} + \mathbf{z}_l \right) - \mathbf{x}_l \right\|_2^2 \Bigg| \widehat{\mathbf{F}} \right] \tag{10.71}$$

$$\text{subject to } \operatorname{tr}\left\{\mathbf{B}_l^H \mathbf{B}_l\right\} = 1, \ \forall l = 1, 2, \ldots, L.$$

Here, the objective function is the expected sum of the errors seen by all the users in the system. The Lagrangian formulation is

$$\begin{aligned}
&L(\mathbf{B}_1, \mathbf{B}_2, \ldots, \mathbf{B}_L, \alpha, \lambda_1, \lambda_2, \ldots, \lambda_L) \\
&= \sum_{l=1}^{L} \mathbb{E}_{\widetilde{\mathbf{F}},\mathbf{z},\mathbf{x}} \left[\left\| \alpha \left(\left(\sum_{j=1}^{L} \mathbf{F}_{lj} \mathbf{B}_j \right) \mathbf{x} + \mathbf{z}_l \right) - \mathbf{x}_l \right\|_2^2 \Bigg| \widehat{\mathbf{F}} \right] + \sum_{l=1}^{L} \lambda_l \left(\operatorname{tr}\{\mathbf{B}_l^H \mathbf{B}_l\} - 1 \right), \\
&= \sum_{l=1}^{L} \operatorname{tr} \mathbb{E}_{\widetilde{\mathbf{F}},\mathbf{z},\mathbf{x}} \left[\left\{ \alpha \left(\left(\sum_{j=1}^{L} \mathbf{F}_{lj} \mathbf{B}_j \right) \mathbf{x} + \mathbf{z}_l \right) - \mathbf{x}_l \right\} \left\{ \alpha \left(\left(\sum_{j=1}^{L} \mathbf{F}_{lj} \mathbf{B}_j \right) \mathbf{x} + \mathbf{z}_l \right) - \mathbf{x}_l \right\}^H \Bigg| \widehat{\mathbf{F}} \right] \\
&\quad + \sum_{l=1}^{L} \lambda_l \left(\operatorname{tr}\{\mathbf{B}_l^H \mathbf{B}_l\} - 1 \right) \\
&= \sum_{l=1}^{L} \operatorname{tr} \mathbb{E}_{\widetilde{\mathbf{F}}} \left[\alpha^2 \left(\sum_{j=1}^{L} \mathbf{F}_{lj} \mathbf{B}_j \right) \left(\sum_{j=1}^{L} \mathbf{F}_{lj} \mathbf{B}_j \right)^H - \alpha \left(\sum_{j=1}^{L} \mathbf{F}_{lj} \mathbf{B}_j \right) \mathbf{I}_l' + \alpha^2 \mathbf{I}_K \right. \\
&\quad \left. - \alpha \mathbf{I}_l'^H \left(\sum_{j=1}^{L} \mathbf{F}_{lj} \mathbf{B}_j \right)^H + \mathbf{I}_K \Bigg| \widehat{\mathbf{F}} \right] + \sum_{l=1}^{L} \lambda_l \left(\operatorname{tr}\{\mathbf{B}_l^H \mathbf{B}_l\} - 1 \right) \\
&= \sum_{l=1}^{L} \operatorname{tr} \left\{ \alpha^2 \left(\sum_{j=1}^{L} \widehat{\mathbf{F}}_{lj} \mathbf{B}_j \right) \left(\sum_{j=1}^{L} \widehat{\mathbf{F}}_{lj} \mathbf{B}_j \right)^H + \alpha^2 \sum_{j=1}^{L} \eta_{lj} \mathbf{B}_j^H \mathbf{B}_j \right. \\
&\quad \left. - \alpha \left(\sum_{j=1}^{L} \widehat{\mathbf{F}}_{lj} \mathbf{B}_j \right) \mathbf{I}_l' - \alpha \mathbf{I}_l'^H \left(\sum_{j=1}^{L} \widehat{\mathbf{F}}_{lj} \mathbf{B}_j \right)^H \right\} + \sum_{l=1}^{L} \lambda_l \left(\operatorname{tr}\{\mathbf{B}_l^H \mathbf{B}_l\} - 1 \right) + (\alpha^2 + 1) KL,
\end{aligned}$$

where

$$\mathbf{I}_l' = \left[\mathbf{0}_{K \times K} \ \mathbf{0}_{K \times K} \ \cdots \underbrace{\mathbf{I}_K}_{l\text{th position}} \cdots \ \mathbf{0}_{K \times K} \right]^T.$$

The following properties are used in the above simplifications: $\mathbb{E}\left[\widetilde{\mathbf{F}}_{jl} \widetilde{\mathbf{F}}_{jl}^H \mid \widehat{\mathbf{F}}_{jl}\right] = \eta_{jl} \mathbf{I}_K$ and $\mathbb{E}\left[\mathbf{x}\mathbf{x}_l^H\right] = \mathbf{I}_l'$. Differentiating the Lagrangian with $\mathbf{B}_j, \alpha$, and equating to zero yields the following three equations:

$$(\alpha^{opt})^2 \left[\sum_{l=1}^{L} \widehat{\mathbf{F}}_{lj}^H \widehat{\mathbf{F}}_{lj} + \left(\sum_{l=1}^{L} \eta_{lj} + \frac{\lambda_j^{opt}}{(\alpha^{opt})^2}\right)\mathbf{I}_M\right]\mathbf{B}_j^{opt} - \alpha^{opt}\sum_{l=1}^{L}\widehat{\mathbf{F}}_{lj}^H \mathbf{I}_l^{'H}$$
$$+ (\alpha^{opt})^2 \sum_{i=1, i\neq j}^{L} \left[\sum_{l=1}^{L} \widehat{\mathbf{F}}_{lj}^H \widehat{\mathbf{F}}_{li}\right]\mathbf{B}_i^{opt} = 0, \quad \forall j = 1, 2, \ldots, L, \tag{10.72}$$

$$\mathrm{tr}\{\mathbf{B}_j^{opt}(\mathbf{B}_j^{opt})^H\} = 1, \quad \forall j = 1, 2, \ldots, L, \tag{10.73}$$

$$\sum_{l=1}^{L} \mathrm{tr}\Bigg\{2\alpha^{opt}\Big(\sum_{j=1}^{L}\widehat{\mathbf{F}}_{lj}\mathbf{B}_j^{opt}\Big)\Big(\sum_{j=1}^{L}\widehat{\mathbf{F}}_{lj}\mathbf{B}_j^{opt}\Big)^H + 2\alpha^{opt}\sum_{j=1}^{L}\eta_{lj}(\mathbf{B}_j^{opt})^H\mathbf{B}_j^{opt}$$
$$-\Big(\sum_{j=1}^{L}\widehat{\mathbf{F}}_{lj}\mathbf{B}_j^{opt}\Big)\mathbf{I}_l^{'} - \big(\mathbf{I}_l^{'}\big)^H\Big(\sum_{j=1}^{L}\widehat{\mathbf{F}}_{lj}\mathbf{B}_j^{opt}\Big)^H\Bigg\} + 2\alpha^{opt}KL = 0. \tag{10.74}$$

Equation (10.72) can be written in the matrix form as

$$\alpha^{opt}(\widehat{\mathbf{F}}^H\widehat{\mathbf{F}} + \mathbf{D})\mathbf{B} = \mathbf{b}, \tag{10.75}$$

where

$$\mathbf{D} = \begin{bmatrix} \mathbf{D}_1 & \mathbf{0}_{M\times M} & \cdots & \mathbf{0}_{M\times M} \\ \mathbf{0}_{M\times M} & \mathbf{D}_2 & \cdots & \mathbf{0}_{M\times M} \\ \vdots & \vdots & \ddots & \vdots \\ \mathbf{0}_{M\times M} & \mathbf{0}_{M\times M} & \cdots & \mathbf{D}_L \end{bmatrix}, \quad \mathbf{D}_i = \left(\sum_{l=1}^{L}\eta_{li} + \frac{\lambda_i^{opt}}{\alpha^{opt2}}\right)\mathbf{I}_M,$$

$$\mathbf{B} = \begin{bmatrix} \mathbf{B}_1^{opt} \\ \mathbf{B}_2^{opt} \\ \vdots \\ \mathbf{B}_L^{opt} \end{bmatrix}, \quad \mathbf{b} = \begin{bmatrix} \sum_{l=1}^{L}\widehat{\mathbf{F}}_{l1}^H\mathbf{I}_l^{'H} \\ \sum_{l=1}^{L}\widehat{\mathbf{F}}_{l2}^H\mathbf{I}_l^{'H} \\ \vdots \\ \sum_{l=1}^{L}\widehat{\mathbf{F}}_{lL}^H\mathbf{I}_l^{'H} \end{bmatrix}.$$

The precoding matrix for all cells can be computed as

$$\mathbf{B} = \frac{1}{\alpha}\left(\widehat{\mathbf{F}}^H\widehat{\mathbf{F}} + \mathbf{D}\right)^{-1}\mathbf{b}. \tag{10.76}$$

$\lambda_1, \lambda_2, \ldots, \lambda_L, \alpha$ can be solved numerically by substituting (10.76) in (10.73) and (10.74).

10.3.4 Performance

Consider a multicell system with $L = 4$ cells and $K = 4$ users in each cell as shown in Fig. 10.14. The average powers at the BS and at the users are taken to

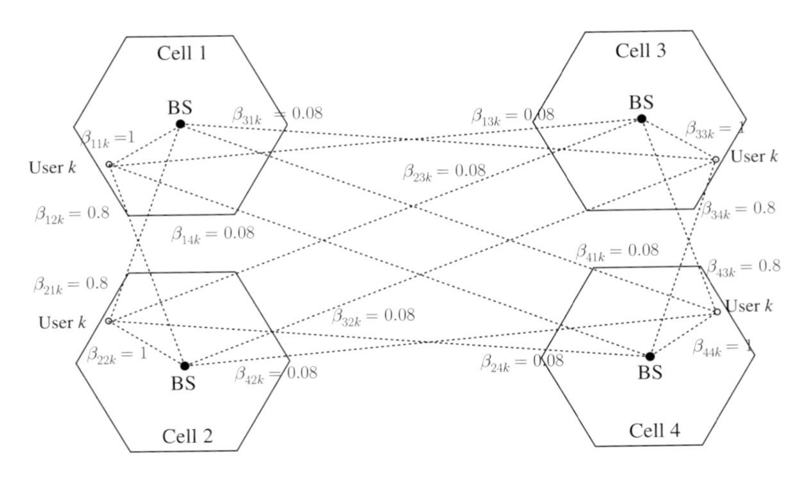

Figure 10.14 Arrangement of cells used in simulation.

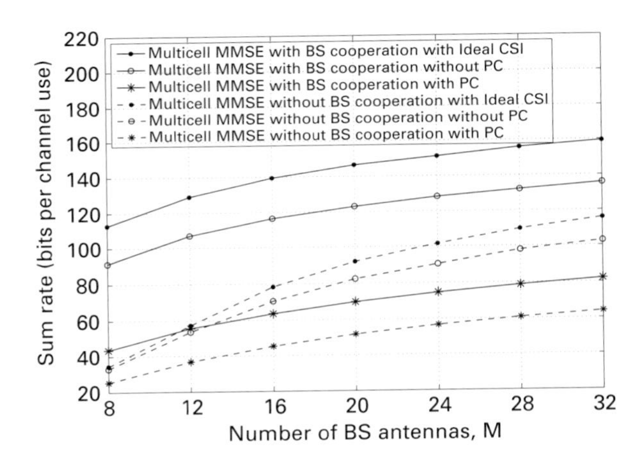

Figure 10.15 Comparison of sum-rates achieved by multicell MMSE precoding with and without BS cooperation in the system with $L = 4$ cells and $K = 4$ users.

be $P_d = 20$ dB and $P_u = 10$ dB, respectively. In cells 1 and 2, orthogonal training sequences are used and they are reused in cells 3 and 4. The propagation factors are taken as follows (see Fig. 10.14): for all k, $\beta_{jlk} = 1$ if $j = l$, $\beta_{jlk} = 0.8$ if $(j, l) \in \{(1, 2), (2, 1), (3, 4), (4, 3)\}$, and $\beta_{jlk} = 0.08$ for all other values of j and l. In Fig. 10.15, the sum of achievable rates of all the users in the system is plotted as a function of the number of BS antennas M for the precoding methods with and without BS cooperation. Achievable sum-rates for the following three cases are shown: (i) with pilot contamination, where orthogonal training sequences are used in two cells and they are reused in the other two cells as described above, (ii) without pilot contamination where orthogonal training sequences are used for all the users in the system, and (iii) the ideal CSI case where it is assumed that BSs have perfect CSI. In case (ii), the CSI is still imperfect because of the estimation error due to MMSE estimation. In Fig. 10.15, the effect of pilot contamination can be clearly seen: the gap between the sum-rate curves for the ideal CSI case and the no pilot contamination case is smaller than the gap between the sum-rate curves of without and with pilot contamination. It can also be seen that precoding with BS cooperation gives greater sum-rates than precoding without BS cooperation.

References

[1] G. Raleigh and J. Cioffi, "Spatio-temporal coding for wireless communication," *IEEE Trans. Commun.*, vol. 44, no. 3, pp. 357–366, Mar. 1998.

[2] L. G. Ordonez, D. P. Palomar, A. P. Zamora, and J. R. Fonollosa, "High-SNR analytical performance of spatial multiplexing MIMO systems with CSI," *IEEE Trans. Signal Process.*, vol. 55, no. 11, pp. 5447–5463, Nov. 2007.

[3] Z. Wang and G. B. Giannakis, "A simple and general parametrization quantifying performance in fading channels," *IEEE Trans. Commun.*, vol. 51, no. 8, pp. 1389–1398, Aug. 2003.

[4] B. Vrigneau, J. Letessier, P. Rostaing, L. Collin, and G. Burel, "Extension of the MIMO precoder based on the minimum euclidean distance: a cross-form matrix," *IEEE J. Select. Topics Signal Process.*, vol. 2, no. 2, pp. 135–146, Apr. 2008.

[5] S. K. Mohammed, E. Viterbo, Y. Hong, and A. Chockalingam, "MIMO precoding with X- and Y-codes," *IEEE Trans. Inform. Theory*, vol. 57, no. 6, pp. 3542–3566, Jun. 2011.

[6] K. P. Srinath and B. S. Rajan, "A low ML-decoding complexity, full-diversity, full-rate MIMO precoder," *IEEE Trans. Signal Process.*, vol. 59, no. 11, pp. 5485–5498, Nov. 2011.

[7] D. Tse and P. Viswanath, *Fundamentals of Wireless Communication.* Cambridge, UK: Cambridge University Press, 2005.

[8] H. Weingarten, Y. Steinberg, and S. Shamai, "The capacity region of the Gaussian multiple-input multiple-output broadcast channel," *IEEE Trans. Inform. Theory*, vol. 52, no. 9, pp. 3936–3964, Sep. 2006.

[9] Q. H. Spencer, C. B. Peel, A. L. Swindlehurst, and M. Haardt, "An introduction to the multi-user MIMO downlink," *IEEE Commun. Mag.*, vol. 42, no. 10, pp. 60–67, Oct. 2004.

[10] D. Gesbert, R. Kountouris, R. W. Heath, C.-B. Chae, and T. Salzer, "Shifting the MIMO paradigm," *IEEE Signal Process. Mag.*, vol. 24, no. 5, pp. 36–46, Sep. 2007.

[11] N. Jindal and A. Goldsmith, "Dirty paper coding vs. TDMA for MIMO broadcast channels," *IEEE Trans. Inform. Theory*, vol. 51, no. 5, pp. 1783–1794, May 2005.

[12] G. Caire and S. Shamai, "On the achievable throughput of a multiantenna Gaussian broadcast channel," *IEEE Trans. Inform. Theory*, vol. 49, no. 7, pp. 1691–1706, Jul. 2003.

[13] M. H. M. Costa, "Writing on dirty-paper," *IEEE Trans. Inform. Theory*, vol. 29, no. 3, pp. 439–441, May 1983.

[14] C. B. Peel, "On dirty-paper coding," *IEEE Signal Process. Mag.*, vol. 20, no. 3, pp. 112–113, May 2003.

[15] S. Vishwanath, N. Jindal, and A. Goldsmith, "Duality, achievable rates and sum-rate capacity of MIMO broadcast channels," *IEEE Trans. Inform. Theory*, vol. 49, no. 10, pp. 2658–2668, Oct. 2003.

[16] P. Viswanath and D. Tse, "Sum capacity of the vector Gaussian broadcast channel and uplink–downlink duality," *IEEE Trans. Inform. Theory*, vol. 49, no. 8, pp. 1912–1921, Aug. 2003.

[17] M. Vu and A. Paulraj, "MIMO wireless linear precoding," *IEEE Signal Process. Mag.*, vol. 24, no. 5, pp. 86–105, Sep. 2007.

[18] Q. H. Spencer, A. L. Swindlehurst, and M. Haardt, "Zero-forcing methods for downlink spatial multiplexing in multiuser MIMO channels," *IEEE Trans. Signal Process.*, vol. 52, no. 2, pp. 461–471, Feb. 2004.

[19] Z. Pan, K.-K. Wong, and T.-S. Ng, "Generalized multiuser orthogonal space division multiplexing," *IEEE Trans. Wireless Commun.*, vol. 3, no. 2, pp. 1969–1973, Nov. 2004.

[20] C.-B. Chae, D. Mazzarese, and R. W. Heath Jr., "Coordinated beamforming for multiuser MIMO systems with limited feedforward," in *Asilomar Conf. on Signals, Systems and Computers*, Pacific Grove, CH, Oct.–Nov. 2006, pp. 1511–1515.

[21] L. Choi and D. Murch, "A transmit preprocessing technique for multiuser MIMO systems using a decomposition approach," *IEEE Trans. Wireless Commun.*, vol. 3, no. 1, pp. 20–24, Jan. 2004.

[22] B. Bandemer, M. Haardt, and S. Visuri, "Linear MMSE multi-user MIMO downlink precoding for users with multiple antennas," in *IEEE PIMRC'2006*, Helsinki, Sep. 2006, pp. 1–5.

[23] J. Zhang, Y. Wu, S. Zhou, and J. Wang, "Joint linear transmitter and receiver design for the downlink of multiuser MIMO systems," *IEEE Commun. Lett.*, vol. 9, no. 11, pp. 991–993, Nov. 2005.

[24] S. Shi, M. Schubert, and H. Boche, "Downlink MMSE transceiver optimization for multiuser MIMO systems: duality and sum-MSE minimization," *IEEE Trans. Signal Process.*, vol. 55, no. 11, pp. 5436–5446, Nov. 2007.

[25] A. Mezghani, M. Joham, R. Hunger, and W. Utschick, "Transceiver design for multi-user MIMO systems," in *WSA2006*, Schtoss Reisenburg, Mar. 2006.

[26] A. M. Khachan, A. J. Tenenbaum, and R. S. Adve, "Linear processing for the downlink in multiuser MIMO systems with multiple data streams," in *IEEE ICC'2008*, Beijing, Jun. 2008, pp. 4113–4118.

[27] A. J. Tenenbaum and R. S. Adve, "Linear processing and sum throughput in the multiuser MIMO downlink," *IEEE Trans. Wireless Commun.*, vol. 8, no. 5, pp. 2652–2661, May 2009.

[28] M. Tomlinson, "New automatic equaliser employing modulo arithmetic," *Electron. Lett.*, vol. 7, no. 5/6, pp. 138–139, Mar. 1971.

[29] H. Harashima and H. Miyakawa, "Matched transmission technique for channels with inter-symbol interference," *IEEE Trans. Commun.*, vol. 20, no. 8, pp. 774–780, Aug. 1972.

[30] A. Mezghani, R. Hunger, M. Joham, and W. Utschick, "Iterative THP transceiver optimization for multi-user MIMO systems based on weighted sum-mse minimization," in *IEEE SPAWC'2006*, Cannes, Jul. 2005, pp. 1–5.

[31] K. Kusume, M. Joham, W. Utschick, and G. Bauch, "Efficient Tomlinson-Harashima precoding for spatial multiplexing on flat MIMO channel," in *IEEE IEEE ICC'2005*, Seoul, May 2005, pp. 2021–2025.

[32] C. B. Peel, B. M. Hochwald, and A. L. Swindlehurst, "A vector-perturbation technique for near-capacity multi-antenna multiuser communication - Part I: Channel inversion and regularization," *IEEE Trans. Commun.*, vol. 53, no. 1, pp. 195–202, Jan. 2005.

[33] ——, "A vector-perturbation technique for near-capacity multi-antenna multiuser communication – Part II: Vector perturbation," *IEEE Trans. Commun.*, vol. 53, no. 3, pp. 537–544, Mar. 2005.

[34] D. Seethaler and G. Matz, "Efficient vector perturbation in multiantenna multiuser systems based on approximate integer relations," in *EUSIPCO'2006*, Florence, Sep. 2004.

[35] S. K. Mohammed, A. Chockalingam, and B. S. Rajan, "A low-complexity precoder for large multiuser MISO systems," in *IEEE VTC'2008*, Singapore, May 2008, pp. 797–801.

[36] T. L. Marzetta, "How much training is required for multiuser MIMO?" in *40th Asilomar Conf. on Signals, Systems and Computers*, Pacific Grove, CA, Oct.–Nov. 2006, pp. 359–363.

[37] J. Jose, A. Ashikhmin, T. Marzetta, and S. Vishwanath, "Pilot contamination and precoding in multi-cell TDD systems," *IEEE Trans. Wireless Commun.*, vol. 10, no. 8, pp. 2640–2651, Aug. 2011.

[38] T. L. Marzetta, "Noncooperative cellular wireless with unlimited numbers of base station antennas," *IEEE Trans. Wireless Commun.*, vol. 9, no. 11, pp. 3590–3600, Nov. 2010.

[39] B. Gopalakrishnan and N. Jindal, "An analysis of pilot contamination on multi-user MIMO cellular systems with many antennas," in *IEEE SPAWC'2011*, San Francisco, CA, Jun. 2011, pp. 381–385.

[40] K. Appaiah, A. Ashikhmin, and T. L. Marzetta, "Pilot contamination reduction in multi-user TDD systems," in *IEEE ICC'2010*, Cape Town, May 2010, pp. 1–5.

[41] B. Hassibi and B. M. Hochwald, "How much training is needed in multiple-antenna wireless links?" *IEEE Trans. Inform. Theory*, vol. 49, no. 4, pp. 951–963, Apr. 2003.

11 MIMO channel models

Channel models play a crucial role in the design and analysis of wireless communication systems. They enable the system designers to analyze the performance of wireless systems and optimize design parameters even before the systems are actually built. They are key ingredients in such design and performance evaluation exercises, which are often carried out through mathematical analysis or computer simulation or a combination of both. A good channel model that accurately captures the real channel behavior is a very valuable tool that can accelerate the development of practical wireless systems. The need for good channel models to aid the design, analysis, and development of MIMO systems in general, and large MIMO systems in particular, is immense. A lot of effort has been directed towards MIMO channel sounding campaigns and MIMO channel modeling. Measurements from these campaigns have aided the formulation of MIMO channel models in wireless standards [1]–[4]. Channel sounding campaigns with large numbers of antennas, in both outdoor and indoor settings, have also appeared, though sparsely, in the literature. Now, with the increasing interest in large MIMO system implementations, there is renewed interest and activity in large MIMO channel sounding. While these channel measurements are expected to yield accurate and realistic models for large MIMO channels, the traditional analytical MIMO channel models which are widely known in the literature, are expected to find continued use. This chapter gives a summary of analytical MIMO channel models and MIMO channel models in current wireless standards, and details of some of the more recent large MIMO channel sounding campaigns.

The number of antenna elements and their geometrical configuration, polarization, and propagation environment influence the real-life MIMO channel characteristics, and hence the corresponding channel models. MIMO channel measurements using linear, circular, planar, and three-dimensional arrays with single or dual polarized antenna elements are common. These measurements can yield stochastic characterization of MIMO channels. They can also be used to validate analytical models. It is typical to plot the amplitude envelope of the MIMO channel coefficients obtained from measurements as normalized pdfs and see if the distributions follow well-known distributions (e.g., Rayleigh, Ricean, log-normal) or other distributions. Another key interest is to capture the spatial characteristics of the MIMO channel in the model. Widely studied MIMO channel models include the Kronecker model [5],[6], the Weichselberger model

[7], the finite scatterers model [8], the maximum entropy model [9], and the virtual channel separation model [10]. Validation of various models is often carried out using data from channel sounding campaigns by extracting the model parameters from measured data, generating synthesized channels by Monte Carlo simulations, and comparing certain metrics from the synthesized channels with those extracted directly from the measurement.

11.1 Analytical channel models

The most commonly used analytical MIMO channel model is the spatially iid frequency non-selective (flat) fading channel model. In this narrowband channel model, the channel gain between any pair of transmit–receive antennas is modeled as a complex Gaussian random variable. This model relies on (*i*) the antenna elements in the transmitter/receiver being spatially well separated, and (*ii*) the presence of a large number of temporally but narrowly separated multipaths (common in a 'rich-scattering' environment), whose combined gain, by the central-limit theorem, can be approximated by a Gaussian random variable. In a pure multipath environment without a line-of-sight (LOS) component, the gains have zero mean and the corresponding amplitude distribution is Rayleigh. If an LOS component exists in addition to the multipath components, then the mean is non-zero and the amplitude distribution is Ricean. In general, the $n_r \times n_t$ channel matrix $\mathbf{H}$ can be considered to be made up of a zero-mean stochastic part $\mathbf{H}_s$ and a deterministic part $\mathbf{H}_d$ according to

$$\mathbf{H} = \sqrt{\frac{1}{1+K}}\mathbf{H}_s + \sqrt{\frac{K}{K+1}}\mathbf{H}_d, \tag{11.1}$$

where K is the Rice factor, defined as the ratio between the powers in the LOS path and the non-LOS paths. $K = 0$ corresponds to a pure multipath channel without a LOS component, and $K = \infty$ corresponds to an unfaded AWGN channel.

11.1.1 Spatial correlation based models

Spatial correlation based analytical models characterize the MIMO channel matrix statistically in terms of the correlations between the entries of the channel matrix. Spatial correlation between the antenna elements at the transmitter/receiver can affect the MIMO capacity and system performance. Therefore, it is of interest to factor in spatial correlation between antenna elements in the MIMO channel models. To this end, focusing on a pure multipath channel without any LOS component, consider $K = 0$ in (11.1), i.e., $\mathbf{H} = \mathbf{H}_s$. Let $\mathbf{h} = \text{vec}(\mathbf{H})$. Defining the $n_t n_r \times n_t n_r$ correlation matrix as $\mathbf{R}_{\mathbf{H}} \triangleq \mathbb{E}[\mathbf{h}\mathbf{h}^H]$, the zero-mean multivariate complex Gaussian distribution of $\mathbf{h}$ can be written as

$$f(\mathbf{h}) = \frac{1}{\pi^{n_t n_r} \det\{\mathbf{R}_{\mathbf{H}}\}} \exp\left(-\mathbf{h}^H \mathbf{R}_{\mathbf{H}}^{-1} \mathbf{h}\right). \tag{11.2}$$

$\mathbf{R_H}$ contains the correlations of all the elements of the channel matrix and describes the spatial statistics. Now, realizations of vector $\mathbf{h}$ with distribution (11.2), and hence the realizations of the channel matrix $\mathbf{H}$, can be obtained by

$$\mathbf{h} = \mathbf{R}_{\mathbf{H}}^{1/2}\mathbf{g}, \tag{11.3}$$

where $\mathbf{R}_{\mathbf{H}}^{1/2}$ is any matrix satisfying $\mathbf{R}_{\mathbf{H}}^{1/2}\mathbf{R}_{\mathbf{H}}^{H^{1/2}} = \mathbf{R_H}$, and $\mathbf{g}$ is an $n_t n_r \times 1$ vector with iid Gaussian entries with zero mean and unit variance. Note that the number of real-valued parameters required to fully specify $\mathbf{R_H}$ is $n_t^2 n_r^2$, which is large for large n_t, n_r. Imposing a certain structure on the correlation matrix can reduce this requirement. Different ways to reduce this requirement lead to different correlation based models which are described below.

The iid model

One of the widely used analytical MIMO channel models is the iid model. In the iid model,

$$\mathbf{R_H} = \rho^2 \mathbf{I}_{n_t n_r}, \tag{11.4}$$

i.e., all entries of $\mathbf{H}$ are uncorrelated and have equal variance ρ^2. A single real-valued parameter ρ fully specifies the channel model in this case. This model corresponds to a spatially white MIMO channel, which, in practice, can occur in rich scattering environments having multipath components uniformly distributed in all directions. The model is attractive due to its simplicity. It has found extensive use in theoretical studies (e.g., information theoretic analysis of MIMO systems) and simulations (e.g., performance evaluation of MIMO systems and algorithms) [11]. The simplicity of the iid model makes it quite attractive for large MIMO system studies.

The Kronecker model

The Kronecker model incorporates spatial correlation between antenna elements. It also assumes that the spatial correlations at the transmitter and receiver are separable. Let $\mathbf{R}_{Tx} = \mathbb{E}[\mathbf{H}^H\mathbf{H}]$ and $\mathbf{R}_{Rx} = \mathbb{E}[\mathbf{H}\mathbf{H}^H]$ denote the transmit correlation matrix and the receive correlation matrix, respectively. Assuming that transmit and receive correlations are separable is equivalent to restricting to correlation matrices that can be written in the Kronecker product form

$$\mathbf{R_H} = \mathbf{R}_{Tx} \otimes \mathbf{R}_{Rx}. \tag{11.5}$$

With this, the vector $\mathbf{h}$ in the Kronecker model becomes

$$\mathbf{h} = (\mathbf{R}_{Tx} \otimes \mathbf{R}_{Rx})^{1/2}\mathbf{g}, \tag{11.6}$$

and

$$\mathbf{H} = \mathbf{R}_{Tx}^{1/2}\mathbf{G}\mathbf{R}_{Rx}^{1/2}, \tag{11.7}$$

where, as before, $\mathbf{g}$ is an $n_t n_r \times 1$ vector with iid Gaussian entries with zero mean and unit variance, and $\mathbf{G}$ is an iid unit variance matrix obtained by performing an inverse vec(.) operation on $\mathbf{g}$. The number of parameters that characterize this model is n_t^2 (parameters in $\mathbf{R}_{Tx}$) plus n_r^2 (parameters in $\mathbf{R}_{Rx}$), unlike $n_t^2 n_r^2$ parameters in the full correlation matrix.

A limitation with the Kronecker model is that it does not take into account the coupling between the direction of departure (DoD) at the transmitter and the direction of arrival (DoA) at the receiver, which is typical in MIMO channels. So the Kronecker model is suitable only in certain environments. In small MIMO systems (e.g., 2×2 and 3×3 indoor MIMO channels at 5.2 GHz and 5.8 GHz), because of the reduced spatial resolution involved, the Kronecker model has been shown to be a good fit to the measured channel [12],[13]. However, application of the Kronecker model to an 8×8 measured channel (e.g., an 8×8 NLOS MIMO channel at 5.2 GHz) has indicated discrepancies in the modeled capacity (the Kronecker model underestimated the capacity extracted from the channel measurements) and the joint spatial DoD–DoA spectra [14]. These discrepancies are likely to be more pronounced in large arrays with high angular resolution [14]. Similar results have been reported in outdoor-to-indoor office MIMO channel measurements in the 5.2 GHz band [15]; while the Kronecker model was found to be a good fit in a 2×8 LOS setup, the fit was found to be not as good in a 16×8 NLOS setup. These observations and results in [15] have served as input to the COST 273 MIMO channel model [16]. The Kronecker model has also been shown to underestimate throughputs in adaptive modulation in MIMO compared to throughputs obtained using measured channel matrices [17].

Ray-tracing methods have been used in [18] to study the suitability of the Kronecker model for use in the two-ring model (Fig. 11.1(a)) and the elliptical model (Fig. 11.1(b)) of scatterer distributions. The two-ring model of scatterers can be viewed as typical of outdoor environments; e.g., in cellular systems where the BS and the mobile are surrounded by different sets of scatterers and there is no LOS between the mobile and the BS. The elliptical model of scatterers can be viewed as typical of indoor environments; e.g., indoor Wi-Fi environments where the transmitter and receiver share the same scatterers [19]. It has been shown that the Kronecker model is suitable for the two-ring model (i.e., suitable for some outdoor environments) and not suitable for the elliptical model (i.e., not suitable for some indoor environments) due to the separability and non-separability of the correlation structures in the former and latter models, respectively.

Despite the limitation of ignoring the coupling between the DoD and DoA at the transmit and receive ends, the Kronecker model has been popularly used in information theoretic capacity analysis and simulation studies [5],[6],[20],[21]. Though the Kronecker model is expected to be increasingly inaccurate for increasing array size and angular resolution, it will still find use in large MIMO system studies because of its simplicity.

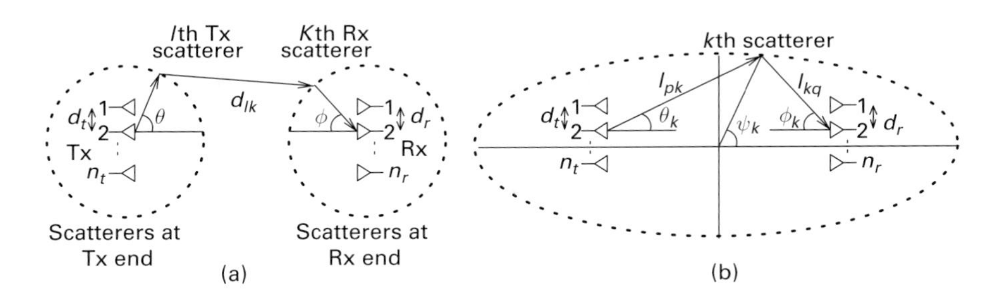

Figure 11.1 Models of scatterer distribution: (a) two-ring model; (b) elliptical model.

The Weichselberger model

The Weichselberger model [2],[7] attempts to overcome the assumption of transmit–receive correlation separability made in the Kronecker model. The DoD spectrum at the transmit end and the DoA spectrum at the receive end are determined by the spatial structure/environment of the channel. Ignoring the coupling between the DoD–DoA spectra amounts to ignoring the effect of a significant part of the spatial structure in the channel. The Weichselberger model takes this coupling into account through a coupling matrix $\mathbf{\Omega}$, which allows joint modeling of the transmit and receive channel correlations. The model uses the eigenvalue decomposition of the transmit and receive correlation matrices,

$$\mathbf{R}_{Tx} = \mathbf{U}_{Tx}\mathbf{\Lambda}_{Tx}\mathbf{U}_{Tx}^H, \tag{11.8}$$

$$\mathbf{R}_{Rx} = \mathbf{U}_{Rx}\mathbf{\Lambda}_{Rx}\mathbf{U}_{Rx}^H, \tag{11.9}$$

where $\mathbf{U}_{Tx}$ and $\mathbf{U}_{Rx}$ are unitary matrices whose columns are the eigenvectors of $\mathbf{R}_{Tx}$ and $\mathbf{R}_{Rx}$, respectively, and $\mathbf{\Lambda}_{Tx}$ and $\mathbf{\Lambda}_{Rx}$ are diagonal matrices with the corresponding eigenvalues. The Weichselberger model is given by

$$\mathbf{H} = \mathbf{U}_{Rx}(\tilde{\mathbf{\Omega}} \odot \mathbf{G})\mathbf{U}_{Tx}^T, \tag{11.10}$$

where $\mathbf{G}$ is an $n_r \times n_t$ iid matrix as defined in the Kronecker model, $\tilde{\mathbf{\Omega}}$ is the element-wise square root of an $n_r \times n_t$ coupling matrix $\mathbf{\Omega}$, and $\odot$ denotes element-wise multiplication. The elements of $\mathbf{\Omega}$ are real-valued and non-negative, and they determine the average power coupling between the transmit and receive eigenmodes. It can be observed that the Weichselberger model in (11.10) becomes the Kronecker model when the coupling matrix is a rank-1 matrix given by

$$\mathbf{\Omega} = \boldsymbol{\lambda}_{Rx}\boldsymbol{\lambda}_{Tx}^T, \tag{11.11}$$

where $\boldsymbol{\lambda}_{Tx}$ and $\boldsymbol{\lambda}_{Rx}$ are vectors containing the eigenvectors of the transmit and receive correlation matrices, respectively.

The Weichselberger model is specified by the transmit and receive eigenmodes ($\mathbf{U}_{Tx}$, $\mathbf{U}_{Rx}$), and the coupling matrix $\mathbf{\Omega}$. The real-valued parameters that characterize the Weichselberger model include: $n_r(n_r - 1)$ parameters for

$\mathbf{U}_{Rx}$, $n_t(n_t-1)$ parameters for $\mathbf{U}_{Tx}$, and $n_r n_t$ parameters for $\mathbf{\Omega}$. In the Weichselberger model, performance metrics like the capacity and the diversity order of the MIMO channel are independent of the transmit and receive eigenmodes, and depend only on the coupling matrix $\mathbf{\Omega}$. The structure of $\mathbf{\Omega}$ can be exploited to design signal processing algorithms to achieve improved capacity, diversity, or beamforming gains [2]. Channel measurements in 2.45 GHz and data analysis have shown that on several channel metrics like capacity, eigenvalue spread, condition numbers, and correlation matrix distance, the Weichselberger model outperformed the Kronecker model in terms of providing a closer match with these metrics obtained from measured data [22]. Outdoor-to-indoor MIMO channel measurements in 5.2 GHz have also shown that the Kronecker model is not well suited and that less restrictive assumptions on the channel as in the Weichselberger model or the full correlation model must be used [15].

11.1.2 Propagation based models

Propagation based analytical models characterize the MIMO channel matrix through propagation parameters. The finite scatterers model, the maximum entropy model, and the virtual channel separation model are examples of propagation based models.

The finite scatterers model

The finite scatterers model is a generic multipath channel model which accounts for each individual multipath component by its path gain and angle of departure (AoD) and angle of arrival (AoR) [8]. The basic premise of the model is that the signal from the transmitter reaches the receiver through a finite number of distinct paths (referred to as multipath components), which are treated according to ray-optical concepts and concepts from the geometric theory of diffraction. Single scattering and double scattering are possible mechanisms in this model. In single scattering, the signal from the transmitter hits a scatterer and bounces to the receiver (paths 1, 2 in Fig. 11.2). In double scattering, the signal bounces off the first scatterer, hits a second scatterer, and bounces to the receiver (paths 3,4,5 in Fig. 11.2). In multiple scattering, the transmitted signal is successively bounced off multiple scatterers until it reaches the receiver. Geometry-based models and ray tracing models are closely related to the finite scatterers model.

In the finite scatterers model, each path is indexed by an integer p, $p = 1, 2, \ldots, P$ where P denotes the finite number of paths. Figure 11.2 shows the finite scatterers model with $P = 5$. Each path has an AoD from the transmit antenna array (denoted by $\phi_{T,p}$ for the pth path), an AoA at the receive antenna array (denoted by $\phi_{R,p}$ for the pth path), a path gain (denoted by ξ_p for the pth path) defined as the ratio of the electric/magnetic field at the location of the receive antenna array to that of the transmit antenna array, and a time delay (denoted by τ_p for the pth path). The model also allows multiple

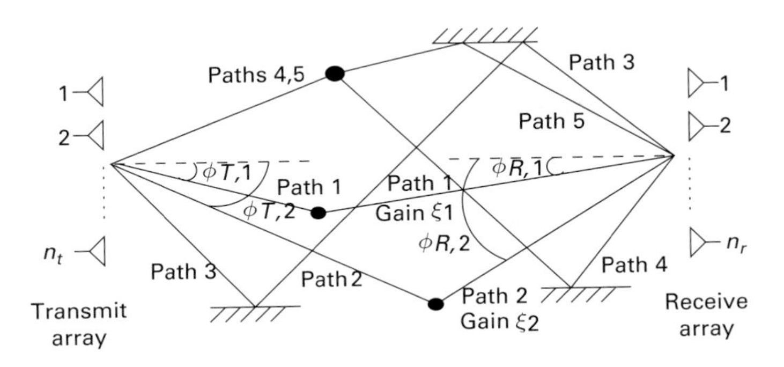

Figure 11.2 The finite scatterers model with $P = 5$.

scattering with split components, where there can be one AoD but two or more AoAs; likewise, there can be multiple AoDs and a single AoA.

Let $\mathbf{s}$ and $\mathbf{r}$ denote the transmit and received signal vectors of size $n_t \times 1$ and $n_r \times 1$, respectively. Let $\boldsymbol{\psi}_p$ denote the $n_r \times 1$ sized response vector at the receive antenna array due to signal with AoA $\phi_{R,p}$. Likewise, let $\boldsymbol{\theta}_p$ denote the $n_t \times 1$ sized steering vector at the transmit side for the signal with AoD $\phi_{T,p}$. The τ_ps can be ignored if the signal bandwidth is narrow compared to the coherence bandwidth of the channel. Assuming narrowband signaling where the τ_ps can be neglected, the received signal vector is given by

$$\mathbf{r} = \sum_{p=1}^{P} \boldsymbol{\psi}_p \xi_p \boldsymbol{\theta}_p^T \mathbf{s} \;=\; \boldsymbol{\Psi}\boldsymbol{\Xi}\boldsymbol{\Theta}^T \mathbf{s}, \tag{11.12}$$

where $\boldsymbol{\Psi} = [\boldsymbol{\psi}_1\, \boldsymbol{\psi}_2\, \cdots\, \boldsymbol{\psi}_P]$ is an $n_r \times P$ matrix, $\boldsymbol{\Theta} = [\boldsymbol{\theta}_1\, \boldsymbol{\theta}_2\, \cdots\, \boldsymbol{\theta}_P]$ is an $n_t \times P$ matrix, and $\boldsymbol{\Xi}$ is a $P \times P$ diagonal matrix with ξ_p as the pth diagonal element. The finite scatterers channel model is then given by

$$\mathbf{H} = \boldsymbol{\Psi}\boldsymbol{\Xi}\boldsymbol{\Theta}^T. \tag{11.13}$$

In this model, the steering and response vectors, $\boldsymbol{\theta}_p$s and $\boldsymbol{\psi}_p$s, incorporate the geometry, directivity, and coupling of the antenna elements. A $\boldsymbol{\psi}_p$ vector for uniform linear antenna array at the receiver with constant inter-element spacing d_r is given by [8]

$$\boldsymbol{\psi}_p = \left\{ \exp\left(j2\pi \frac{n d_r}{\lambda} \sin(\phi_{R,p}) \right),\; n = 0, \ldots, n_r - 1 \right\}^T, \tag{11.14}$$

where λ is the wavelength. Likewise, a $\boldsymbol{\theta}_p$ vector can be defined on the transmit side. Different choices of distributions of AoAs, AoDs, and path gains give different forms of the finite scatterers model. Commonly assumed distributions for AoDs/AoAs are uniform and Laplacian. The cluster model for AoDs/AoAs is also common. Path gains may be Rayleigh distributed or log-normally distributed.

In wideband signaling where the τ_ps cannot be neglected, a tapped delay line representation of the channel model can be written as

$$\mathbf{H}(\tau) = \sum_{l=-\infty}^{\infty} \mathbf{H}_l \delta(\tau - lT_s), \tag{11.15}$$

where $T_s = 1/W$, W is the bandwidth of the signal, and

$$\mathbf{H}_l = \sum_{p=1}^{P} \xi_p \text{sinc}(\tau_p - lT_s) \boldsymbol{\psi}_p \boldsymbol{\theta}_p^T = \boldsymbol{\Psi}(\boldsymbol{\Xi} \odot \mathbf{T}_l)\boldsymbol{\Theta}^T, \tag{11.16}$$

where $\text{sinc}(x) = \sin(\pi x)/\pi x$, $\mathbf{T}_l$ is a $P \times P$ diagonal matrix with $\text{sinc}(\tau_p - lT_s)$ as the pth diagonal element, and $\odot$ denotes element-wise multiplication.

In [23], finite scatterer model parameters are obtained from channel measurements in outdoor urban environments at 2 GHz using two transmit antennas and an eight-antenna uniform circular receive array with approximately half wavelength inter-element spacing. The SAGE algorithm was used to analyze the measured data and determine DoAs and time delays of arrival of each identifiable multipath component. These measurements have suggested that the model parameters, namely, time of arrival, DoA, and amplitude, are independent, the gains, ξ_ps, are log-normally distributed in magnitude and uniformly distributed in phase, the delays, τ_ps, are exponentially distributed, and the distribution of AoAs is approximately uniform.

The maximum entropy model

Maximum entropy models of MIMO channels are based on information theoretic tools [9],[24]–[26]. Introduced in [9], these models use the principle of maximum entropy together with the principle of logical consistency. The model is derived for various levels of channel parameter knowledge, such as knowledge of DoA, DoD, number of scatterers, and powers of the steering directions. In particular, mutual information compliant MIMO channel models are developed. A mutual information compliant model is defined as follows. The mutual information of a MIMO channel with n_t transmit antennas, n_r receive antennas, $n_r \times n_t$ channel matrix $\mathbf{H}$, and SNR ρ is $I_M = \log_2 \det(\mathbf{I}_{n_r} + (\rho/n_t)\mathbf{H}\mathbf{H}^H)$. In many models of $\mathbf{H}$ (e.g., the iid model, the DoA based model, the DoD based model), the cumulative distribution function (CDF) of the mutual information converges to the form [9]

$$F(I_M) = 1 - Q\left(\frac{I_M - n_t\mu}{\sigma}\right), \tag{11.17}$$

where μ and σ depend on the model. A model is said to be mutual information complying if it minimizes the MSE

$$\int_0^\infty |F(I_M) - F_{emp}(I_M)|^2 \, dI_M, \tag{11.18}$$

where $F_{emp}(I_M)$ is the empirical CDF obtained from measurements.

Let s_t and s_r denote the number of scatterers on the transmit and receive sides, respectively. Let $\boldsymbol{\Theta}_{n_t \times s_t}$ and $\boldsymbol{\Psi}_{n_r \times s_r}$ denote the steering and response matrices on the transmit and receive sides, and $\mathbf{P}_t$ and $\mathbf{P}_r$ characterize the respective powers. Let $\boldsymbol{\Omega}_{s_r \times s_t}$ denote the matrix comprising the path gains from the scatterers at the transmit side to the scatterers at the receive side (coupling matrix). Assuming knowledge of the above parameters, the maximum entropy representation of the channel has the following structure [24],[25]:

$$\mathbf{H} = \frac{1}{\sqrt{s_r s_t}} \boldsymbol{\Psi} \mathbf{P}_r^{\frac{1}{2}} (\boldsymbol{\Omega} \odot \mathbf{S}) \mathbf{P}_t^{\frac{1}{2}} \boldsymbol{\Theta}^T, \tag{11.19}$$

where $\mathbf{S}$ is an $s_r \times s_t$ matrix with iid Gaussian entries. The above model is consistent in the sense that models without the knowledge of some of the parameters can be obtained by marginalizing (11.19) with respect to the unknown parameters. Several well-known models can be obtained from (11.19) as special cases through such marginalization. For example, if the channel energy alone is known and other parameters are not known, then the pure iid channel model comes out as a special case by taking $\boldsymbol{\Theta} = \mathbf{F}_{n_t}$, $\boldsymbol{\Psi} = \mathbf{F}_{n_r}$, $\mathbf{P}_r = \mathbf{I}_{n_r}$, and $\mathbf{P}_t = \mathbf{I}_{n_t}$, where $\mathbf{F}_{n_t}$ and $\mathbf{F}_{n_r}$ are $n_t \times n_t$ and $n_r \times n_r$ Fourier matrices, respectively. Likewise, the DoD model in which directions and powers are known only on the transmit side comes out as a special case by taking $\boldsymbol{\Psi} = \mathbf{F}_{n_r}$ and $\mathbf{P}_r = \mathbf{I}_{n_r}$, and the DoA model where directions and and powers are known only on the receive side comes out as a special case by taking $\boldsymbol{\Theta} = \mathbf{F}_{n_t}$ and $\mathbf{P}_t = \mathbf{I}_{n_t}$.

With the above model, the minimization of (11.18) involves an optimization with respect to $\boldsymbol{\Theta}$, $\boldsymbol{\Psi}$, $\mathbf{P}_t$, $\mathbf{P}_r$, and $\boldsymbol{\Omega}$, which is often non-trivial. The simplified model where equal powers and Fourier directions for $\boldsymbol{\Psi}$ and $\boldsymbol{\Theta}$ are used has the form (this simplified model is referred to as the double directional model)

$$\mathbf{H} = \frac{1}{\sqrt{s_t s_r}} \mathbf{F}_{n_r \times s_r} \mathbf{S}_{s_r \times s_t} \mathbf{F}_{s_t \times n_t}, \tag{11.20}$$

where $\mathbf{F}_{n_r \times s_r}$ and $\mathbf{F}_{s_t \times n_t}$ are $n_r \times s_r$ and $s_t \times n_t$ Fourier matrices, respectively. For this simplified model, the optimization in (11.18) is with respect to s_t and s_r, which can be solved [9]. The simplified model in (11.20) becomes the DoD model if $s_r = n_r$, the DoA model if $s_t = n_t$, and the iid model if $s_t = n_t$ and $s_r = n_r$. Measurements in the 2.1 GHz and 5.2 GHz bands in urban environments (e.g., street-grid scenario, open city place, indoor cell site) using an eight element uniform linear array at the transmitter and an 8×4 planar array with 32 antenna elements at the receiver [24],[25] have revealed that (*i*) the mutual information obtained from measurements mostly shows Gaussian behavior, (*ii*) the double directional model in (11.20) accurately fits the data for seven or eight scatterers on the transmit and receive sides (implying it to be a good model for urban environments where number of scatterers is expected to be high), and (*iii*) while the general model (11.19) is good for any number of antennas, the double directional model in (11.20) is good for a large number of antennas like 6×6, 7×7, 8×8. The reasons why the double directional model is good for large number of antennas is that as the number of antennas increases, the resolution increases which enables all the scatterers to be captured, and the Gaussian approximation becomes more realistic.

11.2 Effect of spatial correlation on large MIMO performance: an illustration

The placement of MIMO antenna elements and the propagation conditions witnessed in practice often render the iid fading model inadequate. For example, spatial correlation on the transmit and/or receive side can affect the rank structure of the MIMO channel resulting in degraded channel capacity. The non-LOS (NLOS) correlated MIMO channel model for an outdoor propagation scenario shown in Fig. 11.3 was proposed in [27]. The model explains the existence of 'pinhole' channels which exhibit poor rank properties, even if the spatial fading correlations at both ends of the link are low. In other words, the realization of high MIMO capacity in actual wireless channels is sensitive not only to the fading correlation, but also to the structure of the scattering in the propagation environment.

The channel model in [27] considers linear arrays of antennas at the transmitter and the receiver. The transmitter has n_t omnidirectional transmit antenna elements with inter-element spacing d_t. Likewise, the receiver has n_r omnidirectional receive antenna elements with inter-element spacing d_r. The propagation path between the transmit and receive arrays is obstructed on both sides of the link by a number of significant near-field scatterers (e.g., large objects) referred to as transmit and receive scatterers, which are modeled as omnidirectional ideal scatterers. The maximum ranges of the scatterers from the horizontal axis on the transmit and receive sides are denoted by D_t and D_r, respectively. When omnidirectional antennas are used, D_t and D_r correspond to the transmit and receive scattering radii, respectively. On the receive side, the signal reflected by the scatterers onto the antennas impinges on the array with an angular spread θ_r, which is a function of the distance between the array and the scatterers. Similarly, the angular spread on the transmit side is denoted θ_t. The range between the local scatterers on the transmit and receive sides is denoted by R. It is assumed that the scatterers are located sufficiently far from the antennas that the plane-wave assumption holds. Further, the local scattering condition is assumed, i.e., $D_t \ll R$ and $D_r \ll R$. The number of scatterers on each side, S, is considered to be large enough (typically > 10) for random fading to occur. The complex channel gain matrix of this model is given by [27]

$$\mathbf{H}_c = \frac{1}{\sqrt{S}} \mathbf{R}^{1/2}_{\theta_r,d_r} \mathbf{G}_r \mathbf{R}^{1/2}_{\theta_S,2D_r/S} \mathbf{G}_t \mathbf{R}^{1/2}_{\theta_t,d_t}, \tag{11.21}$$

where $\mathbf{R}_{\theta_t,d_t}$ is the $n_t \times n_t$ transmit correlation matrix, and $\mathbf{R}_{\theta_r,d_r}$ is the $n_r \times n_r$ receive correlation matrix. That is, if $\mathbf{h}$ denotes the received signal vector at the receive antenna array, then $\mathbf{h}$ is modeled as $\mathbf{h} \sim \mathcal{CN}(\mathbf{0}, \mathbf{R}_{\theta_r,d_r})$. Equivalently,

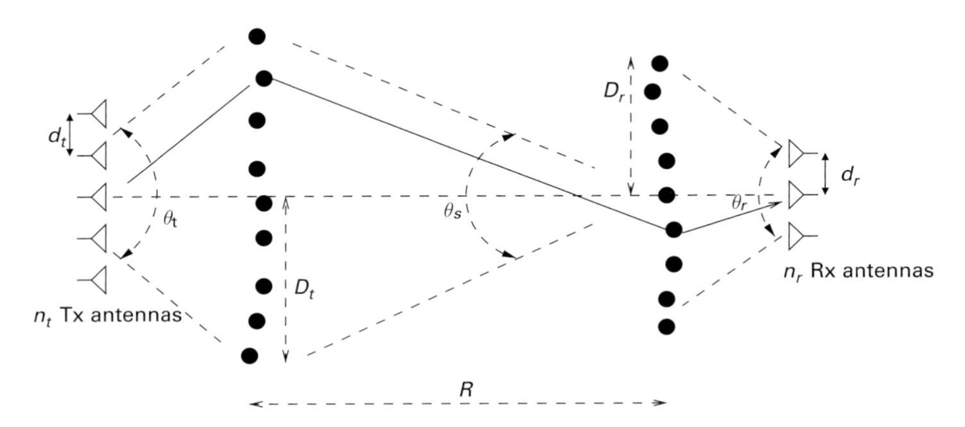

Figure 11.3 Propagation scenario for the analytical MIMO fading channel model in [27].

$\mathbf{h} = \mathbf{R}_{\theta_r,d_r}^{1/2}\mathbf{g}$, where $\mathbf{g} \sim \mathcal{CN}(\mathbf{0}, \mathbf{I}_{n_r})$. A similar definition holds for the transmit correlation matrix $\mathbf{R}_{\theta_t,d_t}$. Accordingly, $\mathbf{G}_t$ in (11.21) is an iid Rayleigh fading matrix of size $S \times n_t$, given by $\mathbf{G}_t = [\mathbf{g}_1\ \mathbf{g}_2\ \cdots\ \mathbf{g}_{n_t}]$, where $\mathbf{g}_n \sim \mathcal{CN}(\mathbf{0}, \mathbf{I}_S)$. Similarly, $\mathbf{G}_r$ is an iid Rayleigh matrix of size $n_r \times S$. The $\mathbf{R}_{\theta_t,d_t}$ and $\mathbf{R}_{\theta_r,d_r}$ matrices control the transmit and receive antenna correlations, respectively. Different assumptions on the statistics of the DoDs/DoAs will yield different expressions for these matrices [19],[27]. The $\mathbf{R}_{\theta_S,2D_r/S}$ matrix in (11.21) is defined as follows. Each scatterer at the transmit side captures the signal from the transmit antennas and radiates it in the form of a plane wave toward the scatterers on the receive side. The receive side scatterers are viewed as an array of S virtual antennas with average spacing $2D_r/S$ and experience an angular spread $\theta_S = 2\tan^{-1}(D_t/R)$. Let the nth transmit antenna's signal captured by the S receive side scatterers be denoted by the vector $\mathbf{y}_n = [y_{n,1}\ y_{n,2}\ \cdots\ y_{n,S}]^T$. Using the approximation that the receive side scatterers form a uniform array of sensors, $\mathbf{y}_n \sim \mathcal{CN}(\mathbf{0}, \mathbf{R}_{\theta_S,2D_r/S})$, or equivalently $\mathbf{y}_n = \mathbf{R}_{\theta_S,2D_r/S}^{1/2}\ \mathbf{g}_n$ with $\mathbf{g}_n \sim \mathcal{CN}(\mathbf{0}, \mathbf{I}_S)$.

11.2.1 Pinhole effect

On the transmit side, the spatial fading correlation between the transmit antennas is governed by the deterministic matrix $\mathbf{R}_{\theta_t,d_t}^{1/2}$, and hence by the local transmit angular spread, the transmit antenna beamwidth, and spacing. Similarly, on the receive side, the fading correlation is controlled by the receive angular spread, the antenna beamwidth, and spacing through $\mathbf{R}_{\theta_r,d_r}^{1/2}$. Even if fading is spatially uncorrelated on both transmit and receive ends (i.e., $\mathbf{R}_{\theta_t,d_t} = \mathbf{I}_{n_t}$ and $\mathbf{R}_{\theta_r,d_r} = \mathbf{I}_{n_r}$), (11.21) shows that it is still possible to have a rank-deficient MIMO channel due to the $\mathbf{R}_{\theta_S,2D_r/S}^{1/2}$ matrix, leading to reduced capacity. Such channels are referred to as pin-hole/key-hole channels [27],[28], where the scattered energy travels through a narrow air pipe between the transmitter and receiver, preventing the channel rank from building up. This happens when the rank of the matrix $\mathbf{R}^{1/2}_{\theta_S,2D_r/S}$ in (11.21) drops, which can be caused by, for example, a large transmit–receive range R or small scattering radii (i.e., small D_t or small D_r or both). It has been also argued that the measurements of [28] showed only rare occurrences of weak pinholes, and that although experimental evidence of pinholes was established in controlled environments in the laboratory in [29],[30], not many true occurrences of pinholes have been reported [31]. Nonetheless, the effect of pinholes on the performance of MIMO systems continues to be widely studied [32]–[35].

11.2.2 Effect of spatial correlation on LAS detector performance

Let us now illustrate the effect of the spatial correlation modeled in (11.21) on the BER performance of the LAS detection algorithm presented in Chapter 5. The parameters that characterize the MIMO channel model in (11.21) are:

- n_t, n_r : the number of transmit and receive (omnidirectional) antennas;
- d_t, d_r: the spacing between antenna elements at the transmit and receive sides;
- R: the distance between the transmit and receive sides;
- D_t, D_r: the transmit and receive scattering radii;
- S: the number of scatterers on each side;
- θ_t, θ_r: the angular spread at the transmit and receiver sides;
- f_c, λ: the carrier frequency, wavelength.

Let us consider the performance of large MIMO systems that employ NO-STBCs. Consider 16×16 and 12×12 full-rate NO-STBCs from CDA with $t = \delta = 1$ [36]. These NO-STBC MIMO systems are suited for practical implementation when fading is quasi-static, i.e., the fading remains constant over one STBC block length. For example, with static terminals in applications like wireless high-definition television (HDTV) distribution, the fading can be slow enough to satisfy the quasi-static fading assumption.

The BER performance of the systems is evaluated for two channel models: (*i*) iid fading channel model and (*ii*) the spatial correlation based channel model in (11.21). In the simulations, the following parameters are used for the channel model in (11.21): $f_c = 5$ GHz, $R = 500$ m, $S = 30$, $D_t = D_r = 20$ m, $\theta_t = \theta_r = 90°$, and $d_t = d_r = 2\lambda/3$. For the considered carrier frequency of 5 GHz, $\lambda = 6$ cm and $d_t = d_r = 4$ cm.

Spatial correlation degrades capacity and BER performance

Figure 11.4 shows the BER performance of a 1-LAS detector in decoding 16×16 NO-STBC with $n_t = n_r = 16$ and 16-QAM. The MMSE output is used as the initial vector for 1-LAS. Uncoded BER and rate-3/4 turbo coded BER (48 bps/Hz spectral efficiency) for iid fading as well as correlated fading are shown. In addition, the theoretical minimum SNRs required to achieve a capacity of 48 bps/Hz in iid as well as correlated fading are also shown. It is seen that the minimum SNR required to achieve a certain capacity (48 bps/Hz) increases for correlated

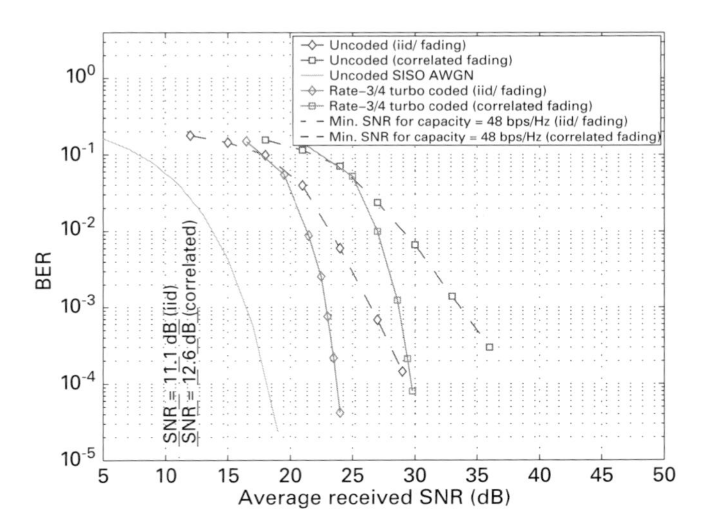

Figure 11.4 Uncoded/coded BER performance of a 1-LAS detector in iid fading and correlated fading in NO-STBC MIMO system with $n_t = n_r = 16$, 16×16 NO-STBC, 16-QAM, rate-3/4 turbo code, 48 bps/Hz.

fading compared to iid fading. From the BER plots in Fig. 11.4, it can be observed that at an uncoded BER of 10^{-3}, the performance in correlated fading degrades by about 7 dB compared to that in iid fading. Likewise, at a rate-3/4 turbo coded BER of 10^{-4}, a performance loss of about 6 dB is observed in correlated fading compared to that in iid fading. In terms of nearness to capacity, the vertical fall of the coded BER for iid fading occurs at about 24 dB SNR, which is about 13 dB away from theoretical minimum required SNR of 11.1 dB. With correlated fading, the detector is observed to perform close to capacity within about 18.5 dB. One way to alleviate such degradation in BER performance due to spatial correlation is to provide more dimensions (i.e., more antennas) on the receive side. The results presented in Fig. 11.5 illustrate this point.

Increasing number of receive antennas improves performance

Figure 11.5 shows the BER performance of a 1-LAS detector in STBC MIMO systems using 12×12 NO-STBC and 16-QAM. The effect of increasing the number of receive antennas for the same physical length of the receiver array ($n_r d_r$) and the same number of transmit antennas (n_t) is illustrated by comparing the performances of the $[n_t = 12, n_r = 12]$ system and the $[n_t = 12, n_r = 18]$ system, keeping $n_r d_r$ fixed at 72 cm and $d_t = d_r$ in both systems. It can be seen that a substantial improvement in uncoded as well as coded BER performance is achieved by increasing n_r beyond n_t. By comparing the performance of $[n_t = 12, n_r = 12]$ and $[n_t = 12, n_r = 18]$ systems, it is observed that the uncoded BER performance with $[n_t = 12, n_r = 18]$ improves by about 17 dB compared to that with $[n_t = 12, n_r = 12]$ at 2×10^{-3} BER. Even the uncoded BER performance with $[n_t = 12, n_r = 18]$ is significantly better than the coded

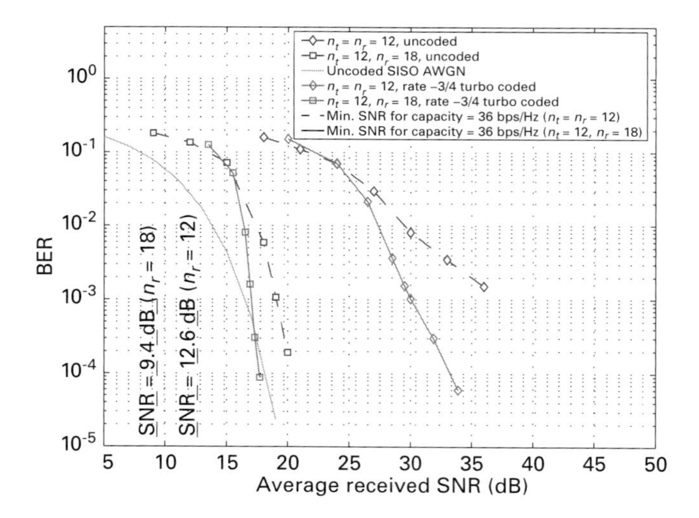

Figure 11.5 BER performance of a large MIMO system with $n_r = n_t$ and $n_r > n_t$ in correlated fading keeping $n_r d_r$ constant and $d_t = d_r$. $n_r d_r = 72$ cm. 12×12 NO-STBC, $n_t = 12$, $n_r = 12, 18$, 16-QAM, rate-3/4 turbo code, 36 bps/Hz.

BER performance with $[n_t = n_r = 12]$ by about 11.5 dB at 10^{-3} BER. With a rate-3/4 turbo code (36 bps/Hz), at a coded BER of 10^{-4}, there is an improvement in performance of about 13 dB with $[n_t = 12, n_r = 18]$ compared to $[n_t = n_r = 12]$. With $[n_t = 12, n_r = 18]$, the vertical fall of coded BER is such that it is about 8 dB from the theoretical minimum SNR needed to achieve capacity. Therefore, it is seen that the BER performance loss that occurs due to spatial correlation can be alleviated by using more receive antennas.

At this point, it is appropriate to note that spatial correlation need not always be harmful. For example, transmit correlation in MIMO fading can be exploited by using non-isotropic inputs (precoding) based on knowledge of the channel correlation matrices. While [37]–[39] have proposed correlation-exploiting precoders for orthogonal/quasi-orthogonal small MIMO systems in correlated Rayleigh/Ricean fading channels, such precoders for large MIMO systems remain to be studied.

11.3 Standardized channel models

A key advantage of channel models is that they establish reproducible channel conditions, which is very useful in system development and testing. To exploit this advantage in the development of new wireless systems, parameter sets for channel models representative of different target scenarios are defined by standards bodies. Such reference MIMO channel models are defined in wireless standards like WiFi (IEEE 802.11) and 3GPP (LTE/LTE-A).

WiFi refers to a collection of standards for implementing indoor wireless local area networks (WLAN) in the 2.4 GHz and 5 GHz frequency bands for unlicensed use. Earlier standards in the WiFi family, including IEEE 802.11b/11g/11a, use single-antenna terminals and access points. Multiantenna techniques were adopted in later standards including IEEE 802.11n/11ac in order to significantly increase the data rates compared to the rates in the 802.11b/g/a standards. IEEE 802.11n supports up to 600 Mbps through spatial multiplexing of up to four data streams simultaneously in the same frequency using up to four antennas. IEEE 802.11ac aims to support multigigabit rates (up to 3.6 Gbps) using up to eight data streams, and allows multiuser MIMO configuration. IEEE 802.11ac is also referred to as 5G WiFi or gigabit WiFi.

3GPP refers to a family of standards for third generation (3G) mobile radio communication and beyond. Channel modeling efforts in 2G and 3G mobile radio systems under the European research initiative 'cooperation in science and technology (COST)' have resulted in COST channel models for different radio environments including micro-, macro-, and pico-cell scenarios [40]. These initiatives defined channel models (e.g., COST 259, COST 273 models) that include the directional characteristics of radio propagation, and are therefore suitable for simulation of smart antennas and MIMO systems. In particular, the spatial channel model (SCM) was developed by 3GPP as a common reference model for evaluating different MIMO techniques in the 2 GHz band in outdoor environments [41]. The system bandwidth in the SCM model is 5 MHz. The issue of this narrow bandwidth was addressed in the WINNER channel model [42], where the SCM model was extended for a 100 MHz system bandwidth as well as a 5 GHz center frequency. This extended model is referred to as the SCM-Extended (SCME) model. The WINNER model is very general and covers many scenarios. It is a multicluster model similar to SCM. Cluster based models are quite relevant because practical measurements show clusters, and clusters reduce the number of parameters considerably. Many standard MIMO channel models rely on clusters (e.g., IEEE 802.11n, 3GPP-SCM, COST 273, WINNER II). In the WINNER model, clusters are placed to generate given azimuth power spectra at the transmit side and at the receive side. Each cluster has 20 multipath components. Eighteen different scenarios are parameterized by large numbers of measurements in outdoor, indoor, outdoor-to-indoor, with and without LOS, and high-speed scenarios. A simplified version of the SCME model has been adopted for standardization of the 3GPP long-term evolution [43],[44].

11.3.1 Models in IEEE 802.11 WiFi

WiFi standards like IEEE 802.11n and IEEE 802.11ac have adopted MIMO techniques for indoor communications in the 2.4 GHz and 5 GHz unlicensed bands. The 11n standard defines MIMO configurations with up to four antennas, the more recent 11ac standard defines MIMO configurations with up to eight antennas. Reference MIMO channel models for these standards are defined under the Task Group IEEE 802.11n (TGn) channel model framework with a focus on

Table 11.1. *TGn channel models A–F*

Parameters	A	B	C	D	E	F
Avg. 1st wall dist. (m)	5	5	5	10	20	30
RMS delay spread (ns)	0	15	30	50	100	150
Maximum delay (ns)	0	80	200	390	730	1050
Number of taps	1	9	14	18	18	18
Number of clusters	N/A	2	2	3	4	6
K (dB) LOS/NLOS	$0/-\infty$	$0/-\infty$	$0/-\infty$	$3/-\infty$	$6/-\infty$	$6/-\infty$

MIMO WLANs [45]. Six models (TGn models A–F) are defined. The models assume linear antenna arrays at the transmitter and receiver with 1/2, 1, and 4 wavelength spacing between antenna elements.

To model the frequency selectivity/delay spread characteristics of the channel, multicluster tapped delay line models with different numbers of taps are defined. Table 11.1 shows the parameter sets that define the delay spread characteristics for the TGn models A–F, which are meant to reflect the characteristics of the modeled environment. Note that TGn model A is the frequency-flat fading model. The multicluster model is based on the cluster model developed by Saleh and Valenzuela [46]. Depending on the model, the number of clusters varies from 2 to 6, which is chosen to be consistent with several experimentally determined results reported in the literature. Power, angular spread, AoA, and AoD values are assigned to each tap and cluster using statistical methods that agree with experimentally determined values reported in the literature. Cluster angular spread has been experimentally found to be in the range 20°–40°, and the mean AoA has been found to be random with a uniform distribution. The power, angular spread at the transmitter and receiver, and AoA and AoD for each tap and cluster for models A–F are tabulated in Appendix C of [46]. For a given antenna configuration, the channel matrix $\mathbf{H}$ of size $n_r \times n_t$ for each tap can be determined with the knowledge of each tap power, AS, and AoA/AoD. Modeling of this channel matrix also considers transmit and receive correlation matrices, following the approach in [47],[48], where the correlation matrix for each tap is based on the power angular spectrum (PAS) with angular spread being the second moment of PAS.

The temporal fading characteristics of indoor wireless channels are quite different from those of outdoor mobile channels. In indoor wireless systems, the transmitter and receiver are often stationary with people moving in between. Whereas, in outdoor mobile systems, user terminals often move at different speeds through an environment. Therefore, the Doppler bandwidth in indoor wireless channels can be significantly smaller than the Doppler bandwidths in outdoor mobile channels. To model the time selectivity of the indoor wireless channel, a bell-shaped Doppler spectrum is defined in the WiFi standard. The Doppler spectrum assumes reflectors moving in the environment at 1.2 km/h speed. This corresponds to about 6 Hz Doppler in the 5 GHz band and 3 Hz in the 2.4 GHz band.

Channel models for polarized antennas are also defined. Reference [46] refers to a weblink [49], where a Matlab implementation of the TGn MIMO channel models with appropriate antenna correlation properties is available for download.

11.3.2 Models in 3GPP/LTE

3GPP defines two types of SCM in [43] for MIMO simulations, one for calibration purposes and another for system simulation and performance comparison purposes. In the model for calibration, each resolvable path in the model is characterized by its own spatial parameters including angular spread, AoA/AoD, and power azimuth spectrum. All paths are assumed to be independent at both the BS and mobile station sides. Antenna element spacings of $0.5\lambda, 4\lambda$, and 10λ are defined on the BS side, and a spacing of 0.5λ is defined at the mobile station side. The per-path Doppler spectrum is defined as a function of the direction of travel and the per-path PAS and AoA at the mobile station. The average complex correlation and magnitude of the complex correlation between BS antennas and between mobile station antennas are also defined for different antenna spacings, angular spread, and AoA.

The model for system simulation defines the methodology and parameters for generating the spatial and temporal channel coefficients between a given BS and mobile station for use in system level simulations. For an M-element BS/mobile station array and an N-element mobile station/BS array, the channel coefficients for one of L multipath components are defined by an $M \times N$ matrix of complex amplitudes. Each channel matrix varies as a function of time due to the movement of the mobile station. The overall procedure for generating the channel matrices consists of three basic steps: (i) specify an environment – either suburban macro, urban macro, or urban micro, (ii) obtain the parameters to be used in simulations, associated with that environment, and (iii) generate the channel coefficients based on the parameters. Practical antennas in hand held mobile devices require antenna spacings much less than 0.5λ. So polarized antennas are likely to be a way to implement multiple antennas. A cross-polarized model is defined to address this scenario.

In the LTE standard, propagation conditions are defined under static propagation conditions and multipath fading propagation conditions [44]. The multipath propagation conditions in MIMO scenarios consist of (i) a set of correlation matrices defining the correlation between the BS (referred to as eNodeB in LTE) and user equipment (UE) antennas, (ii) a delay profile in the form of a tapped delay line, characterized by a number of taps at fixed positions, the rms delay spread, and the maximum delay spanned by the taps, and (iii) a combination of channel model parameters that include the delay profile and the Doppler spectrum characterized by a spectrum shape and maximum Doppler frequency. The MIMO channel correlation matrices apply for the antenna configuration using uniform linear arrays at both eNodeB and UE. Correlation matrices for the eNodeB ($\mathbf{R}_{eNB}$) are defined for one antenna, two antennas, and four antennas, in terms of

Table 11.2. *Delay profiles in LTE channel models*

Models	No. of taps	Delay spread (rms, ns)	Max. delay (span, ns)	Max. Doppler (Hz)
EPA	7	45	410	5, 70, 300
EVA	9	357	2510	5, 70, 300
ETU	9	991	5000	5, 70, 300

a parameter α. Likewise, correlation matrices for the UE ($\mathbf{R}_{UE}$) are defined for one antenna, two antennas, and four antennas, in terms of another parameter β. Spatial correlation matrices ($\mathbf{R}_{spat} = \mathbf{R}_{eNB} \otimes \mathbf{R}_{UE}$) are defined to characterize the spatial correlation between the antennas at the eNodeB and UE in terms of the parameters α and β. These correlation matrices are defined for the cases of 1×2, 2×2, 4×2, and 4×4 MIMO configurations. For cases with more antennas at either eNodeB or UE or both, the channel spatial correlation matrix can still be expressed as the Kronecker product of $\mathbf{R}_{eNB}$ and $\mathbf{R}_{UE}$ according to $\mathbf{R}_{spat} = \mathbf{R}_{eNB} \otimes \mathbf{R}_{UE}$. The (α, β) values for low, medium, and high correlation levels are defined to be $(0, 0)$, $(0.3, 0.9)$, and $(0.9, 0.9)$, respectively. The spatial correlation matrix for the low correlation case with $\alpha = \beta = 0$ for an $n \times m$ MIMO configuration is nothing but $\mathbf{I}_{nm}$. MIMO correlation matrices using cross-polarized antennas are also defined, in which case the $\mathbf{R}_{spat}$ is given by $\mathbf{R}_{spat} = \mathbf{P}(\mathbf{R}_{eNB} \otimes \mathbf{\Gamma} \otimes \mathbf{R}_{UE})\mathbf{P}^T$, where $\mathbf{R}_{eNB}$ is the correlation matrix at the eNodeB with the same polarization, $\mathbf{R}_{UE}$ is the correlation matrix at the UE with the same polarization, $\mathbf{\Gamma}$ is a polarization correlation matrix defined to be a function of a parameter γ, and $\mathbf{P}$ is a permutation matrix. $\mathbf{R}_{eNB}$, $\mathbf{R}_{UE}$ matrices for a two-antenna transmitter using one pair of cross-polarized antenna elements, a four-antenna transmitter using two pairs of cross-polarized antenna elements, and an eight-antenna transmitter using four pairs of cross-polarized antenna elements are defined. For the high-correlation case, the values of α, β, γ are 0.9, 0.9, 0.3, respectively.

The delay profiles are defined for low, medium, and high delay spread environments, representative of extended pedestrian A (EPA), extended vehicular A model (EVA), and extended typical urban (ETU) models, respectively. Likewise, the time variation of the channel is defined through low (5 Hz), medium (70 Hz), and high (300 Hz) Doppler frequencies. Table 11.2 shows the delay profiles and maximum Doppler frequencies, and Table 11.3 shows the power delay profiles for the EPA, EVA, and ETU scenarios.

11.4 Large MIMO channel measurement campaigns

MIMO channel measurements using a large number of antennas have been reported in the literature. With the growing interest in the practical implementation of large MIMO systems, more and more channel sounding measurements

Table 11.3. *Power delay profiles in LTE channel models*

EPA		EVA		ETU	
Excess tap delay (ns)	Relative power (dB)	Excess tap delay (ns)	Relative power (dB)	Excess tap delay (ns)	Relative power (dB)
0	0.0	0	0.0	0	−1.0
30	−1.0	30	−1.5	50	−1.0
70	−2.0	150	−1.4	120	−1.0
90	−3.0	310	−3.6	200	0.0
110	−8.0	370	−0.6	230	0.0
190	−17.2	710	−9.1	500	0.0
410	−20.8	1090	−7.0	1600	−3.0
		1730	−12.0	2300	−5.0
		2510	−16.9	5000	−7.0

using large antenna arrays are being reported. Indoor and outdoor measurements in the 2 GHz and 5 GHz bands are common. Some key requirements of MIMO channel measurement systems include: (i) good angular resolution to distinguish DoA and DoD, (ii) polarization discrimination capability to determine the usefulness of orthogonal polarizations as parallel channels, and (iii) the capability to record the channel continuously to investigate the multipath behavior as the user terminal moves, which allows the use of the Doppler domain in the signal analysis. Some of the large MIMO channel sounding campaigns and measurements reported in the literature are summarized in this subsection.

Early indoor measurements with 12×15 MIMO

Early MIMO channel measurements using a large number of antenna elements include those reported in [50],[51] using 12 transmit antennas and 15 receive antennas in the 1.95 GHz band with 30 kHz system bandwidth (narrowband MIMO) in indoor environments under strong and weak LOS conditions. The antenna elements were mounted on 2 ft $\times$ 2 ft panels, and were either vertically or horizontally polarized and arranged in alternate polarizations on a 4×4 grid, separated by $\lambda/2$ ($\approx$ 8 cm). Results from these measurements highlighted the effect of spatial correlation on the MIMO capacity. The 12×15 system measurement results showed that the system capacity does not scale linearly with the number of elements (as one would have expected in a rich scattering environment) which indicates that the signals are correlated, limiting the effective rank of the channel matrix. Subsequently, the effect of antenna polarization on the MIMO capacity was studied in a 4×4 MIMO configuration [52], noting that a system of size 12×15 may not be practical due to complexity and cost (reflecting the technology landscape and views at that time).

Figure 11.6 Sixteen-element antenna array in a user terminal (laptop). Urban outdoor MIMO channel sounding in 2.11 GHz. Photo source: [53].

Early urban outdoor measurements with 16×16 MIMO

Outdoor narrowband MIMO measurements in the 2.11 GHz band using 16 transmit antennas and 16 receive antennas were carried out in Manhattan, and the results were reported in [53]. The BS antenna array was a horizontal linear array of eight pairs of antennas, each pair consisting of a vertically and horizontally polarized radiating slot element. The length of the entire BS array was 3 m, which is 20λ at 2 GHz. The user terminal (laptop) array was rectangular with approximately $\lambda/2$ spacing between elements with alternating polarizations and oriented in the vertical plane (see Fig. 11.6). Spatial correlation model parameters were derived from the measured data. High capacities (80% of the fully scattering, i.e., iid, channel capacity) were found to be achieved. The AoA and Doppler spread were also derived from the temporal correlation of field components [54]. Results showed that the AoA at the mobile station was not uniformly distributed, which, in turn, resulted in approximately twice the correlation time than the predicted values from the Jakes model [55]. The measured median coherence time was at least a few seconds for stationary channels and 90 ms at the pedestrian speed of 3 km/hr. The measured median rms angular spread at the mobile was 22.5° for horizontally polarized antennas and 25.5° for vertically polarized antennas.

32×64 indoor measurements

Indoor MIMO channel sounding experiments in the 5.3 GHz band with 120 MHz bandwidth under LOS and NLOS scenarios (representative of WLAN environments) using dual polarized 64-element cylindrical antenna structure and a 21-element semi-spherical antenna structure were reported in [56],[57]. Channel measurements between one transmitter (semi-spherical mount) and two receivers (one semi-spherical mount and one cylindrical mount) each having 32 channels were carried out, essentially acquiring two 32×32 channel matrices or equivalently one effective channel matrix of size 64×32. The path loss (exponents 2.2,

8.2 for Tx-Rx1 and 1.5, 9.7 for Tx-Rx2) and power delay profile (maximum delay of about 450 ns) characteristics, and MIMO capacity results obtained from these measurements were reported.

16 × 16 indoor measurements

Contributions to IEEE 802.11 task group TGac towards 802.11ac channel modeling have considered indoor channel sounding measurements with 8×8 and 16×16 MIMO configurations in the 5.17 GHz band under LOS and NLOS settings [58]. The eight-element antenna was a polarized slot antenna array with $\lambda/2$ separation between slot pairs. The 16-element antenna was a linear dipole antenna array with $\lambda/2$ separation between the elements. Results from measurements suggest that TGn channel models can be used for 11ac if the system bandwidth is less than 100 MHz. For bandwidths more than 100 MHz, the channel tap spacing may have to be reduced to 1 ns instead of the 10 ns spacing in 11n and wider bandwidth channel measurements are needed to derive channel models for such large bandwidths.

16 × 32 indoor measurements

In [59], wideband indoor channel sounding measurements were carried out for the 16×32 MIMO configuration in the 5.8 GHz band with 100 MHz bandwidth to investigate both the spatial as well as the temporal characteristics of the channel. The antennas used at the transmit and receive ends were planar arrays of monopoles. The monopoles were arranged in an 8×12 rectangular grid (see Fig. 11.7). Sixteen monopoles (4×4 grid) were used at the transmitter, and 32 monopoles (4×8 grid) were used at the receiver. The monopoles were about 0.3λ in length and spaced about $\lambda/2$ apart. Measurements were carried out at different locations in the same building, e.g., open laboratory, room-to-room, basement, building level crossing. At 10 dB SNR, the statistics of the obtained capacities showed mean capacities in the range 32–51 bps/Hz. The Doppler bandwidth was found to be within about ± 2 Hz, which was comparable to the expected ± 0.6 Hz; the receive array was moved at a speed of about 1/30 m/s, which corresponds to a maximum Doppler of 0.6 Hz at 5.8 GHz carrier frequency. The excess Doppler was attributed to other changes in the environment, such as people walking by, etc. The coherence bandwidths of the channels were found to be in the range 5–28 MHz.

24 × 24 and 36 × 36 MIMO cubes

Compact antenna designs are important in large MIMO systems, where individual antennas need to be densely packed in a compact volume with low mutual coupling between various antenna ports. Popular compact antenna design approaches in mobile communications include patch antennas and planar inverted F antennas. Another interesting approach to compact MIMO antenna design

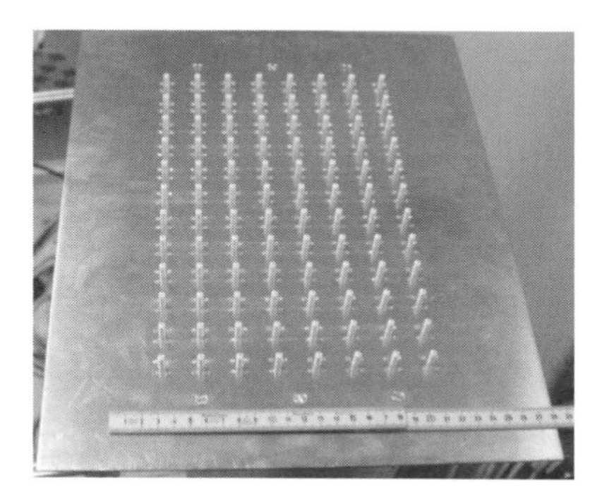

Figure 11.7 Planar monopole array in an 8×12 rectangular grid. 16×32 indoor wideband measurements in 5.8 GHz. Photo source: [59].

is the MIMO cube [60]. The MIMO cube takes advantage of spatial and polarization diversities in a compact volume. In [61], 24-port and 36-port MIMO cube geometries have been proposed and tested (see Fig. 11.8).

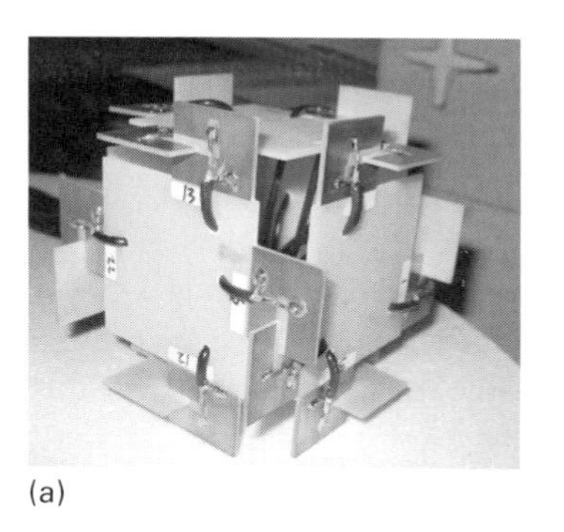

(a)

(b)

Figure 11.8 24×24 and 36×36 channel sounding using MIMO cubes in 2.7 GHz: (a) 24-antenna MIMO cube (80 mm $\times$ 80 mm $\times$ 80 mm); (b) 36-antenna MIMO cube (120 mm $\times$ 120 mm $\times$ 120 mm). Photo source: [61].

In the 24-port cube, twelve pairs of $\lambda/4$ slot antennas are distributed on each edge of an 80 mm $\times$ 80 mm $\times$ 80 mm cube providing low mutual coupling at 2.7 GHz operating frequency. The 36-port design consists of a combination of 24 $\lambda/2$ slot antennas and 12 $\lambda/4$ slot antennas built on a 120 mm $\times$ 120 mm $\times$ 120 mm cube with an operating frequency of 2.82 GHz. Due to inadequate test equipment/hardware for activating and testing all the 36 ports simultaneously, channel measurements were carried out in 4×4 configurations in an indoor environment for a total of around 100 combinations of four ports from the total of 36 ports. Measurement results indicated that the mutual coupling in the MIMO cube is low enough not to affect the performance significantly. At an SNR of 20 dB, the estimated channel capacity of the 36-port MIMO cube was 159 bps/Hz compared to the 36×36 iid MIMO capacity of 197 bps/Hz.

Figure 11.9 Sixteen-element UCA. 8×16 outdoor-to-indoor office measurements in 5.2 GHz. Photo source: [62].

8×16 outdoor-to-indoor measurements

A MIMO channel measurement campaign for an outdoor-to-indoor office scenario in the 5.2 GHz band with 120 MHz system bandwidth has been reported in [62]. The transmit antenna was an eight-element dual polarized uniform linear array (ULA) of patch elements with $\lambda/2$ spacing. The receive antenna was a 16-element uniform circular array (UCA) of vertically polarized monopole elements (see Fig. 11.9). Various parameters including DoA, DoD, distributions of rms directional spreads and delay spreads, and complex amplitudes were derived from the measured data. The angular dispersion at the outdoor link end was found to be small; the mean direction spread was in the range of 0.09–0.24. At the indoor link end, the angular dispersion was much larger (mean direction spreads in the range of 0.69–0.82). The delay spread was measured to be in the range of 5–25 ns. The DoA spectrum was found to depend noticeably on the DoD. Using the ergodic channel capacity as a metric, the performances of the Kronecker, virtual channel representation, and Weichselberger models for this outdoor-to-indoor scenario were compared. The Kronecker model was found not to be suitable due to the breakdown of the DoA–DoD decoupling assumptions. The Weichselberger model was found to provide a better fit to the measured capacity for both the LOS and NLOS scenarios.

Measurements with 128-element antenna arrays

An emerging architecture for mobile communication beyond 4G is one where the BS is equipped with tens to hundreds of antennas and each user terminal is equipped with one or more antennas. While theoretical and signal processing algorithm related studies for such large-scale systems assume well-known channel models (e.g., the iid model, the pin-hole model), sufficient validation of these models through real-world channel measurements is essential. The literature on channel measurements with such large antenna arrays is still limited.

(a)

(b)

Figure 11.10 128-element antenna arrays. Measurements in 2.6 GHz: (a) Cylindrical patch array (b) Planar patch array. Photo source: [63].

Large MIMO measurement campaigns have started to address this subject [63]–[65].

In [63], experiments with 128-element antenna arrays in the 2.6 GHz band with 50 MHz bandwidth were reported. The receive antenna array was a cylindrical patch array with 16 dual polarized antennas in each ring and four such rings stacked on top of each other, giving a total of 128 antenna ports (see Fig. 11.10). The diameter and height of the cylinder were 29.4 cm and 28.3 cm, respectively. The spacing between adjacent antenna elements was $\lambda/2$ (about 6 cm at a carrier frequency of 2.6 GHz). The transmit antenna array consisted of a planar patch array having two rows with eight dual polarized antennas in each row, giving a total of 32 antenna ports. Measurements were carried out in a residential area in an outdoor-to-indoor setting (representing a scenario with one indoor BS and multiple outdoor users). The first antenna in the 32-antenna transmit array was selected to represent a single-antenna user terminal. The BS receive antenna array was placed upstairs in a house in a particular street, and several (ten) outdoor locations in the next two streets were chosen as sites for the user terminal transmitter. The two-user scenario was tested by selecting two different locations among the chosen locations for keeping the user transmitter. The measured data in a two-user scenario in which the two user transmitters were kept at sites near to each other showed that the channel correlation is higher in the measured channels than in the iid channels. This is because the two user locations were close and probably had common scatterers that made the channels similar. Also, measurements made by varying the number of receive antennas at the indoor BS indicated that the average correlation decreased as the number of receive antennas was increased. Overall, the results indicated that most of the theoretical benefits of large-scale MIMO could be realized over the measured channels also.

Another measurement campaign was carried out in the 2.6 GHz band using a 50 MHz bandwidth employing a 128-element BS receive linear array with $\lambda/2$ spacing (hence, the array is 7.3 m long) and 36 different single-antenna user transmitter positions, 26 of them in LOS conditions and the remaining 10 in

NLOS conditions, and the results were presented in [64]. Parameters like the Rice factor, the received power levels over the array, the antenna correlation, and the eigenvalue distributions were analyzed. It has been remarked that the propagation conditions, from a large-array point of view, are actually better than expected. It has also been observed that the near-field effects and the non-stationarities over the array help to decorrelate the channel for different users, thereby providing favorable channel conditions with stable channels and low interference for the considered single antenna users.

An outdoor measurement campaign with large-scale antenna arrays in the 2.6 GHz band with a 20 MHz bandwidth was reported in [65]. A virtual antenna array (a rotating antenna array with 16 angular positions to emulate a cylindrical array of 112 elements) was mounted on the top of a large building at a height of about 20 m. Two mobile single-antenna receivers were 2 m apart on top of a car. The measurement positions were selected to provide a good mix of different channel conditions (LOS and NLOS) which can be considered as representative for a residential urban area. The channel capacity and achievable sum-rates with linear precoding were estimated using the measured data. Different metrics including the correlation coefficient and condition number were analyzed to find to what extent channel orthogonality between different terminals can be established by scaling up the number of transmit antennas. The results indicated that in spite of significant differences between the iid and the measured channels, a large fraction of the theoretical performance gains of large antenna arrays could be achieved in practice.

11.5 Compact antenna arrays

For a given antenna aperture constraint, increasing the number of antenna elements decreases the amount of spacing and thus increases the correlation. As a rule of thumb, the inter-element spacing should be not less than $\lambda/2$ to successfully decorrelate the incoming waves. This spacing may not be always possible in large MIMO systems. Compact antenna arrays are antenna arrays with inter-element spacing less than $\lambda/2$. They play an important role in large MIMO systems because of the limited space available for mounting antenna elements in communication terminals and the detrimental effects of mutual coupling in antenna arrays. Mutual coupling not only affects the MIMO channel capacity but also affects the array's radiation efficiency. The coupling becomes more pronounced as the antenna spacing decreases. Compact arrays designed to preserve MIMO channel capacity are crucial in large MIMO systems. Designing antenna arrays which are compact yet demonstrate acceptable mutual coupling and radiation efficiency is challenging.

One approach to designing antenna arrays that preserve MIMO channel capacity is to use matching networks. Conjugate matching networks and load matching networks have been developed to combat coupling-induced correlation and

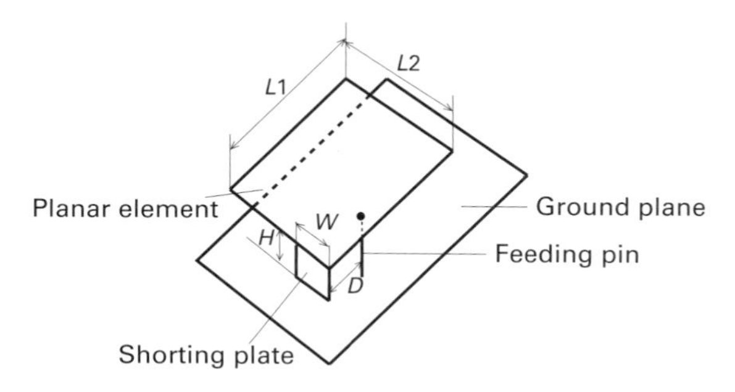

Figure 11.11 PIFA.

impedance mismatch. These are narrowband methods. Designing and realizing matching networks for multiband and wideband MIMO transceivers is more difficult. Also, a conjugate matching network, though mathematically tractable, may be difficult to realize as a compact physical circuit. Therefore, the realization of compact antenna arrays that are designed to preserve MIMO channel capacity without the need for matching networks is of interest. One approach that suits well for this objective is the PIFA approach.

11.5.1 PIFA

PIFA was introduced in 1987 [66]. It has good attributes like a low profile, good radiation characteristics, and a wide bandwidth. Because of these attributes, PIFA has emerged as one of the most promising low profile antenna design approaches. A broad range of applications use PIFA as the basic antenna element. These include mobile phones, wireless sensors, radio-frequency identification (RFID), ultra wideband (UWB) and MIMO systems, and wearable devices, covering several frequency bands of interest in outdoor and indoor applications; e.g., GSM (890–960 MHz), PCS (1850–1990 MHz), Bluetooth (2.4–2.48 GHz), DVB-H (UHF: 470–862 MHz; L: 1452–1492 MHz), WiFi (2.4–2.485 GHz; 5.16–5.5 GHz).

The name PIFA originates from the linear inverted F antennas (IFA), which are wire structures above a ground plane, forming an F shape. IFA is a variant of the monopole where the top section is folded down so as to be parallel to the ground plane. PIFA can be considered as a kind of IFA with the wire radiator element replaced by a plate to enhance the bandwidth. A PIFA typically consists of (*i*) a ground plane, (*ii*) a rectangular planar element of length $L1$ and width $L2$, placed above the ground plane at a certain height H, (*iii*) a short-circuit plate of width W (typically of a narrower width than that of the short side $L2$ of the planar element), and (*iv*) a feed pin placed at a distance D along the long side of the planar element, as shown in Fig. 11.11. PIFA characteristics are affected by a number of parameters including the dimensions of the ground plane, length, width, height, and position of the planar element (top plate), the positions and widths of shorting plate and feed pin/plate.

When $W = L2$, the shorting plate spans the entire short side of the planar element. In this case, the PIFA is resonant (i.e., has the maximum radiation efficiency) when

$$L1 = \frac{\lambda}{4}, \quad \text{if } W = L2. \tag{11.22}$$

The above relation between the resonant length and the shorting plate width can be explained by considering how a $\lambda/4$ patch antenna radiates. A $\lambda/4$ patch antenna needs a quarter-wavelength of space between the edge and the shorting area. If $W = L2$, then the distance from one edge to the short is simply $L1$, which gives (11.22).

When $W = 0$ (or $W \ll L2$), the shorting plate becomes a shorting pin. In this case, the PIFA is resonant at

$$L1 + L2 = \frac{\lambda}{4} \quad \text{if } W = 0. \tag{11.23}$$

Then, since it is the fringing fields along the edge that give rise to radiation in microstrip antennas, the length from the open-circuited radiating edge (the far edge in Fig. 11.11) to the shorting pin is on average equal to $L1 + L2$. This can be seen by measuring the distance from any point on the far edge of the PIFA to the shorting pin. The clockwise and counter-clockwise paths always add up to $2(L1+L2)$. So, on average, resonance will occur when the path length $(L1+L2)$ for a single path is a quarter-wavelength.

In general, the resonant length of a PIFA as a function of its parameters can be approximated as

$$L1 + L2 - W = \frac{\lambda}{4}. \tag{11.24}$$

For example, a PIFA with $L1 = 2.5$ cm, $L2 = 1.5$ cm, $W = 1$ cm, and an air dielectric between the ground plane and planar element will resonate at 2.5 GHz. The width of the feed plate plays an important role in broadening the antenna bandwidth. The length of the PIFA antenna can be further reduced by capacitive loading, e.g., by adding a capacitance between the feed point and the open edge. The capacitive load can be produced by adding a plate (parallel to the ground) to produce a parallel plate capacitor. While this capacitive loading approach can reduce the resonance length from $\lambda/4$ to less than $\lambda/8$, it comes at the cost of some loss in radiation efficiency and bandwidth.

11.5.2 PIFAs as elements in compact arrays

A PIFA is relatively robust to influence from another nearby PIFA because of the radiating element's low profile and proximity to the ground plane. This makes it a good candidate for use in a compact array. The design and testing of compact arrays using multiple PIFAs (dual and quad PIFAs) with an inter-PIFA spacing of 27 mm in 2.4–2.5 GHz ($< \lambda/4$ spacing) have been reported in [67]. Measurement results there have shown that a compact PIFA array's eigenvalue distributions are nearly as close to the ideal iid Gaussian channel's distributions as those of the 1λ dipole uniform linear array, despite the fact that the PIFA array was subject to the same or worse propagation-induced correlation. Compact antenna arrays with PIFA as the basic element are a promising approach in large MIMO systems.

11.5.3 MIMO cubes

Another interesting compact antenna array approach suited for large MIMO systems is the MIMO cube [61],[68],[69]. A cube is an attractive structure for building multiple antennas with low mutual couplings between antenna ports, because any two adjacent faces in a cube are perpendicular to each other. In addition, any two opposite faces in a cube have the farthest separation compared with other three-dimensional structures with the same volume. An antenna cube, therefore, can take advantage of spatial and polarization orthogonality to implement a large number of antennas within a constrained volume.

A compact arrangement of 12 dipoles placed on the 12 edges of a cube is shown in Fig. 11.12. Instead of using dipoles on the edges, slots on the edges [70] or slot-type dipoles on the edges [71],[72] can be used in order to improve the coupling performance at high frequencies. Quarter-wave length resonators are advantageous because they occupy less space than other types of resonators. These resonators can be quarter-wave slots or PIFAs.

For large MIMO systems, radiating elements can be built on a cube's edges, faces, and/or corners. An $80\,\text{mm} \times 80\,\text{mm} \times 80\,\text{mm}$ MIMO cube consisting of 24 quarter-wave slot antennas with two antennas placed on each edge of the cube was presented in [61]; see Fig. 11.8(a). A measured worst isolation between ports of -18 dB at 2.7 GHz and a slot antenna gain of 4.4 dBi have been reported. In addition to the edge mounted antennas, more antennas can be mounted on the faces of the cube. Taking this approach, a 36-antenna MIMO cube with 24 antennas placed on the edges and 12 more antennas placed on the faces of the cube in a volume of $120\,\text{mm} \times 120\,\text{mm} \times 120\,\text{mm}$ was also presented [61]; see Fig. 11.8(b). The edge antennas are 12 half-wave slot antenna pairs, each pair interlaced perpendicularly on each edge. The 12 antennas on the faces (2 antennas on each face) were quarter-wave slot antennas printed with 45° rotation relative to the edges of the cube. This helped to reduce the mutual couplings among the ports. Measured port isolations were better than -20 dB and the antenna gains were 4.3 dBi for half-wave slot antennas and 4.2 dBi for quarter-wave slot antennas, respectively, at 2.8 GHz. Even more antennas can be accommodated by using pairs of PIFA and quarter-wave slot antennas stacked closely together to form a set of nearly colocated antennas to achieve polarization diversity. A 48-antenna MIMO based on this approach was reported in [73]. Three colocated PIFA and quarter-wave slot antenna pairs were arranged orthogonally and spaced at the eight corners of the cube (total of 48 antennas) in a volume of $103\,\text{mm} \times 103\,\text{mm} \times 103\,\text{mm}$. The resonant frequency was 2.4 GHz and the port isolations were below -20 dB.

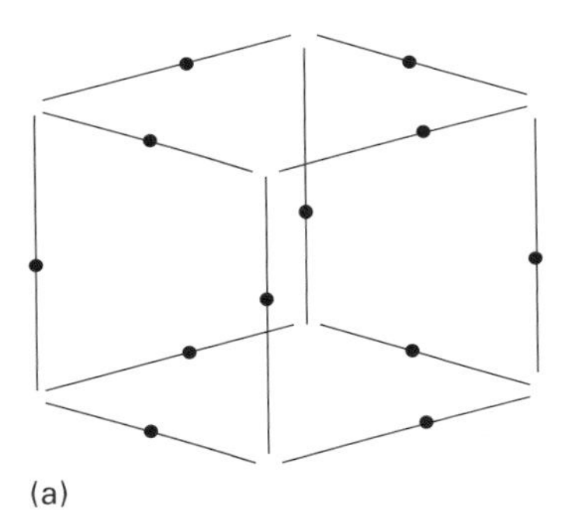
(a)

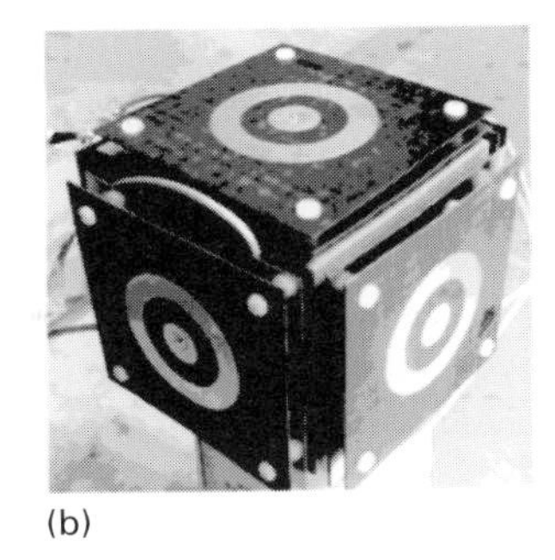
(b)

Figure 11.12 MIMO cube antennas: (a) 12 dipoles arranged along the edges of a cube. (b) 18-antenna MIMO cube. Photo source: [74].

Another compact antenna MIMO cube with 18 antennas mounted in a cube of size $0.76\lambda \times 0.76\lambda \times 0.76\lambda$ was presented in [74]; see Fig. 11.12(b). For the considered frequency band of 2.4–2.8 GHz, a cube of size 94 mm × 94 mm × 94 mm was used. The conformal and low profile tri-polarization antenna [75] which makes full use of the polarization domain was used as the basic antenna element in the MIMO cube. On each face of the cube, a three-port tri-polarization antenna was mounted. A ring patch which functioned as two independent orthogonal polarized antennas, and a disk-loaded monopole constituted the tri-polarization antenna. A metal backing configuration was adopted for these antennas, so the ground of all antennas formed a well-shielded Faraday cage, where other functional circuits could be housed. Indoor measurements in a laboratory environment were made using this MIMO cube and good results were reported.

In summary, MIMO cube developments and experiments suggest that the antenna cube is an attractive approach to support a large number of antennas in a small volume with high isolation of antenna ports.

References

[1] H. Ozcelik, N. Czink, and E. Bonek, "What makes a good MIMO channel model?" in *IEEE VTC'2005 Spring*, Stockholm, vol. 1, Sep. 2005, pp. 156–160.

[2] W. Weichselberger, Spatial structure of multiple antenna radio channels – a signal processing viewpoint. Ph.D dissertation, Technische Universitt Wien, Vienna, Austria, Dec. 2003.

[3] H. Ozcelik, Indoor MIMO channel models. Ph.D dissertation, Institutfr Nachrichtentechnik, Technische Universitt Wien, Vienna, Austria, Dec. 2004.

[4] P. Almers, E. Bonek, A. Burr, *et al.*, "Survey of channel and radio propagation models for wireless MIMO systems," *EURASIP J. Wireless Commun. and Networking*, pp. volume 2007, article–ID 19 070, 19 pages, 2007.

[5] D.-S. Shiu, G. J. Foschini, M. J. Gans, and J. M. Kahn, "Fading correlation and its effect on the capacity of multielement antenna systems," *IEEE Trans. Commun.*, vol. 48, no. 3, pp. 502–513, Mar. 2000.

[6] J. P. Kermoal, L. Schumacher, K. I. Pedersen, P. E. Mogensen, and F. Frederiksen, "A stochastic MIMO radio channel model with experimental validation," *IEEE J. Sel. Areas Commun.*, vol. 20, no. 6, pp. 1211–1226, Jun. 2002.

[7] W. Weichselberger, M. Herdin, H. Ozcelik, and E. Bonek, "A stochastic MIMO channel model with joint correlation of both link ends," *IEEE Trans. Wireless Commun.*, vol. 5, no. 1, pp. 90–100, Jan. 2006.

[8] A. G. Burr, "Capacity bounds and estimates for the finite scatterers MIMO wireless channel," *IEEE J. Sel. Areas in Commun.*, vol. 21, no. 5, pp. 812–818, May 2003.

[9] M. Debbah and R. R. Muller, "MIMO channel modeling and the principle of maximum entropy," *IEEE Trans. Inform. Theory*, vol. 51, no. 5, pp. 1667–1690, May 2005.

[10] A. M. Sayeed, "Deconstructing multiantenna fading channels," *IEEE Trans. Signal Process.*, vol. 50, no. 10, pp. 2563–2579, Oct. 2002.

[11] D. Tse and P. Viswanath, *Fundamentals of Wireless Communication*. Cambridge, UK: Cambridge University, 2005.

[12] K. Yu, M. Bengtsson, B. Ottersten, *et al.*, "Second order statistics of NLOS indoor MIMO channels based on 5.2 GHz measurements," in *IEEE GLOBECOM'2001*, San Antonio, TX, Nov. 2001, pp. 156–160.

[13] R. Stridh, K. Yu, B. Ottersten, and P. Karlsson, "MIMO channel capacity and modeling issues on a measured indoor radio channel at 5.8 GHz," *IEEE Trans. Wireless Commun.*, vol. 4, no. 3, pp. 895–903, May 2005.

[14] H. Ozcelik, M. Herdin, W. Weichselberger, J. Wallace, and E. Bonek, "Deficiencies of 'Kronecker' MIMO radio channel model," *Electronics Lett.*, vol. 39, pp. 1209–1210, Aug. 2003.

[15] S. Wyne, A. F. Molisch, P. Almers, *et al.*, "Outdoor-to-indoor office MIMO measurements and analysis at 5.2 GHz," *IEEE Trans. Veh. Tech.*, vol. 57, no. 3, pp. 1374–1386, May 2008.

[16] L. Correia, *COST 273 Final Report: Towards Mobile Broadband Multimedia Networks*. Amsterdam, Netherlands: Elsevier, 2006.

[17] L. Wood and W. S. Hodgkiss, "Impact of channel models on adaptive M-QAM modulation for MIMO systems," in *IEEE WCNC'2008*, Las vegas, NV, Apr. 2008, pp. 1316–1321.

[18] H. Tong and S. A. Zekavat, "On the suitable environments of the Kronecker product form in MIMO channel modeling," in *IEEE WCNC'2008*, Las Vegas, NV, Apr. 2008, pp. 780–784.

[19] R. Ertel, P. Cardieri, K. Sowerby, T. Rappaport, and J. Reed, "Overview of spatial channel models for antenna array communication systems," *IEEE Pers. Commun.*, vol. 5, no. 1, pp. 10–22, Feb. 1998.

[20] D. Chizhik, F. Rashid-Farrokhi, J. Ling, and A. Lozano, "Effect of antenna separation on the capacity of BLAST in correlated channels," *IEEE Commun. Lett.*, vol. 4, no. 11, pp. 337–339, Nov. 2000.

[21] A. A. Abouda, H. M. El-Sallabi, L. Vuokko, and S. G. Haggman, "Performance of stochastic Kronecker MIMO radio channel model in urban microcells," in *IEEE PIMRC'2006*, Helsinki, Sep. 2006, pp. 1–5.

[22] L. C. Wood and W. S. Hodgkiss, "MIMO channel models and performance metrics," in *IEEE GLOBECOM'2007*, Washington, DC, Nov. 2007, pp. 3740–3744.

[23] S. E. Foo, M. A. Beach, and A. G. Burr, "Wideband outdoor MIMO channel model derived from directional channel measurements at 2 GHz," in *WPMC'2004*, Abano Terme, Sep. 2004.

[24] M. Debbah and R. Muller, "Capacity complying MIMO channel models," in *Asilomar Conf. on Signals, Systems and Computers*, Pacific Grove, CA, vol. 2, Nov. 2003, pp. 1815–1819.

[25] M. Debbah, R. Muller, H. Hofstetter, and P. Lehne, "Validation of mutual information complying MIMO channel models," online http://citeseerx.ist.psu.edu/viewdoc/summary?doi=10.1.1.2.1829, 2004.

[26] B. T. Maharaj, L. P. Linde, and J. W. Wallace, "MIMO channel modelling: the Kronecker model and maximum entropy," in *IEEE WCNC'2007*, Hong Kong, Mar. 2007, pp. 1911–1914.

[27] D. Gesbert, H. Bolcskei, D. A. Gore, and A. J. Paulraj, "Outdoor MIMO wireless channels: models and performance prediction," *IEEE Trans. Commun.*, vol. 50, no. 12, pp. 1926–1934, Dec. 2002.

[28] D. Chizhik, G. J. Foschini, and R. A. Valenzuela, "Capacities of multi-element transmit and receive antennas: correlations and keyholes," *Electronics Lett.*, vol. 37, pp. 1099–1100, Jun. 2000.

[29] P. Almers, F. Tufvesson, and A. F. Molisch, "Measurement of keyhole efffects in a wireless multiple-input multiple-output (MIMO) channel," *IEEE Commun. Lett.*, vol. 7, no. 8, pp. 373–375, Aug. 2003.

[30] ——, "Keyhole effect in MIMO wireless channels: measurements and theory," *IEEE Trans. Wireless Commun.*, vol. 5, no. 12, pp. 3596–3604, May 2006.

[31] E. Bonek, M. Herdin, W. Weichselberger, and H. Ozcelik, "MIMO - study propagation first!" in *IEEE ISSPIT'2003*, Darmstadt, Dec. 2003, pp. 150–153.

[32] G. Levin and S. Loyka, "On the outage capacity distribution of correlated keyhole MIMO channels," *IEEE Trans. Inform. Theory*, vol. 54, no. 7, pp. 3232–3245, Jul. 2008.

[33] A. Nezampour, A. Nasri, and R. Schober, "Asymptotic analysis of space-time codes in generalized keyhole fading channels," *IEEE Trans. Wireless Commun.*, vol. 10, no. 6, pp. 1863–1873, Jun. 2011.

[34] C. Zhong, S. Jin, K.-K. Wong, and M. R. McKay, "Ergodic mutual information analysis for multi-keyhole MIMO channels," *IEEE Trans. Wireless Commun.*, vol. 10, no. 6, pp. 1754–1763, Jun. 2011.

[35] T. Q. Duong, H. A. Suraweera, T. A. Tsiftsis, H.-J. Zepernick, and A. Nallanathan, "Keyhole effect in dual-hop MIMO AF relay transmission with space-time block codes," *IEEE Trans. Commun.*, vol. 60, no. 12, pp. 3683–3693, Dec. 2012.

[36] B. A. Sethuraman, B. S. Rajan, and V. Shashidhar, "Full-diversity high-rate space-time block codes from division algebras," *IEEE Trans. Inform. Theory*, vol. 49, no. 10, pp. 2596–2616, Oct. 2003.

[37] M. Vu and A. J. Paulraj, "Optimal linear precoders for MIMO wireless correlated channels with nonzero mean in spacetime coded systems," *IEEE Trans. Signal Process.*, vol. 54, no. 6, pp. 2318–2332, Jun. 2006.

[38] H. R. Bahrami and T. Le-Ngoc, "Precoder design based on correlation matrices for MIMO systems," *IEEE Trans. Wireless Commun.*, vol. 5, no. 12, pp. 3579–3587, Dec. 2006.

[39] K. T. Phan, S. A. Vorobyov, and C. Tellambura, "Precoder design for space-time coded systems with correlated Rayleigh fading channels using convex optimization," *IEEE Trans. Signal Process.*, vol. 57, no. 2, pp. 814–819, Feb. 2009.

[40] A. F. Molisch, H. Asplund, H. Heddergott, M. Steinbauer, and T. Zwick, "The COST259 directional channel model - part i: Overview and methodology," *IEEE Trans. Wireless Commun.*, vol. 5, no. 12, pp. 3421–3433, Dec. 2006.

[41] "Spatial channel model for multiple input multiple output (MIMO) simulations," 3GPP-3GPP2 Spatial Channel Model Ad-hoc Group; 3GPP TR 25.996, v6.1.0, 2003–09.

[42] M. Narandic, C. Schneider, R. Thoma, *et al.*, Comparison of SCM, SCME, and WINNER channel models, *IEEE VTC'2007 Spring*, Dublin, pp. 413–417, Apr. 2007.

[43] "Spatial channel model for multiple input multiple output (MIMO) simulations (Release 11)," 3GPP Technical Specification Group Radio Access Network; 3GPP TR 25.996, v11.0.0, 2012-09.

[44] "User equipment (UE) radio transmission and reception (Release 10)," 3GPP Technical Specification Group Radio Access Network; 3GPP TS 36.101, v10.6.0, 2012-03.

[45] V. Erceg, "TGn channel models," IEEE P802.11 wireless LANs: doc.: IEEE 802.11-03/940r43, May 2004.

[46] A. A. M. Saleh and R. A. Valenzuela, "A statistical model for indoor multipath propagation," *IEEE J. Sel. Areas in Commun.*, vol. 5, no. 2, pp. 128–137, Oct. 1987.

[47] J. P. Kermoal, L. Schumacher, P. E. Mogensen, and K. I. Pedersen, "Experimental investigation of correlation properties of MIMO radio channels for indoor picocell scenario," in *IEEE VTC'2000*, Boston, MA, Sep. 2000, pp. 14–21.

[48] L. Schumacher, K. I. Pedersen, and P. E. Mogensen, "From antenna spacings to theoretical capacities - guidelines for simulating MIMO systems," in *IEEE PIMRC'2002*, Cannes, Sep. 2002, pp. 587–592.

[49] L. Schumacher, "LAN MIMO channel Matlab program," http://www.info.fundp.ac.be/∼lsc/Research.

[50] P. Kyritsi, P. W. Wolniansky, and R. A. Valenzuela, "Indoor BLAST measurements: capacity of multi-element antenna systems," in *Multi-Access, Mobility and Teletraffic for Wireless Commun.*, vol. 5, Dec. 2000, pp. 49–60.

[51] P. Kyritsi, D. C. Cox, R. A. Valenzuela, and P. W. Wolniansky, "Correlation analysis based on MIMO channel measurements in an indoor environment," *IEEE J. Sel. Areas Commun.*, vol. 21, no. 5, pp. 713–720, Jun. 2003.

[52] ——, "Effect of antenna polarization on the capacity of a multiple element system in an indoor environment," *IEEE J. Sel. Areas Commun.*, vol. 20, no. 6, pp. 1227–1239, Aug. 2002.

[53] D. Chizhik, J. Ling, P. W. Wolniansky, *et al.*, "Multiple-input-multiple-output measurements and modeling in Manhattan," *IEEE J. Sel. Areas Commun.*, vol. 21, no. 3, pp. 321–331, Apr. 2003.

[54] H. Xu, M. Gans, D. Chizhik, *et al.*, "Spatial and temporal variations of MIMO channels and impacts on capacity," in *IEEE ICC'2002*, vol. 1, Apr.-May 2002, pp. 262–266.

[55] W. C. Jakes, Ed., *Microwave Mobile Communications*. New York, NY: IEEE Press, 1974.

[56] J. Koivunen, Characterisation of MIMO propagation channel in multi-link scenarios. Master's thesis, Helsinki University of Technology, Finland, Dec. 2007.

[57] J. Koivunen, P. Almers, V.-M. Kolmonen, *et al.*, "Dynamic multi-link indoor MIMO measurements at 5.3 GHz," in *IEEE EuCAP'2007*, Edinburgh, Nov. 2007, pp. 1–6.

[58] G. Breit, H. Sampath, V. K. Jones, *et al.*, "802.11ac channel modeling," doc. IEEE 802.11-09/0088r0, submission to Task Group TGac, 19 Jan. 2009.

[59] J. O. Nielsen, J. B. Andersen, P. C. F. Eggers, *et al.*, "Measurements of indoor 16 × 32 wideband MIMO channels at 5.8 GHz," in *IEEE ISSSTA'2004*, Sydney, Aug.–Sep. 2004, pp. 864–868.

[60] B. N. Getu and J. B. Andersen, "The MIMO cube - a compact MIMO antenna," *IEEE Trans. Wireless Commun.*, vol. 4, no. 3, pp. 1136–1141, May 2005.

[61] C.-Y. Chiu, J.-B. Yan, and R. D. Murch, "24-port and 36-port antenna cubes suitable for MIMO wireless communications,," *IEEE Trans. Antennas and Propagation*, vol. 56, no. 4, pp. 1170–1176, Apr. 2008.

[62] S. Wyne, A. F. Molisch, P. Almers, *et al.*, "Outdoor-to-indoor office MIMO measurements and analysis at 5.2 GHz," *IEEE Trans. Veh. Tech.*, vol. 57, no. 3, pp. 1374–1386, May 2008.

[63] X. Gao, O. Edfors, F. Rusek, and F. Tufvesson, "Linear pre-coding performance in measured very-large MIMO channels," in *IEEE VTC'2011 Fall*, San Francisco, CA, Sep. 2011, pp. 1–5.

[64] S. Payami and F. Tufvesson, "Channel measurements and analysis for very large array systems at 2.6 GHz," in *European Conf. Antennas and Prop. (EUCAP'2012)*, Prague, Mar. 2012, pp. 433–437.

[65] J. Hoydis, C. Hoek, T. Wild, and S. ten Brink, "Channel measurements for large antenna arrays," in *Intl. Symp. on wireless Commun. Sys. (ISWCS)*, Paris, Aug. 2012, pp. 811–815.

[66] T. Taga and K. Tsunekawa, "Performance analysis of a built-in planar inverted F antenna," *IEEE J. Sel. Areas Commun.*, vol. 5, no. 5, pp. 921–929, Jun. 1987.

[67] D. W. Browne, M. Manteghi, M. P. Fitz, and Y. Rahmat-Samii, "Experiments with compact antenna arrays for MIMO radio communications," *IEEE Trans. Antennas and Propagation*, vol. 11, no. 54, pp. 3239–3250, Nov. 2006.

[68] J. B. Andersen and B. N. Getu, "The MIMO cube - a compact MIMO antenna," in *Intl. Symp. on Wireless Personal Multimedia Commun.*, vol. 1, Oct. 2002, pp. 112–114.

[69] ——, "The MIMO cube - a compact MIMO antenna," *IEEE Trans. Wireless Commun.*, vol. 4, no. 3, pp. 1136–1141, May 2005.

[70] J. X. Yun and R. G. Vaughan, "Slot MIMO cube," in *IEEE Antennas and Propagation Society Intl. Symp.*, Toronto, Jul. 2010, pp. 1–4.

[71] A. Nemeth, L. Sziics, and L. Nagy, "MIMO cube formed of slot dipoles," in *IST Mobile and Wireless Commun. Summit*, Budapest, Jul. 2007, pp. 1–5.

[72] L. Nagy, "Modified MIMO cube for enhanced channel capacity," *Intl J. Antennas and Propagation*, vol. 2012, Article ID 734896, 10 pages. doi:10.1155/2012/734896.

[73] C.-Y. Chiu and R. D. Murch, "Overview of multiple antenna designs for handheld devices and base stations," in *Intl Workshop on Antenna Technology (iWAT'2011)*, Hong Kong, Mar. 2011, pp. 74–77.

[74] J. Zheng, X. Gao, and Z. Feng, "A compact eighteen-port antenna cube for MIMO systems," *IEEE Trans. Antennas and Propagation.*, vol. 60, no. 2, pp. 445–455, Feb. 2012.

[75] X. Gao, H. Zhong, Z. Zhang, Z. Feng, and M. F. Iskander, "Low-profile planar tri-polarization antenna for WLAN communications," *IEEE Trans. Antennas and Prop. Lett.*, vol. 9, pp. 83–86, Feb. 2010.

12 Large MIMO testbeds

As in any new or emerging technology, demonstrators, testbeds, and prototypes play an important role in the development of large MIMO systems. The terms demonstrators, testbeds, and prototypes are often used loosely and interchangeably to refer to practical proof-of-concept-like implementations. The following broad definitions from [1],[2] bring out some key differences between them.

- A *demonstrator* is meant primarily to showcase and demonstrate technology to customers. Generally, it involves implementation of a new idea, concept, or standard that has been already established and has been finalized to some extent. Therefore, the requirements on functionality and design time are more important than scalability.
- A *testbed* is meant for research in general. It is a platform that allows the testing or verification of new algorithms or ideas under real-world conditions. Therefore, testbeds are expected to be more modular, scalable, and extendable.
- A *prototype* is meant to be serve as the initial realization of a research idea or a standard in real time, as a reference, a proof-of-concept, or a platform for future developments and improvements. It is often intended to evolve a prototype into a product.

From these definitions, one can see that testbeds and prototypes can play crucial roles in the research and development phase. While prototypes need necessarily to operate in real time, a testbed can be a real-time testbed or a non-real-time (offline) testbed depending on the available resources in comparison with the real-time computation need.

The first laboratory demonstration of a MIMO wireless communication system was reported in the late 1990s by the Bell Labs [3], where a laboratory prototype of an 8×12 V-BLAST MIMO system using eight transmit antennas and twelve receive antennas was demonstrated in an indoor laboratory/office environment. The distance between the transmitter and the receiver was about 12 m. It was a narrowband system which operated at a carrier frequency of 1.9 GHz and a bandwidth of 30 kHz. The antenna arrays consisted of $\lambda/2$ wire dipoles mounted in various arrangements; receive dipoles were mounted on the surface of a metallic hemisphere approximately 20 cm in diameter, and transmit dipoles were mounted on a flat metal sheet in a rectangular array configuration with about $\lambda/2$ inter-element spacing. The system employed spatial multiplexing,

16-QAM modulation, and ZF-SIC detection. A transmission rate of 777.6 kbps was achieved using 30 kHz bandwidth, demonstrating a high spectral efficiency of 25.92 bps/Hz.

Subsequent to the V-BLAST demonstration in 1998, over the years, several MIMO testbeds and prototypes were designed and built in many research laboratories and universities around the world. They were built with different purposes and scenarios in mind: e.g., to demonstrate MIMO gains of different MIMO configurations and signaling, to study and demonstrate multiuser MIMO communication, to provide hardware platforms for MIMO signal processing algorithms development and testing, to address education and standardization requirements and objectives [4]–[11].

A majority of the MIMO testbeds and prototypes built and reported have adopted MIMO configurations with only a small number of antennas. Many of them use 2–4 antennas, and some of them use up to eight antennas. Large MIMO testbeds and prototypes that use tens of antennas have started to emerge. This chapter summarizes the state-of-the-art in large MIMO testbeds/prototypes. In particular, four large MIMO testbeds/prototypes with the number of antennas ranging from 12 to 64 developed at different research laboratories and universities, and reported in the literature, are presented in the sections below. While these large MIMO testbeds/prototypes may be the first few of their kind, more such large MIMO testbeds/prototypes are expected to be built and reported in the years to come.

12.1 12×12 point-to-point MIMO system

A point-to-point MIMO system with 12 transmit and 12 receive antennas operating at 4.635 GHz carrier frequency with a 101.4 MHz bandwidth was reported in [12]. A data rate of 5 Gbps has been demonstrated, which, at a 101.4 MHz system bandwidth, corresponds to about 50 bps/Hz spectral efficiency. The system used spatial multiplexing of 12 data streams, MIMO-OFDM signaling using 1536 subcarriers with a subcarrier spacing of 65.919 kHz, turbo coding with rate 8/9 and constraint length 4, and 64-QAM modulation. A complexity-reduced ML detection with QR decomposition and the M-algorithm was used for signal detection at the receiver. The channel gains on each subcarrier were estimated using two-dimensional MMSE channel estimation.

The system was field tested in an outdoor urban environment. The 12 transmit antennas at the BS were mounted in a linear array configuration on the rooftop of a building at a height of about 26 m. The inter-antenna spacing was about 11λ (about 70 cm for the carrier frequency of 4.635 GHz). The 12 receive antennas at the mobile station (MS) were mounted on top of a van at a height of about 3.5 m. The receive antennas were dipole antennas mounted in a linear array configuration with an inter-antenna spacing of about 3.1λ (20 cm for a 4.635 MHz carrier frequency). Measurements were taken when the van was driven around at an average speed of about 10 km/hr with a majority of the measurements made under NLOS conditions. The van's route was such that the distance between the transmitter and the receiver was between 150 m and 200 m. Field measurement results demonstrated that a transmission rate of 4.915 Gbps with an average packet error rate below 10^{-2} was achieved at most of the locations along the measurement route.

12.2 8 × 16 point-to-point MIMO system at 10 Gbps rate

A 10 Gbps wireless mobile transmission at a carrier frequency of 11 GHz with a 400 MHz bandwidth (i.e., 25 bps/Hz spectral efficiency) has been demonstrated in an 8 × 16 MIMO system [13]. Spatial multiplexing with 64-QAM modulation and turbo coding was used. Eight transmit antennas were mounted on top of a van and 16 receive antennas were used at the BS. Transmission at a rate of 10 Gbps was carried out from the van. Tests and measurements were conducted when the van was driven around at about 9 km/hr in an urban environment. This experiment is expected to help pave the way for future super-high-bit-rate mobile communication.

12.3 16 × 16 multiuser MIMO system

A 16 × 16 multiuser MIMO testbed in an indoor environment was reported in [14]. The testbed was configured to have 16 transmit antennas at the access point and multiple user terminals. Each user terminal can have one or more receive antennas such that the total number of receive antennas across all user terminals is 16. An example configuration for which experimental results have been reported is four user terminals with four antennas each, as shown in Fig. 12.1. The system can have a maximum bandwidth of 100 MHz and the RF carrier frequency is 4.85 GHz. The system uses convolutional coding with constraint length 7, bit interleaving in space and frequency domains, adaptive modulation, and OFDM signaling using 64-point IFFT. The transmit chain and receive chain block diagrams of the testbed are shown in Fig. 12.2. The allowed coding rates include 1/2, 2/3, 3/4, 5/6, and 7/8, and the modulation alphabets include 4-QAM, 16-QAM, 64-QAM, 256-QAM, and 1024-QAM. The receiver sensitivity ranges from −20 dBm to −70 dBm. A transmission frame format based on the one in IEEE 802.11a standard, extended to enable multiuser MIMO-OFDM transmission, was adopted. A downlink beamforming scheme was implemented in the multiuser MIMO testbed. The channel matrices were estimated at the receiver from known preamble symbols embedded in the frame, and the data symbols were decoded using the ZF algorithm. The channel matrices on all subcarriers were initially fed back to the transmitter, and were used for beamforming.

The testbed was deployed inside a room of size 17.6 m × 12.1 m × 3 m. Sleeve

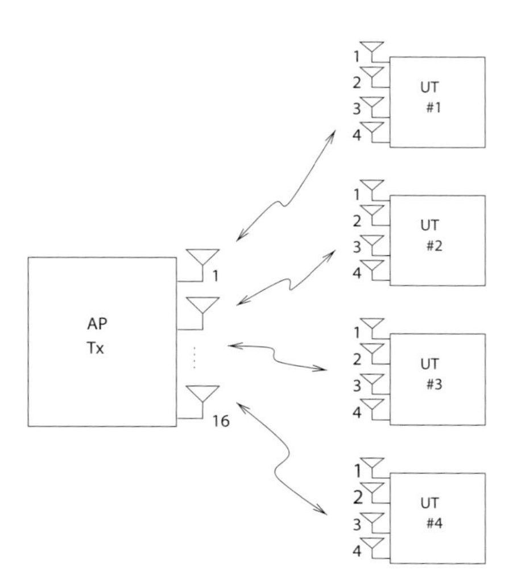

Figure 12.1 16×16 multiuser MIMO testbed at 4.85 GHz in an indoor environment [14]. (AP: access point; UT: user terminal.)

linear array antennas with an inter-element spacing of λ at the transmitter and $\lambda/2$ at the receiver were used. The heights of the transmitter and receiver antennas were 1.7 m and 0.7 m, respectively. The maximum transmission power per antenna was -6 dBm. Using this testbed, spectral efficiencies achieved in the multiuser MIMO downlink transmission scheme for different combinations of the number of UTs (n_u) and the number of receive antennas per UT (n_r) have been evaluated for 16 transmit antennas at the AP ($n_t = 16$) at a 20 MHz bandwidth. For example, results for ($n_u = 2, n_r = 8$), ($n_u = 4, n_r = 4$), ($n_u = 8, n_r = 2$) are presented in [14]. Results have shown that the downlink beamforming scheme achieved a spectral efficiency of 43.5 bps/Hz (870 Mbps) at 31 dB SNR and 50 bps/Hz (1 Gbps) at 36 dB SNR. These achieved spectral efficiencies are quite impressive compared with those achieved in current wireless systems and standards.

12.4 64×15 multiuser MIMO system (Argos)

The design, realization, and evaluation of a multiuser MIMO system that serves several single-antenna user terminals through a BS with a large number ($\gg 10$) of antennas was reported in [15]. The system is called *Argos* – the name of a 100-eyed giant in Greek mythology. This study reported results from an Argos prototype system with 64 antennas at the BS capable of serving 15 user terminals simultaneously. Built using off-the-shelf WARP boards [16] in a modular fashion, the Argos prototype system employs TDD and multiuser beamforming (MUBF) to send independent data streams to multiple user terminals simultaneously.

Linear precoding techniques are used for MUBF. These techniques include ZF MUBF and conjugate MUBF. Let $\mathbf{x}$ denote the n_u-length vector consisting of the

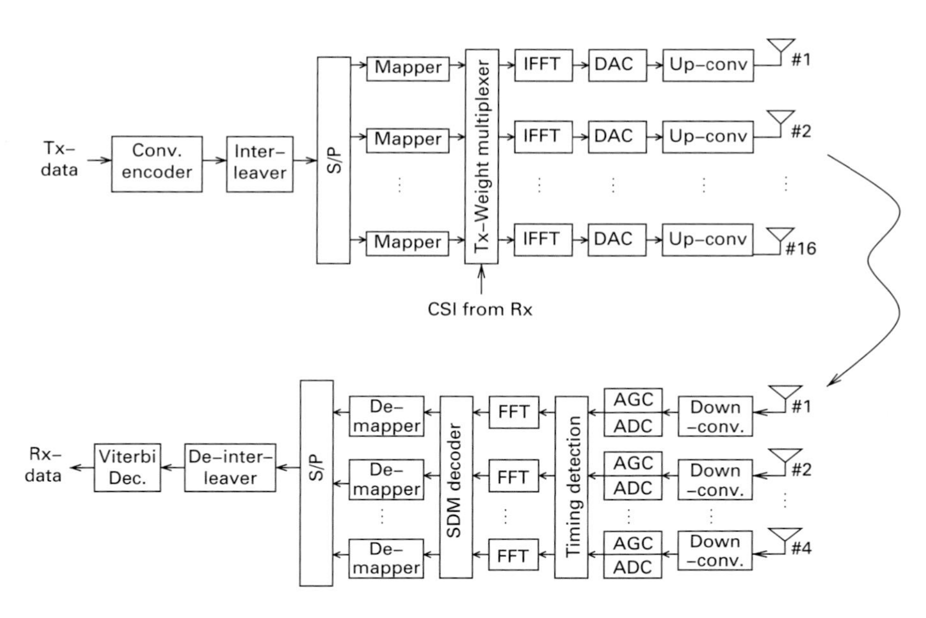

Figure 12.2 Transmit and receive chains in a 16 × 16 multiuser MIMO testbed [14] with $n_t = 16$ and $n_u = n_r = 4$.

data symbols meant for n_u user terminals on the downlink. The linear precoder generates an n_t-length vector $\mathbf{s}$ from $\mathbf{x}$ for the n_t BS antennas to transmit. It obtains $\mathbf{s}$ by multiplying $\mathbf{x}$ by an $n_t \times n_u$ precoding matrix $\mathbf{P}$, such that $\mathbf{s} = \mathbf{P}\mathbf{x}$. The entries of the matrix $\mathbf{P}$ are the beamforming weights, obtained using knowledge of the CSI. Let $\mathbf{H}$ denote the $n_t \times n_u$ channel matrix between the n_t BS antennas and n_u user terminals. In conjugate beamforming, the beamforming weights are the complex conjugates of the CSI, i.e., the precoding matrix $\mathbf{P} = c\mathbf{H}^*$, where c is a normalizing constant and $\mathbf{H}^*$ is the complex conjugate of $\mathbf{H}$. In ZF MUBF, the precoding matrix is $\mathbf{P} = \mathbf{H}^*\left(\mathbf{H}^T\mathbf{H}^*\right)^{-1}$, which forces the inter-user interference to zero. While conjugate MUBF has the advantage of significantly lower complexity, ZF MUBF achieves a much better performance.

The Argos BS includes 16 WARP boards, each board acting as four radios with daughter cards and four antennas. The radios operate at the 2.4 or 5 GHz ISM bands with a 20 MHz bandwidth and 625 kHz subcarrier spacing. The 64 antennas are compactly placed on a custom rack-mount platform. The system achieves 85 bps/Hz spectral efficiency using ZF MUBF. With the computationally less demanding conjugate MUBF, the system achieves a spectral efficiency of 38 bps/Hz. Adoption of TDD allows the estimated CSI on the uplink to be used for downlink beamforming (due to channel reciprocity). This avoids the need for CSI feedback from the user terminals. A scalable channel estimation architecture that computes the full CSI at the BS using only n_u uplink pilots, independent of the number of BS antennas, is employed. The computation of the beamforming weights is carried out locally at each antenna to avoid the data-transfer overhead incurred in a centralized approach. In addition, local power scaling at each antenna is used as an approximate substitute for global power scaling. These features contribute to increased scalability in the number of BS antennas.

The experimental setup included 64-antenna BS hardware (mounted on a movable rack so that the BS can be placed at various indoor locations) and 15 user terminals distributed in different locations inside a building. Tests were carried out in both LOS and non-LOS channel conditions between the BS and the user terminals. Received signal-to-interference noise ratios (SINRs) were measured at various user terminals and the sum capacity was obtained by adding the individual capacities towards the user terminals. Experiments in this indoor set up have revealed that, fixing the number of user terminals at 15, the measured sum capacity using ZF MUBF and conjugate MUBF increased as the number of active BS antennas was increased from 15 to 64. Likewise, fixing the number of BS antennas at 64 and increasing the number of active user terminals from 1 to 15 also increased the sum capacity, approximately linearly with the number of user terminals. In the maximum configuration of 64 BS antennas and 15 user terminals, ZF MUBF achieved 85 bps/Hz and conjugate MUBF achieved 35 bps/Hz, which are significantly higher spectral efficiencies than those achieved in current wireless standards.

12.5 32×14 multiuser MIMO system (Ngara)

Hardware realization and experiments on a multiuser MIMO system with a large number of antennas at a central access point and single-antenna stationary user terminals to provide high-speed internet access in rural areas have been reported in [17],[18]. The rural wireless broadband access demonstrator is called *Ngara*, which is a word of the Aboriginal Darug people of Australia meaning "to listen, hear, and think." The system consists of a central access point unit and multiple user terminal units. The access point provides internet access to the user terminals through over-the-air communication between the user terminals and the access point in FDD mode (see Fig. 12.3). The RF carrier frequencies on the uplink (user-terminal-to-access-point link) and the downlink (access-point-to-user-terminal link) are 806 MHz and 638 MHz, respectively. The system bandwidth is 14 MHz. The choice of this frequency plan was made to enable the future use of the system in the digital dividend frequency band of 694–820 MHz for bi-directional wireless communication. The downlink frequency is close to that of the digital TV broadcasting. Use of these lower carrier frequencies (very high frequency (VHF) and ultra high frequency (UHF)) has the advantage of covering a larger area. The air interface allows the use of 54 different OFDMA subchannels and up to 32 SDMA beams. Time and frequency synchronization in the system is achieved by the use of global positioning system (GPS) receivers at the access point and user terminals. The system achieves a spectral efficiency of 67 bps/Hz.

The access point is equipped with a UCA of up to 32 antenna el-

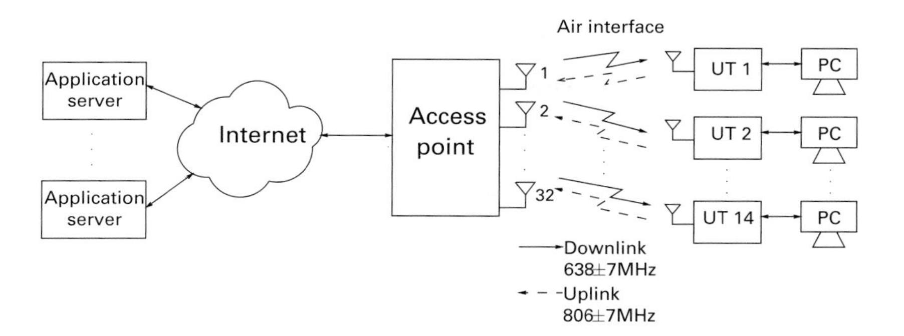

Figure 12.3 Ngara 32 × 14 multiuser MIMO testbed for high-speed internet access in rural areas [18]. (UT: user terminal.)

ements in order to provide a uniform azimuthal coverage (see Fig. 12.4). The user terminal is provided with a single antenna, typically mounted outdoors (e.g., on a rooftop) so that GPS signal is always available for synchronization. In order to reduce the computational load for signal processing at the AP, ZF detection on the uplink and ZF precoding on the downlink have been chosen. Since a LOS component is typical in rural environments, the performance of ZF precoding based multiuser MIMO downlink when the access point is equipped with a UCA in an LOS environment was studied in [19].

User terminal unit

Each user terminal has a hardware unit consisting of the user terminal antenna, RF diplexer, transmit and receive radio frequency/intermediate frequency (RF/IF) chains, signal processing and clocking blocks, and data and programming interfaces. On the receive side, the received RF signal at 638 MHz is down converted to 70 MHz, analog-to-digital converted using a 12-bit analog-to-digital converter (ADC), and fed to a Xylinx Virtex-6 FPGA for baseband processing. The field programmable gate array (FPGA) interfaces with the user personal computer (PC) through a 1 Gbps Ethernet (1GbE) interface. On the transmit side, the FPGA output is digital-to-analog converted using a 16-bit digital-to-analog converter (DAC) at 70 MHz, up converted to 806 MHz RF frequency, and amplified using the power amplifier for transmission. Each user terminal has a GPS receiver which provides accurate timing and a 10 MHz frequency reference to the signal processing modules. The user terminal MAC is also implemented in the FPGA.

Access point unit

The access point unit hardware serves up to 32 antennas using RF/IF chains provisioned for up to 32 parallel channels in a modular fashion. Baseband processing functions are performed by multiple Virtex-6 FPGAs. Three front-end

Figure 12.4 Ngara 32 × 14 multiuser MIMO testbed in the UHF band. Photo source: [18].

FPGAs interface with ADCs and DACs of the different channels and perform time-domain/frequency-domain processing – front-end FPGAs 0 and 1 serve 12 channels each and front-end FPGA 2 serves the remaining 8 channels. Three back-end FPGAs – one for transmit and two for receive – perform information bit processing functions. For example, LDPC decoding functions are carried out by the two back-end receive FPGAs. Another major computational load in baseband processing comes from the matrix inversions needed for ZF detection on the uplink and ZF precoding on the downlink. Since there are 54 subcarriers in the system, for 14 user terminals and 32 antennas at the access point, inversion of 3456 matrices, each matrix of size 14×32, is needed. Through efficient matrix inversion implementations, all these inversions are implemented in one Virtex-6 FPGA. One more FPGA is used to implement the MAC functions. There is a total of nine FPGAs. The FPGAs are configured and controlled through the USB2 interface. The USBs from all the FPGAs are connected to a USB hub so that the controlling PC/laptop can connect to all the FPGAs via a single USB cable. The MAC unit interfaces with an Ethernet virtual LAN (VLAN) trunking switch through a 10GbE Ethernet interface. The Ethernet VLAN switch, on the other end, supports multiple 1GbE Ethernet ports, logically one for each UT. This design provides for flexible multicast and quality of service which can be implemented by standard Ethernet switches located on the AP network.

Demonstration

The Ngara demonstrator was tested in the laboratory environment shown in Fig. 12.4. The user terminals were located uniformly on the two sides of a rectangular room, and the access point was located at the center. Thirty-two vertically polarized folded dipole antennas with an inter-element separation of about 0.4λ form a UCA. While tests with 18 user terminals were carried out using offline signal processing, tests with 14 user terminals were carried out in real time and these demonstrated simultaneous real-time video streaming from 14 user terminals. In the real-time tests with 14 user terminals, each user terminal was connected to a NetBook or a video streamer via an Ethernet cable. The source of the streamed video was a DVD quality MPEG2 file. The total Ethernet throughput per user terminal was approximately 25 Mbps (though the full capacity of the user terminal was 41.18 Mbps). Each user terminal then converted the Ethernet packets into uplink wireless packets and sent them to the access point. The access point decoded the wireless packets and passed the corresponding Ethernet packets to the Ethernet VLAN switch via the 10GbE Ethernet port. The total Ethernet throughput at the access point was approximately 350 Mbps. At the access point, 14 MacBooks were connected to the Ethernet VLAN switch to display the videos streamed from the user terminals. The results demonstrated very good quality video transfer.

12.6 Summary

As can be seen, the spectral efficiencies achieved in all of the large MIMO testbeds presented in this chapter are about an order higher than those in current wireless standards. Results from all these testbeds clearly demonstrate that the high-spectral-efficiency potential promised by large MIMO systems can indeed be realized in practice. Of course, more work is needed to make large MIMO systems commercially viable. Towards this end, more and more testbeds and prototypes in different configurations, in different frequency bands, in different environments are expected to be reported in the years to come. Improved and efficient large MIMO signal processing algorithms and architectures will be devised and tried out in these testbeds. These large MIMO testbed experiences will naturally become valuable inputs for defining the next generation wireless standards like 5G and beyond.

To take large MIMO systems forward in a big way from here, development of low power application-specific integrated circuits (ASICs) for large MIMO signal processing, highly integrated RF/mixed signal ICs, single chip ADCs/DACs with more and more converters built in, and compact and conformal antenna arrays needs investment and focused efforts. Identifying application scenarios that can exploit large MIMO benefits (including scenarios where user terminals can also have a large number of antennas – e.g., TVs, laptops, note pads, tablets, smart phones) and devising suitable large MIMO architectures, algorithms, low cost implementation approaches, and solutions for those scenarios will be rewarding.

References

[1] A. Burg and M. Rupp, Demonstrators and testbeds, In *Smart Antennas: State of the Art*, T. Kaiser, A. Bourcloux, H. Boche, *et al.*, Eds., EURASIP Book Series on Signal Processing and Communications, Vol. 3. New York, NY: Hindawi Publishing Corporation, 2005.

[2] M. Rupp, C. Mehlfuhrer, S. Caban, *et al.*, "Testbeds and rapid prototyping in wireless system design," *EURASIP Newsletter*, vol. 17, no. 3, pp. 32–50, Sep. 2006.

[3] P. W. Wolniansky, G. J. Foschini, G. D. Golden, and R. A. Valenzuela, "V-BLAST: an architecture for realizing very high data rates over the rich-scattering wireless channel," in *URSI Intl Symp. Signals, Systems and Electronics (ISSSE)*, Sept.–Oct. 1998, pp. 295–300.

[4] R. M. Rao, W. Zhu, *et al.*, "Multi-antenna testbeds for research and education in wireless communications," *IEEE Commun. Mag.*, pp. 72–81, Dec. 2004.

[5] S. Caban, C. Mehlfuhrer, R. Langwieser, A. L. Scholtz, and M. Rupp, "Vienna MIMO testbed," *EURASIP J. Appl. Signal Process.*, pp. 1–13, volume 2006, Article ID 54868, DOI 10.1155/ASP/2006/54868.

[6] P. Goud Jr., R. Hang, D. Truhachev, and C. Schlegel, "A portable MIMO testbed and selected channel measurements," *EURASIP J. Appl. Signal Process.*, pp. 1–11, volume 2006, Article ID 51490, DOI 10.1155/ASP/2006/51490.

[7] J. A. Garcia-Naya, M. Gonzalez-Lopez, and L. Castedo, "An overview of MIMO testbed technology," in *ISIVC'208*, Bilbao, Jul. 2008.

[8] K. Kim and M. Torlak, "Rapid prototyping of a cost effective and flexible 4 × 4 MIMO testbed," in *IEEE Sensor Array and Multichannel Signal Processing Workshop (SAM'2008)*, Darmstadt, Jul. 2008, pp. 5–8.

[9] P. Luethi, M. Wenk, T. Koch, N. Felber, W. Fichtner, and M. Lerjen, "Multi-user MIMO testbed," in *WinTech'08*, San Francisco, CA, Sep. 2008, pp. 4584–4589.

[10] N. Jalden, S. Bergman, P. Zetterberg, B. Ottersten, and K. Werner, "Cross layer implementation of a multi-user MIMO test-bed," in *IEEE WCNC'2010*, Sydney, Apr. 2010.

[11] P. Chambers, X. Hong, Z. Chen, C.-X. Wang, M. Beach, and H. Haas, "The UC4G wireless MIMO testbed," in *IEEE GLOBECOM'2012*, Anaheim, CA, Dec. 2012, pp. 4584–4589.

[12] H. Taoka and K. Higuchi, "Field experiments on 5-Gbit/s ultra-highspeed packet transmission using MIMO multiplexing in broadband packet radio access," *NTT DoCoMo Tech. J.*, vol. 9, no. 2, pp. 25–31, Sep. 2007.

[13] NTT DOCOMO press release, "DOCOMO and Tokyo insitute of Techology achieve world's first 10 Gbps packet transmission in outdoor experiment – paving the way for super-high-bit-rate mobile communication," 27 Feb. 2013 http://www.nttdocomo.co.jp/english/info/media_center/pr/2013/0227_00.html. (accessed 26 July 2013).

[14] K. Nishimori, R. Kudo, N. Honma, Y. Takatori, and M. Mizoguchi, "16 × 16 multiuser MIMO testbed employing simple adaptive modulation scheme," in *IEEE VTC'2009 Spring*, Barcelona, Apr. 2009, pp. 1–5.

[15] H. C. Shepard, N. Anand, L. Li, *et al.*, "Argos: Practical many-antenna base stations," in *MobiCom'2012*, Istanbul, Aug. 2012.

[16] P. Murphy, A. Sabharwal and B. Aazhang, "Design of WARP: a wireless open-access research platform", in *European Signal Processing conference (EUSIPCO' 2006)*, Florence, Sep. 2006.

[17] H. Suzuki, I. B. Collings, D. Hayman, J. Pathikulangara, Z. Chen, and R. Kendall, "Large-scale multiple antenna fixed wireless systems for rural areas," in *IEEE PIMRC'2012*, Sydney, Sep. 2012, pp. 1622–1627.

[18] H. Suzuki, R. Kendall, K. Anderson, *et al.*, "Highly spectrally efficient Ngara rural wireless broadband access demonstrator," in *Intl Symp. on Commun. and Inform. Tech. (ISCIT'2012)*, Gold Coast, Oct. 2012, pp. 914–919.
[19] H. Suzuki, D. B. Hayman, J. Pathikulangara, *et al.*, "Design criteria of uniform circular array for multi-user MIMO in rural areas," in *IEEE WCNC'2010*, Sydney, Apr. 2010, pp. 1–6.

Author index

Subject index